Green Photosynthetic Bacteria

Green Photosynthetic Bacteria

Edited by

J. M. Olson
Odense University
Odense, Denmark

J. G. Ormerod
University of Oslo
Oslo, Norway

J. Amesz
State University of Leiden
Leiden, The Netherlands

E. Stackebrandt
Christian-Albrechts-Universität
Kiel, Federal Republic of Germany

and

H. G. Trüper
Rheinische Friedrich-Wilhelms-Universität
Bonn, Federal Republic of Germany

PLENUM PRESS • NEW YORK AND LONDON

Library of Congress Cataloging in Publication Data

EMBO Workshop on Green Photosynthetic Bacteria (1987: Nyborg, Denmark)
 Green photosynthetic bacteria.

 "Based on the proceedings of the EMBO Workshop on Green Photosynthetic Bacteria, held
August 19–21, 1987, in Nyborg, Denmark" — T.p. verso.
 Includes bibliographical references and index.
 1. Bacteria, Photosynthetic — Congresses. I. Olson, John M. II. European Molecular Biology
Organization. III. Title.
QR88.5.E52   1987   589.9   88-12419
ISBN-13: 978-1-4612-8296-9   e-ISBN-13: 978-1-4613-1021-1
DOI: 10.1007/978-1-4613-1021-1

Based on the Proceedings of the EMBO Workshop on Green Photosynthetic Bacteria,
held August 19–21, 1987, in Nyborg, Denmark

© 1988 Plenum Press, New York
Softcover reprint of the hardcover 1st edition 1988
A Division of Plenum Publishing Corporation
233 Spring Street, New York, N.Y. 10013

PREFACE

This volume grew out of the EMBO Workshop on Green Photosynthetic Bacteria held from 19 to 21 August 1987 in Nyborg, Denmark. The papers and summaries which appear in this volume were presented either orally or as posters at the workshop. In accordance with EMBO rules publication in the workshop volume was entirely optional, and a few participants in the workshop elected not to submit papers or summaries.

The workshop was organized by the editors and financed by the European Molecular Biology Organization (EMBO), the Danish Natural Science Research Council, the Carlsberg Foundation, and the Commission of the European Communities. We are happy for the opportunity to thank these organizations publicly for their support.

The edited manuscripts were written by Lisbeth Johansen using an IBM PC with the program Dantekst and a Mannesmann 290 (Tally) printer.

Caroline Olson helped with the editing and the preparation of the index.

J. M. Olson

CONTENTS

PART III. REACTION CENTERS

PART IV. ELECTRON TRANSPORT

PART V. CARBON, HYDROGEN, NITROGEN AND SULFUR METABOLISM

PART VI. PHYLOGENY AND TAXONOMY

PART VII. PHYSIOLOGY AND ECOLOGY

PART VIII. GENETICS

x

INTRODUCTION

J. M. Olson

Institute of Biochemistry
Odense University
DK-5230 Odense M, Denmark

Green photosynthetic bacteria have traditionally included only green sulfur bacteria (Chlorobiaceae) and green filamentous bacteria (Chloroflexaceae), but in this volume are included also the recently-discovered heliobacteria (Gram-positive line), whose reaction centers are strikingly similar to those of the green sulfur bacteria.

Two important papers on reaction centers (not included in this volume) have been published recently. Nitschke et al. (1987) have discovered 2 early electron acceptors (Fe-S centers) in the reaction center of Chlorobium limicola, and Shiozawa et al. (1987) have shown that the reaction center of Chloroflexus aurantiacus contains 2 (rather than 3) polypeptides of M_r = 24 and 24.5 kDa respectively.

This volume begins with papers describing structures on the molecular and macromolecular level and events on a picosecond time scale. It ends with papers describing organisms and populations, and events on time scales of hours, days and weeks. The scope of the volume covers biophysics, biochemistry, physiology, ecology, taxonomy and phylogeny. There is only one paper (Ormerod, 1988) dealing with genetics, an area which ought to receive much more attention in the near future.

An annoying problem in terminology has plagued researchers working with green bacteria since Olson (1980) used the term "baseplate" for two different structures in the same article. The term was used first to denote the 2-dimensional crystalline array of BChl a-protein (809 nm) between the chlorosome core and the cytoplasmic membrane in Cb. limicola f. thiosulfatophilum. (This is the sense in which the term had been used by Staehelin et al. (1980).) The term "baseplate" was also used by Olson (1980) to denote that part of the chlorosome (Cf. aurantiacus) which contains the 792-nm BChl a. (This is the sense in which the term was later used by Feick and Fuller (1984).) Chloroflexus does not contain a crystalline array of BChl a-protein (809 nm) between the chlorosome and cytoplasmic membrane as in Chlorobium, but Chlorobium does contain a 794-nm BChl a in the chlorosome like the 792-nm BChl a of Chloroflexus (Gerola and Olson, 1986).

In this volume Wullink and van Bruggen (1988) use the term "baseplate" in the original sense of Staehelin et al. (1980), while Blankenship et al. (1988) and Oelze and Golecki (1988) use the term in the sense of Feick and Fuller (1984). In the editor's opinion the term "baseplate" should probably be used in the future only to denote that part of the chlorosome (Chloroflexus or Chlorobium) in contact with the "attachment site" on the cytoplasmic membrane. The term "attachement site" might then be used for the 2-dimensional crystal of BChl a-protein in Chlorobium and Prosthecochloris

and the special sites containing the B808-866 BChl a-protein in the
cytoplasmic membrane of Chloroflexus.

Non-standard abbreviations used in this volume include BChl
(bacteriochlorophyll), Chl (chlorophyll) and Cyt (cytochrome).

REFERENCES

Blankenship, R. E., Brune, D. C., Freeman, J. M., King, G. H., McManus, J.
    D., Nozawa, T., and Wittmershaus, B. P., 1988, Energy trapping and
    electron transfer in Chloroflexus aurantiacus, This volume.
Feick, R. G., and Fuller, R. C., 1984, Topography of the photosynthetic
    apparatus of Chloroflexus aurantiacus, Biochemistry, 23:3693.
Gerola, P. D., and Olson, J. M., 1986, A new bacteriochlorophyll a-protein
    complex associated with chlorosomes of green sulfur bacteria,
    Biochim. Biophys. Acta, 848:69.
Oelze, J., and Golecki, J. R., 1988, Growth rate and the control of
    development of the photosynthetic apparatus in Chloroflexus
    aurantiacus as studied on the basis of cytoplasmic membrane structure
    and chlorosome size, This volume.
Olson, J. M., 1980, Chlorophyll organization in green photosynthetic
    bacteria, Biochim. Biophys. Acta, 594:33.
Ormerod, J. G., 1988, Natural genetic transformation in Chlorobium, This
    volume.
Nitschke, W., Feiker, U., Lockau, W., and Hauska, G., 1987, The photosystem
    of the green sulfur bacterium Chlorobium limicola contains 2 early
    electron acceptors similar to photosystem-I, FEBS Lett., 218:283.
Shiozawa, J. A., Lottspeich, F., and Feick, R., 1987, The photochemical
    reaction center of Chloroflexus aurantiacus is composed of two
    structurally similar polypeptides, Eur. J. Biochem., 167:595.
Staehelin, L. A., Golecki, J. R., and Drews, G., 1980, Supramolecular
    organization of chlorosomes (Chlorobium vesicles) and of their
    membrane attachment sites in Chlorobium limicola, Biochim. Biophys.
    Acta, 589:30.
Wullink, W., and van Bruggen, E. F. J., 1988, Structural studies on
    chlorosomes from Prosthecochloris aestuarii, This volume.

STRUCTURAL STUDIES ON CHLOROSOMES FROM PROSTHECOCHLORIS AESTUARII

W. Wullink and E.F.J. van Bruggen

Biochemisch Laboratorium
Rijksuniversiteit Groningen
The Netherlands

INTRODUCTION

In 1964 an electron microscope study of thin sections of several strains of green bacteria belonging to the genus Chlorobium revealed that these organisms always contained vesicle-like elements connected with the cytoplasmic membrane. Upon fractionation these elements turned out to be associated with a fraction that contained a very high specific chlorophyll content. These observations led to the assumption that these vesicles were the site of the photosynthetic apparatus of green bacteria (Cohen-Bazire, 1964). In 1980 a structural model of these vesicles, then called chlorosomes, in Chlorobium limicola was proposed, based on freeze-fracture studies (Staehelin et al., 1980). In this study evidence was presented for the occurrence of a crystalline baseplate consisting of BChl a-protein between the BChl c-protein in the chlorosome core and the reaction centres in the cytoplasmic membrane. The observation that the chlorosome contains some BChl a-protein, not identical to that present in the baseplate (Gerola and Olson, 1986), led to a slightly modified model (Fig. 1). The baseplate BChl a-protein from the related bacterium, Prosthecochloris aestuarii, has been crystallized and its structure determined by X-ray techniques at 0.19-nm resolution (Fenna et al., 1974; Tronrud et al., 1986).

In this study we show that P. aestuarii contains chlorosomes similar in size and shape to those found in Chlorobium. These chlorosomes also are attached to a crystalline baseplate and contain rod-shaped elements. Membrane-embedded particles are probably reaction-centre complexes and associated proteins. One of the ultimate goals of this study is to make a 3-dimensional reconstruction of the structure of the reaction-centre complexes. Since objects with periodicity offer the best possibilities for computer reconstruction, we aimed to obtain the reaction-centre complexes in regular arrays. Considering the possibility of isolating the membrane-embedded reaction-centre complexes and associated proteins in regular arrays when isolated in connection with the crystalline baseplate, we decided to try to isolate whole chlorosomes attached to small patches of the cytoplasmic membrane. For this purpose we applied a mild fixation treatment or a treatment with a reversible crosslinker to the whole P. aestuarii cells before disrupting them. Fractions obtained after partial purification were analysed by electron microscopy, using negative staining techniques, and polyacrylamide gel electrophoresis (PAGE).

MATERIALS AND METHODS

The study was carried out with P. aestuarii cells grown anaerobically in the laboratory of Dr. J. Amesz (R.U., Leiden) according to the method described by Holt et al. (1966).

Fixation and Embedding for Sectioning

The cells were either fixed in culture medium by adding 25% (v/v) glutaraldehyde to a final concentration of 1% and incubating 120 min at 20-25°C followed by postfixation with 1% OsO_4 (w/v) in sodium phosphate buffer (pH 7.2) for 90 min or by washing in aqua dest and resuspending in 1.5% $KMnO_4$ for 20 min followed by repeatedly washing with aqua dest, until the washing fluid remained colourless. In both cases the samples were additionally contrasted by incubation for 60 min in 1% uranyl acetate. After dehydration with increasing concentrations of ethanol, the samples were embedded in SPURR embedding resin (Spurr, 1969).

Freeze-Fracturing

Aliquots of about 5 µl of a concentrated cell suspension in 10 mM sodium phosphate buffer (pH 7.4) containing 10 mM sodium ascorbate (phosphate-ascorbate buffer) and captured between two copper discs, were quickly frozen in N_2 sludge and subsequently mounted in a fracturing device under liquid N_2. Freeze-fracture replicas were prepared at -130°C in a Biotech apparatus using standard procedures. The replicas were cleaned with commercial bleach.

Fixation Prior to Chlorosome Isolation

The bacterial cells were fixed in culture medium by adding 8% formaldehyde in phosphate-ascorbate buffer to a final concentration of 2%. After 30 min of stirring at 20-25°C the cells were centrifuged for 10 min at 4,600 x g, resuspended in a half volume of 2% formaldehyde in phosphate-ascorbate buffer and stirred for 60 min. Eventually the cells were

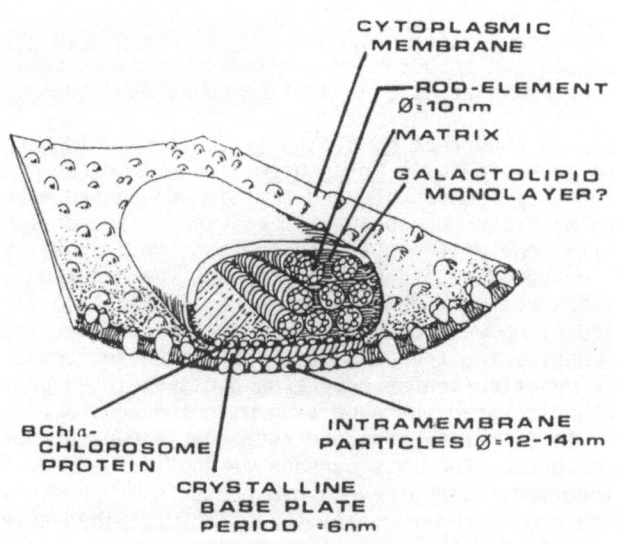

Fig. 1. Model of a Chlorobium chlorosomes (Staehelin et al., 1980; Gerola and Olson, 1986).

pelletted and resuspended in phosphate-ascorbate buffer and kept at 4°C.

Crosslinking Prior to Chlorosome Isolation

The bacterial cells were harvested by centrifuging and resuspended in one volume of a 5 mM sodium phosphate buffer (pH 7.0). The resulting protein concentration was 3 mg/ml as measured by the method of Bradford (1976). Dithiobis(succinimidylpropionate) (DSP), a crosslinker containing a reducible disulfide bond, that can make connections over a distance up to 1.25 nm (Lomant and Fairbanks, 1976), was dissolved in a minimal amount of dimethyl sulphoxide (DMSO) and added to two volumes of 5 mM sodium phosphate buffer (pH 7.0). When the cell suspension was mixed with the DPS solution, the resulting DSP concentration was 1.33 mg/ml. The mixture was kept at 4°C overnight. The crosslinking reaction was stopped by adding one tenth of the mixture's volume of freshly prepared, ice-cold ammonium acetate (1 M).

Isolation of Chlorosomes

After fixation or crosslinking, the cells were fragmented by sonication on melting ice (15 times for 1 min, separated by 1.5 min, using a Branson sonifier B15 at 50 W). The suspension of fragmented cells was centrifuged for 20 min at 18,000 x g. The resulting supernatant was centrifuged for 120 min at 48,000 x g (adapted from Gerola and Olson, 1986), and the pellet was resuspended in 10 mM cacodylate buffer (pH 6.5). To remove small debris from this sample it was centrifuged through a 20% sucrose cushion. The pellet was resuspended in and dialysed against cacodylate buffer and then was stored at 4°C.

Uncrosslinking

Partial uncrosslinking was carried out by adding to a volume of sample one equal volume of 10 mM Tris-HCl buffer (pH 8.5) containing 2.5 to 100 mM dithiothreitol (DTT) and incubating for 30 min at 37°C (Lomant and Fairbanks, 1976). Total uncrosslinking was achieved by adding a β-mercapto-ethanol containing standard denaturation mix (Laemmli, 1970) prior to PAGE.

Preparation of Different Samples from Isolated Chlorosomes

Formaldehyde-fixed chlorosomes were homogenized in a Potter-Elvehjem homogenizer (Potter, 1955) or disrupted by osmotic shock by first equilibrating the sample in 1.5 M sucrose followed by 30-fold dilution in 10 mM cacodylate buffer (pH 6.5). DSP-crosslinked chlorosomes were solubilized by incubation with 0.75% n-octyl β-D-glucopyranoside (β-OG) or 0.05% Triton X-100. Chlorosomes uncrosslinked by DTT-treatment were fractionated by pipetting the sample (1 ml) on top of a 20%-sucrose cushion in cacodylate buffer (5 ml) and subsequent centrifugation for 60 min at 30,000 x g. A volume of about 900 µl was withdrawn from the top and marked TOP. The rest of the supernatant was poured into another centrifuge tube, and the pellet was resuspended in 10 mM cacodylate buffer (pH 6.5) and marked PELLET 1. The supernatant was centrifuged again for 60 min at 40,000 x g and the resulting supernatant was separated from the pellet, which was resuspended in 10 mM cacodylate buffer (pH 6.5) and marked PELLET 2. All samples were stored at 4°C.

Electron Microscopy

Thin sections were mounted on bare copper grids (400 mesh) and stained with lead citrate (Reynolds, 1963). Freeze-fracture replicas were mounted on bare copper grids (400 mesh). For other preparations we used formvar/coal-coated copper grids (400 mesh) that had been subjected to glow discharge at 300 V in 0.4 atm air for 15 s just before use. Specimens were

prepared using the 2-step droplet technique with an aqueous solution of 1%
uranyl acetate as a negative stain. The electron microscopes used were a
JEOL JEM 1200 EX, a Philips EM 300 and a Philips EM 201, usually at a
magnification (on the negative) of 40,000 to 50,000 x. Electron micrographs
were made on either Kodak SO-163 sheets or Kodak Fine Grain Release Positive
(FGRP) 35-mm negative film.

Gel Electrophoresis

PAGE was carried out on 8% polyacrylamide slab gels, with a 4%
polyacrylamide stacking gel. Prior to application to the gel the samples
were either incubated for 3 min at 100°C in a denaturing mix containing
β-mercaptoethanol and sodium dodecyl sulfate (SDS) (Laemmli, 1970) or 30 min
at 37°C in the same mix without β-mercaptoethanol. Sometimes the samples
were just mixed with SDS without β-mercapto ethanol and applied to the
gel immediately. After electrophoresis the gels were stained in a solution
of 0.1% Coomassie Brilliant Blue (CBB) in 45% H_2O, 45% methanol and 10%
acetic acid for 2 h followed by destaining for at least 24 h in 85% H_2O,
7.5% methanol and 7.5% acetic acid.

RESULTS AND DISCUSSION

Thin Sectioning

Thin sections of glutaraldehyde/OsO$_4$-fixed cells show the chlorosomes
as grey oblong bodies attached to the inner cytoplasmic membrane (see Fig.
2A, arrowheads). In the KMnO$_4$-fixed cells the chlorosomes appear as very
heavily stained bodies (see Fig. 2B). In both cases no internal structure of
the chlorosomes is visible. The overall sizes found (40-60 x 60-175 nm) are
similar to those found in Chlorobium (Cohen-Bazire et al., 1964; Staehelin
et al., 1980).

Freeze-Fracturing

The freeze-fracture replicas show that P. aestuarii chlorosomes are

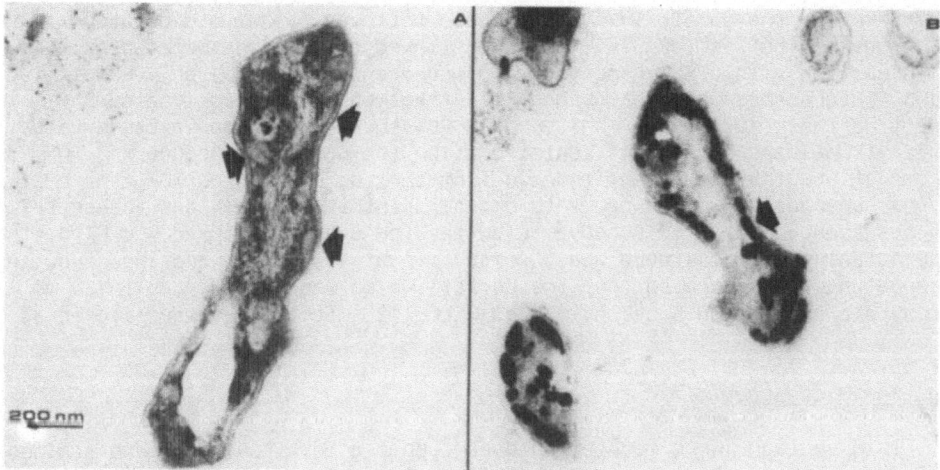

Fig. 2. Thin sections of P. aestuarii cells fixed with glutaraldehyde/OsO$_4$
(A) and thin sections of P. aestuarii cells fixed with KMnO$_4$ (B).
Note the chlorosomes indicated with arrowheads.

6

attached to a crystalline baseplate with a 6.5-nm periodicity determined by optical diffraction (Fig. 3A, arrow). Also some rod-shaped elements with a 11-nm diameter are visible (Fig. 3B, encircled area) as well as more or less circular particles (Ø ≈ 10 nm) grouped together in an area of about the size of a chlorosome attachment site (Fig. 3C, encircled area). These observations are comparable with those on Cb. limicola (Staehelin et al., 1980).

Negative Staining

Purified chlorosomes were negatively stained with uranyl acetate. The electron micrographs showed that small patches of membrane were attached to most of the chlorosome cores. These membranes turned out to contain membrane-embedded protein complexes, probably reaction-centre complexes together with associated proteins. These looked like the complexes observed in isolates from some Chlorobium strains (Cruden and Stanier, 1970). To our disappointment the complexes found were not present in regular arrays, so they still did not offer the most desirable starting point for computer 3-dimensional reconstruction. The protein complexes, which appeared to consist of at least 4 subunits, were estimated to be 16.7 ± 1.8 nm in diameter; the subunits measured about 8 nm (Fig. 4A,B). Side views of the chlorosomes showed their connections to the membrane (Fig. 4C,D).

Disrupting the formaldehyde-fixed chlorosomes by osmotic shock or homogenization released rod-shaped elements 11 nm in diameter and up to about the same length as an intact chlorosome. Sometimes a 3-nm striation perpendicular to the main axis of the rod as well as an accumulation of stain along the axis of the rod were visible, indicating that the rods might consist of ring-shaped subunits (Ø ≈ 11 nm, width ≈ 3 nm) (Fig. 5A,B).

Solubilisation in β-OG or Triton X-100 of DSP-crosslinked chlorosomes also resulted in the release of rod-shaped elements as well as their probable component subunits mentioned above (Fig. 6B,C). Sometimes bare chlorosomal cores were visible (possibly indicating resistance against

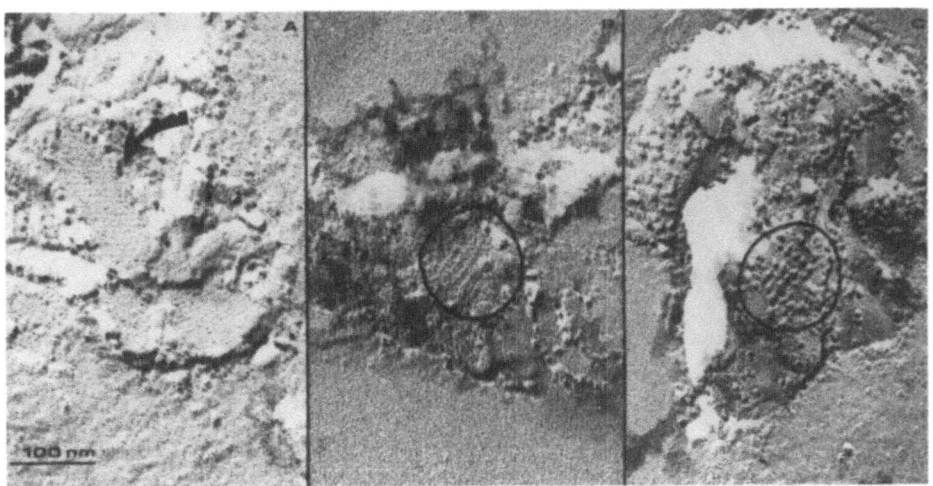

Fig. 3. Freeze-fracture replicas of P. aestuarii showing (A) crystalline baseplates (6.5-nm periodicity) (arrow), (B) rod-shaped elements (Ø ≈ 11 nm) (encircled area), and (C) possible attachment site in cytoplasmic inner membrane with embedded particles (encircled area).

solubilization), but no membrane fragments with embedded protein complexes remained to be seen.

Uncrosslinking of DSP-crosslinked chlorosomes with variable amounts of DTT revealed naked chlorosome cores that appear to disintegrate into thread-like structures. Membrane fragments appeared to contain small protein particles (Ø ≈ 6 nm) and free protein particles about equivalent in size and shape to the protein complex subunits visible in Fig. 4A and B but also quite similar to the trimer of BChl a-baseplate protein (Fenna et al., 1974). Even without DTT uncrosslinking occurs, as judged by electron microscopy. We tried to purify the three different classes of objects mentioned above, using the fractionation method described under MATERIALS AND METHODS. The result is visualized in Fig. 7, in which (A) represents the fraction marked TOP, that consists almost exclusively of free protein particles (Ø ≈ 8.5 nm); (B) represents the fraction marked PELLET 1,

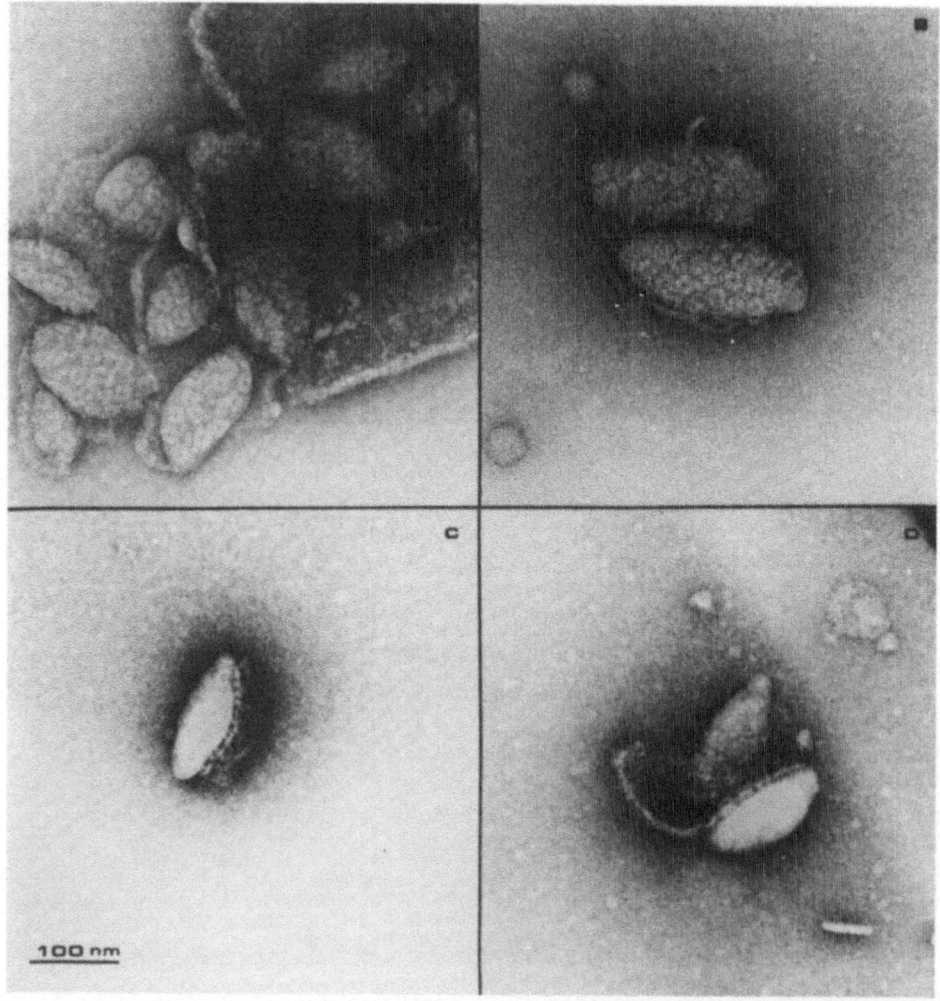

Fig. 4. P. aestuarii chlorosomes negatively stained with uranyl acetate. (A,B) Top views showing membrane-embedded complexes. (C,D) Side views showing the attachment of the membrane patches to the chlorosome core.

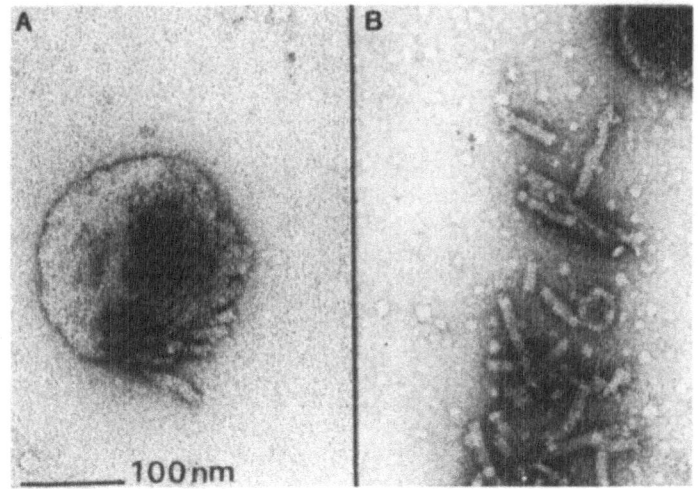

Fig. 5. Rod-shaped elements released from fixed _P. aestuarii_ chlorosomes
after osmotic shock (A) and pottering (B). Negatively stained with
1% uranyl acetate.

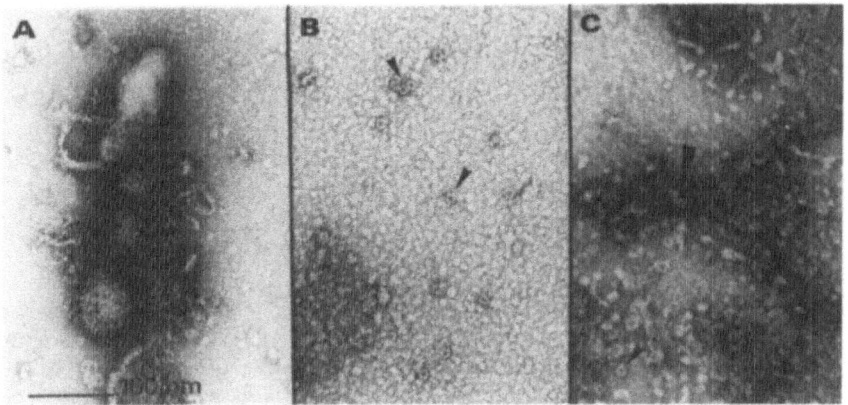

Fig. 6. The result of treatment of DSP-crosslinked _P. aestuarii_ chlorosomes
with 5 mM DTT (A), 0.75% β-OG (B), 0.05% Triton X-100 (C).
Negatively stained with 1% uranyl acetate. Note the rods (large
arrowhead) and the ring-shaped elements (small arrowheads).

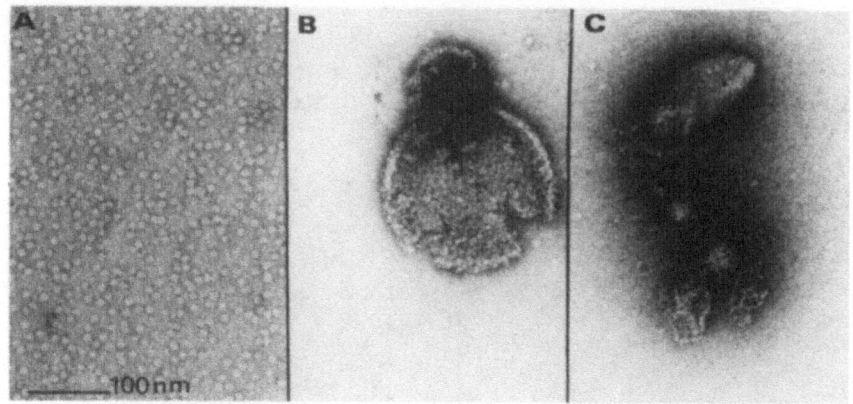

Fig. 7. Different fractions of P. aestuarii chlorosomes uncrosslinked with
DTT (5 mM). (A) the fraction marked TOP, (B) PELLET 1, (C) PELLET
2. Negatively stained with 1% uranyl acetate.

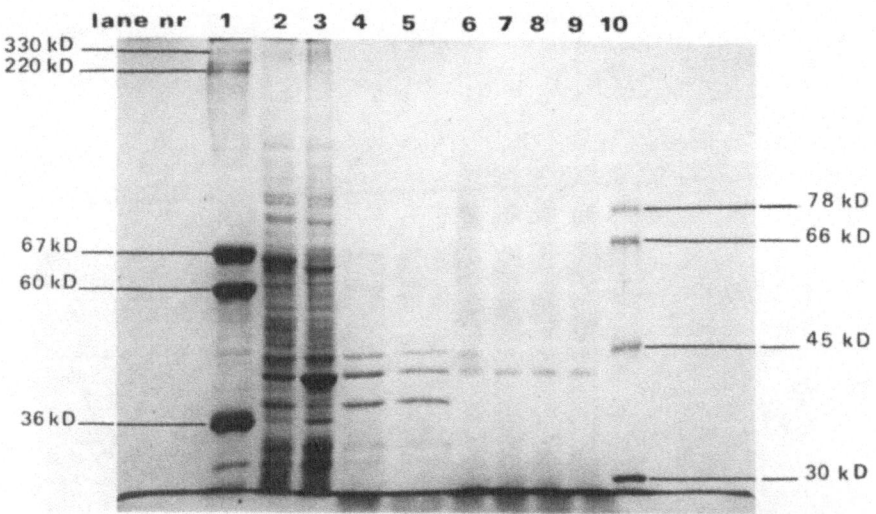

Fig. 8. SDS-PAGE, 8% polyacrylamide. Lanes 1 and 10 contain weight markers.
Lane 2 contains Complex I. Lane 3 contains PP. Lane 4 contains
purified total chlorosome fraction. Lanes 5 and 6 contain the total
chlorosome fraction after uncrosslinking treatment (30 min, 37°C),
with no DTT present. Lanes 7, 8 and 9 contain the chlorosome
fraction after uncrosslinking treatment (30 min, 37°C) with 25, 50
and 100 mM DTT respectively. Before applying them to the gel samples
1-5 and 10 were treated with denaturing mix including
β-mercaptoethanol (3 min, 100°C) (Laemmli, 1970), samples 6-9 were
treated with denaturing mix without β-mercaptoethanol (30 min,
37°C). kD = kDa.

consisting of membrane patches with 6-nm particles and free 8.5-nm particles and (C) represents the fraction marked PELLET 2, that consists mainly of chlorosome cores and thread-like structures. Often some 6-nm particles attached to the chlorosome core are visible.

Polyacrylamide Gel Electrophoresis

Fig. 8 shows the polypeptide composition of purified DSP-crosslinked chlorosomes after uncrosslinking with either β-mercaptoethanol or different concentrations of DTT compared with two functionally characterized samples from P. aestuarii: pigment-protein complex (PP) and Complex I (Swarthoff, 1982). The β-mercaptoethanol uncrosslinked chlorosome samples show only 3 major bands of about 40, 42 and 45 kDa respectively (lanes 4 and 5). The 42-kDa band is comparable with the dominant band of the PP preparation (lane 3), which contains the baseplate protein subunit (Swarthoff, 1982). Some minor bands of about 32 and 34 kDa as well as some ranging from 58 to 70 kDa are also visible. The DTT uncrosslinked samples show clear bands only at 42 and 45 kDa, with smears at about 30, 50 and 80 kDa (lanes 7-9). The sample that was treated like the ones uncrosslinked by DTT, but with no DTT present (lane 6) shows fairly clear 42- and 45-kDa bands, which makes it quite similar to the samples treated with DTT. There is also a vague band of about 100 kDa (perhaps representing a dimer of a polypeptide present in the 40-, 42- or 45-kDa band). (For more details on the treatment of the samples before application to the gel, see the legends to Fig. 8.)

Fig. 9 shows the effect of uncrosslinking with different (low) DTT concentrations and the difference between application of the samples

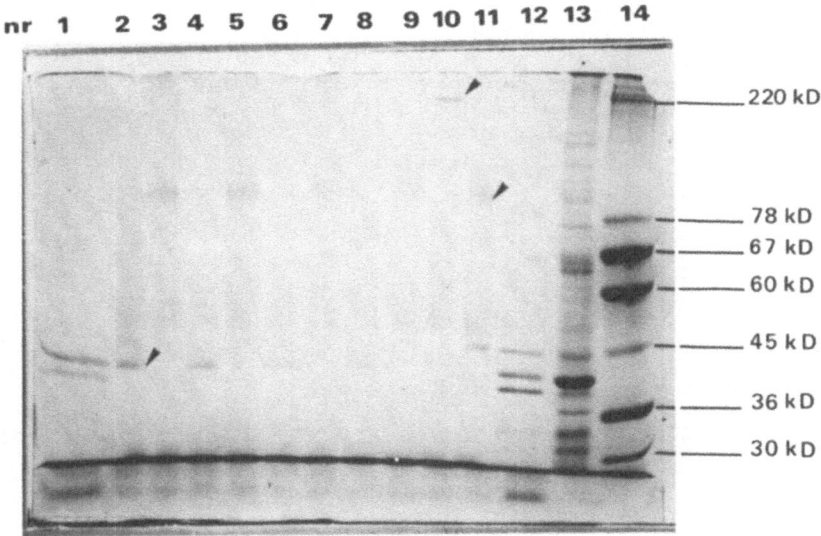

Fig. 9. SDS-PAGE, 8% polyacrylamide. Lanes 1-12 contain the purified total chlorosome fraction. Lane 13 contains PP. Lane 14 contains weight markers. Before applying them to the gel the samples in lanes 1 and 12-14 were heated at 100°C for 3 min in denaturing mix containing β-mercaptoethanol. The samples in lanes 1, 2-3, 4-5, 6-7, 8-9 and 10-11 were submitted to uncrosslinking treatment with 100, 20, 10, 5, 2.5 and 0 mM DTT respectively. The samples in lanes 1, 2, 4, 6, 8 and 10 were incubated for 30 min at 37°C in a denaturing mix without β-mercaptoethanol before application to the gel; the samples in lanes 3, 5, 7, 9 and 11 were applied to the gel immediately. kD = kDa.

to the gel immediately after uncrosslinking and application after a prolonged incubation in denaturation mix without β-mercaptoethanol. In the first case the polypeptide patterns appear to be rather similar (lanes 3, 5, 7 and 9); they all show a vague band of about 100 kDa, but only the sample treated with no DTT at all (lane 11) shows a clear 45-kDa band. The sample treated with 2.5 mM DTT (lane 9) shows an indication of a band of about 220 kDa. In the second case there appears to be evidence that treatment with increasing concentrations of DTT brings about a gradual transformation of a 220-kDa (arrowhead) tetrameric, pentameric or hexameric form to a 42-kDa (arrowhead) (monomeric) form of one polypeptide (lanes 2, 4, 6, 8 and 10). In fact also a 100-kDa (dimeric) form appears as a vague band in lane 2. (For details see the legends to the figure.)

Fig. 10 shows the polypeptide patterns of different fractions of purified uncrosslinked (5 mM DTT) chlorosomes. The fraction marked TOP (lane 7) contains only one clear band, namely the 42-kDa band, possibly originating from the BChl a-baseplate protein, but not migrating to exactly the same spot as the comparable band of the PP sample (lane 8). The fractions marked PELLET 1 (lane 6) and PELLET 2 (lane 5) are very similar; they both show the normal purified chlorosome pattern with a reduced intensity of the 42-kDa band. Lane 2 shows the pattern of a purified chlorosome fraction that has been treated with 2 M KSCN after isolation. Note that the band corresponding to the 42-kDa band from other samples (lanes 4-7 and 9), migrates somewhat further, almost to exactly the same spot as the major band of the PP sample (lane 8) does. (For more details see the legends to the figure.)

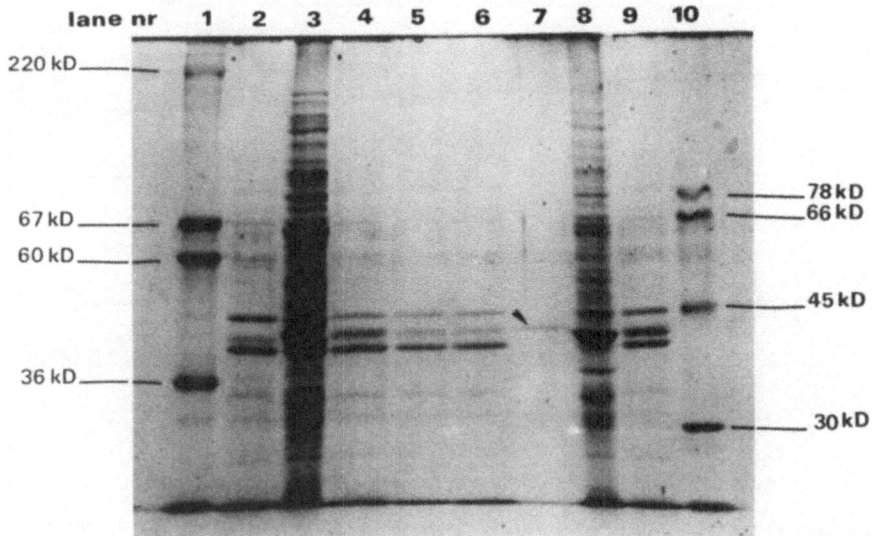

Fig. 10. SDS-PAGE, 8% polyacrylamide. Lanes 1 and 10 contain weight markers. Lane 2 contains a purified chlorosome sample treated with 2 M KSCN. Lane 3 contains Complex 1. Lanes 4-7 contain uncrosslinked (5 mM DTT) chlorosome samples. Lane 4 contains the unfractionated sample. Lanes 5-7 contain the fractions marked PELLET 2, PELLET 1 and TOP respectively. Lane 8 contains PP. Lane 9 contains a purified chlorosome sample. Before application to the gel all samples were treated with a denaturing mix containing β-mercaptoethanol (3 min, 100°C). kD = kDa.

Concluding Remarks

This study makes clear that P. aestuarii chlorosomes are quite similar to those found in different Chlorobium strains (Cohen-Bazire et al., 1964; Staehelin et al., 1980; Gerola and Olson, 1986). Protein complexes embedded in patches of membrane attached to the chlorosomes give an SDS-PAGE polypeptide pattern consisting of only 3 major bands (40, 42 and 45 kDa), which is somewhat surprising, considering the complexity of the total chlorosome unit. The 42-kDa band is at least partly coming from a 8.5-nm particle that has been purified (Fig. 7A). This particle could be the BChl a-baseplate protein studied already by X-ray techniques (Fenna et al., 1974; Tronrud et al., 1986), but it also is similar to the subunits of the membrane-embedded protein complexes visualized in Fig. 4A,B. This observation is in agreement with the proposal (based on linear dichroism studies) that the BChl a-baseplate protein is very similar to a non-baseplate BChl a-protein associated with the reaction-centre (Swarthoff et al., 1980). The origin of the other two major bands and their relation to the reaction-centre complexes, the chlorosome rods and the 6-nm particles remain to be cleared up.

ACKNOWLEDGEMENTS

The authors wish to thank Dr. J. Amesz and coworkers for the gift of P. aestuarii cells and fractions, Drs. M. J. de Hoop for advice and help with the sectioning experiments and Mr. F. Dijk for help and advice with the freeze-fracturing experiments. This investigation was supported by the Netherlands Foundation for Chemical Research (SON) and financed by the Netherlands Organisation for the advancement of Pure Research (ZWO).

Note added in proof: Gerola and Olson (1986) showed by SDS-PAGE that purified chlorosomes (not attached to baseplates or membrane fragments) contained major proteins in a band at approx. 4-5 kDa (M_r). This was the only band visualized by Coomassie Brilliant Blue staining. Six other bands at about 9, 20, 26, 27, 32 and 38 kDa (M_r) were visualized only after silver staining. The absence of a protein band at 4-5 kDa (M_r) in our gels might be due to the fact that we used 8% gels, whereas Gerola and Olson (1986) used 16% gels.

REFERENCES

Bradford, M. M., 1976, A rapid and sensitive method for the quantitation of microgram quantities of protein utilizing the principle of protein-dye binding, Anal. Biochem., 72:248.

Cohen-Bazire, G., Pfennig, N., and Kunisawa, R., 1964, The fine structure of green bacteria, J. Cell Biol., 22:207.

Cruden, D. L., and Stanier, R. Y., 1970, The characterization of chlorobium vesicles and membranes isolated from green bacteria, Arch. Microbiol., 72:115.

Fenna, R. E., Matthews, B. W., Olson, J. M., and Shaw, E. K., 1974, Structure of a bacteriochlorophyll-protein from the green photosynthetic bacterium Chlorobium limicola; crystallographic evidence for a trimer, J. Mol. Biol., 84:231.

Gerola, P. D., and Olson, J. M., 1986, A new bacteriochlorophyll a-protein complex associated with chlorosomes of green sulfur bacteria, Biochim. Biophys. Acta, 848, 69.

Holt, S. C., Conti, S. F., and Fuller, R. C., 1966, Photosynthetic apparatus in the green bacterium Chloropseudomonas ethylicum, J. Bact., 91:311.

Laemmli, U. K., 1970, Cleavage of structural proteins during the assembly of the head of bacteriophage T4, _Nature_ (London), 227:680.

Lomant, A. J., and Fairbanks, G., 1976, Chemical probes of extended biological structures: synthesis and properties of the cleavable cross-linking reagent [^{35}S] dithiobis(succinimidyl propionate), _J. Mol. Biol_., 104:243.

Potter, V. R., 1955, Tissue homogenates, _in_: "Methods in Enzymology Vol. 1," S. P. Colowick and N. O. Kaplan, eds., Academic Press, New York.

Reynolds, E. S., 1963, The use of lead citrate at high pH as an electron opaque stain in electron microscopy, _J. Cell Biol_., 71:136.

Swarthoff, T., de Grooth, B. G., Meiburg, R. F., Rijgersberg, C. P., and Amesz, J., 1980, Orientation of pigments and pigment-protein complexes in the green photosynthetic bacterium _Prosthecochloris aestuarii_, _Biochim. Biophys. Acta_, 593:51.

Swarthoff, T., 1982, The photosynthetic apparatus of a green bacterium, _Ph.D. Thesis_, University of Leiden.

Spurr, A. R., 1969, A low-viscosity epoxy resin embedding medium for electron microscopy, _J. Ultrastruc. Res_., 26:31.

Staehelin, L. A., Golecki, J. R., and Drews, G., 1980, Supramolecular organisation of chlorosomes (chlorobium vesicles) and their membrane attachment sites in _Chlorobium limicola_, _Biochim. Biophys. Acta_, 589:30.

Tronrud, D. E., Schmid, M. F., and Matthews, B. W., 1986, Structure and x-ray aminoacid sequence of a bacteriochlorophyll _a_ protein from _Prosthecochloris aestuarii_ refined at 1.9 Å resolution, _J. Mol. Biol_., 188:443.

SPIN LABEL STUDIES ON CHLOROSOMES FROM GREEN BACTERIA

R.P. Cox, M.T. Jensen, M. Miller and J.P. Pedersen

Institute of Biochemistry
Odense University, Campusvej 55
DK-5230 Odense M, Denmark

INTRODUCTION

The light-harvesting pigments of phototrophic bacteria belonging to the green sulfur bacteria (Chlorobiaceae) and the green filamentous bacteria (Chloroflexaceae) are localised in chlorosomes (Sprague and Varga, 1986). These are flattened cylinders attached to the cellular membrane. They comprise a core of hydrophobic BChl c-binding proteins and an envelope of lipid arranged as a monolayer (Staehelin et al., 1978, 1980). Such a monolayer is an unusual structure in biological systems and a comparison of its biophysical properties with natural and artificial lipid bilayers is of interest. We report here a study of the dynamics of the chlorosome lipid phase using electron spin resonance (ESR) spectrometry of spin labels. We have studied the mesophile Chlorobium and the thermophile Chloroflexus and made comparative measurements at both of their respective temperatures of growth.

METHODS

Bacteria

Chlorobium limicola f. thiosulfatophilum, strain 6230 (Tassajara) was grown at 30°C in the mineral salts of Medium 2 described by Pfennig and Trüper (1981) without NH₄Cl, and supplemented with 0.7 g/l ammonium acetate and 1 g/l sodium thiosulphate. Cultures were grown in completely filled 1.0 litre bottles at a distance of 25 cm from a 60 W incandescent lamp. Chloroflexus aurantiacus strain OK-70-fl was kindly provided by Prof. J.R. Ormerod, University of Oslo, and was grown at 55°C in D medium salts (Castenholz and Pierson, 1981) supplemented with 5 g/l sodium acetate trihydrate, 2 g/l casamino acids (Difco), 1 g/l yeast extract and 0.6 g/l NH₄Cl and buffered with 10 mM Tris-HCl pH 8.0 (J.R. Ormerod, personal communication). Cells were grown in 1.0 litre stoppered bottles completely filled with medium. The bottles were incubated on a rotary shaker at 150 rev/min and illuminated with a single 18 W fluorescent tube (distance of closest approach, 2.0 cm).

Chlorosome Preparation

Chlorosomes were prepared by breaking cells in a French Press in the presence of a high concentration of salt, followed by density gradient centrifugation as described by Gerola and Olson (1986) except that 2.5 M NaBr was used instead of NaSCN (P.D. Gerola, personal communication). No detergents are used in this procedure. The chlorosome fraction was extensively dialysed against 10 mM potassium phosphate buffer pH 7.0 to remove ascorbate which would reduce the nitroxide spin labels used.

Spin Label Measurements

Nitroxide spin labels were obtained from Sigma or Molecular Probes (Junction City, Oregon, U.S.A.).

Chlorosomes were labelled by adding an ethanolic solution of the nitroxide to a small glass tube and evaporating the solvent in a stream of argon. The chlorosome suspension was then added and the tube agitated on a vibratory mixer. In all cases the suspension contained chlorosomes (A 745 = 150, approximately equivalent to 1.5 mg BChl c/ml), 50 µg/ml spin label, 30 mM tris-oxalato chromium III as a "spin-broadening agent", 2 mM potassium ferricyanide to counteract reduction of the spin label, and 9 mM potassium phosphate buffer pH 7.0.

Suspensions of spin-labelled liposomes were prepared from dipalmitoyl phosphatidyl choline (Sigma) in a similar way, except that the phospholipid dissolved in methanol was added to the spin label before evaporation. Liposomes were formed by ultrasonication for a short period.

ESR Spectra were taken using a Varian E-104A spectrometer. The sample was contained in glass capillary tubes sealed with modelling wax and held in a quartz ESR tube. Equilibration of the spin label with the sample was checked by running 3 successive spectra.

RESULTS

Dynamics of Localised Spin Labels

One type of information about membrane lipid dynamics is provided by the ESR spectra of spin-labelled fatty acids with a nitroxide group close to the carboxylic acid group, such as 5-doxyl stearate and 7-doxyl stearate. These are presumed to become arranged with the acid group at the aqueous/lipid interface and the nitroxide group a short distance into the lipid phase. The motion of the nitroxide free radical will then be mainly rotation in a plane parallel to the surface of the lipid phase. A typical ESR spectrum of this type is shown in Fig. 1.

Under certain conditions the ESR spectrum can be analysed to give an "order parameter" for this motion. In the present case, this was only possible for the measurements at 30°C. The results are shown in Table 1.

In the case of the measurements at 55°C, the order parameters could not be determined from the ESR spectrum. An alternative empirical measure of mobility is the distance between the two extreme peaks of the spectrum. These values are also shown in Table 1.

These values are within the range found for artificial bilayer membranes. The extrema splitting found for 5-doxyl stearate in liposomes made from dipalmitoyl phosphatidyl choline was 61 Gauss at 25°C. This is below the phase-transition temperature, and the lipid will be in the gel phase. The corresponding value in the liquid crystalline ronfiguration at 50°C was 41 Gauss.

Table 1. ESR Spectral Parameters of Doxyl Stearates in
          Chlorosomes. Order parameters were calculated
          using the formula given by Marsh (1981). 5-DS,
          5-doxyl stearate; 7-DS, 7 doxyl stearate.

Bacterium	Temp. (°C)	Spin Label	Order Parameter	Extrema Splitting (Gauss)
Chlorobium	30	5-DS	0.64	55
	30	7-DS	0.61	57
	55	5-DS	*	(45)+
	55	7-DS	*	*
Chloroflexus	30	5-DS	0.83	59
	30	7-DS	0.81	59
	55	5-DS	*	45
	55	7-DS	*	*

* Spin label was too mobile to allow the relevant parameter
  to be determined from the ESR spectrum.
+ Estimate from spectrum with very small high-field peak.

Dynamics of Delocalised Spin Labels

The motion of the nitroxide group in spin-labelled fatty acids with the
doxyl group some distance from the acid group approaches that of free
movement in all three dimensions. Such mobility is also found with
hydrophobic spin labels without an "anchoring" group, such as 5-doxyl
decane, which is expected to become distributed throughout the lipid phase.
The ESR spectra of such probes can be analysed to give "apparent rotational
correlation times". Although the quantitative significance of such values is
questionable at best for heterogeneous biological membranes, and the
necessary assumptions break down at values greater than about 1.0 ns (Marsh,
1981), such values do at least provide a basis for semi-quantitative
comparison of different probes and materials.

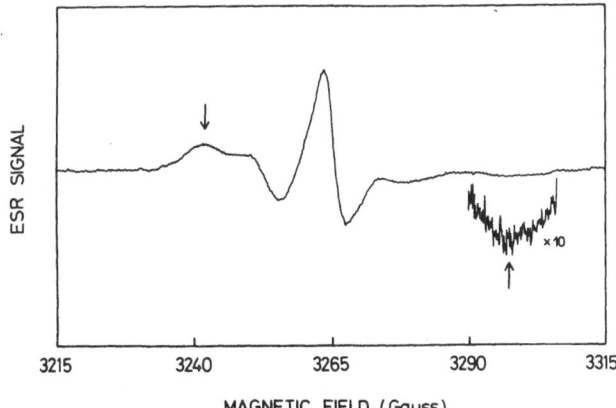

Fig. 1. ESR spectrum (30°C) of Chlorobium chlorosomes labelled with 5-doxyl
        stearate. Microwave frequency was 9.175 GHz, microwave power 10 mW,
        and modulation amplitude 1.0 Gauss. The two arrows show the
        positions of the peaks used in determining the extrema splittings.

Table 2. Apparent Rotational Correlation Times for Spin Labels
in Chlorosomes and Artificial Bilayer Membranes.
Correlation times (τ_s) were calculated from the
formula given by Marsh (1981). 5-DD, 5-doxyl decane.
7-DS, 12-DS, 16-DS; 7-, 12-, and 16-doxyl stearates.
DPPC, dipalmitoyl phosphatidyl choline.

Sample	Bacterium	Spin Label	Correlation time	
			30°C	55°C
Chlorosomes	Chlorobium	12-DS	*	1.40
		16-DS	1.22	0.65
		5-DD	1.27	0.69
Chlorosomes	Chloroflexus	7-DS	*	1.88
		12-DS	*	1.24
		16-DS	2.26	0.68
		5-DD	2.64	0.61
DPPC Liposomes		12-DS	ND	0.69
		16-DS	1.87	0.34
		5-DD	0.98	0.15

*Spin label was insufficiently mobile to allow the correlation
time to be determined from the formula used. ND, not determined

A typical spectrum of a spin label showing rapid motion of this type is
shown in Fig. 2.

The values of the apparent rotational correlation times obtained for a
series of spin labels in chlorosomes from the two types of bacteria at 30°C
and 55°C are shown in Table 2.

These results suggest the presence of a gradient of increasing fluidity
from the surface of the envelope towards the interior. Such a gradient is
also observed in bilayer membranes (Quinn, 1981).

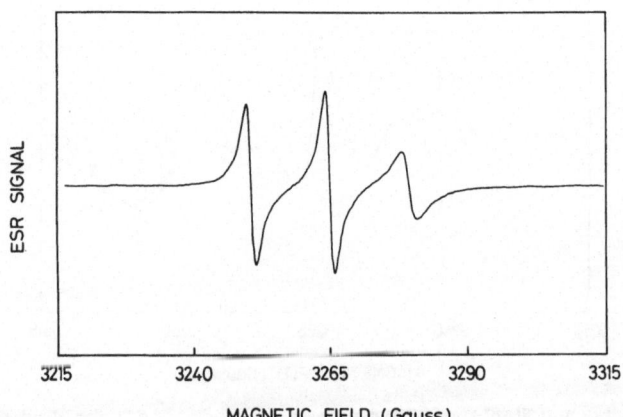

Fig. 2. ESR spectrum (55°C) of Chloroflexus chlorosomes labelled with
16-doxyl stearate. Microwave frequency was 9.175 GHz, microwave
power 10 mW, and modulation amplitude 0.5 Gauss.

The results of similar measurements of apparent correlation times with liposomes of dipalmitoyl phosphatidylcholine are also shown in Table 2. A comparison suggests that chlorosomes from _Chloroflexus_ are more rigid than the artificial bilayer membrane at the same temperature. This is particularly the case when 5-doxyl decane is considered. It is less valuable to compare the _Chlorobium_ chlorosomes with the liposomes at 30°C, since the liposomes are in the gel phase at this temperature. However, the results with 5-doxyl decane suggest that the chlorosomes have a high degree of rigidity under these conditions.

A comparison of the relative mobility of the freely distributed 5-doxyl decane and the anchored 16-doxyl stearate is also of interest. The relatively low mobility of the freely distributed probe in both types of chlorosome might be the result of reduced flexibility of the hydrocarbon chains due to interaction with the chlorophyll-binding proteins.

Temperature Dependence of Chlorosome Fluidity

The possible existence in chlorosomes of phase changes corresponding to those observed in artificial phospholipid bilayer membranes was investigated by measuring the temperature-dependence of fluidity at the growth temperature and below. Such studies are best made with a probe with a low apparent rotational correlation time, such that the maximal value does not exceed the upper limit of validity of the equations used in the calculation (Marsh, 1981). The best choice for this type of measurement in chlorosomes was found to be the small spin-label tetramethyl piperidine N-oxide (Tempo). This is more water-soluble than the doxyl spin labels, and high concentrations are needed to get a satisfactorily large signal from the probe in the lipid phase. It was also found necessary to increase the concentration of the spin-broadening agent tris-oxalato chromium (III) to 55 mM and to correct the ESR spectra for the small residual contribution due to the aqueous probe.

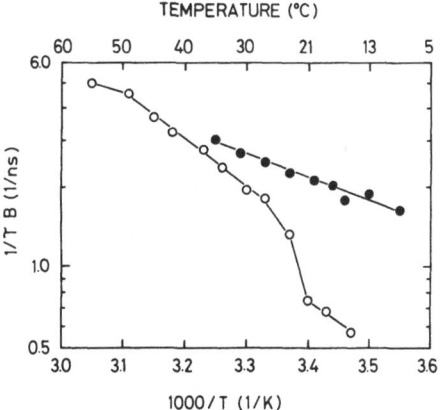

Fig. 3. Arrhenius plot of the mobility of Tempo in chlorosomes from
Chloroflexus (o) and _Chlorobium_ (o). Chlorosomes at a concentration
corresponding to A (745) = 100 for _Chlorobium_ and A (745) = 130 for
Chloroflexus were suspended in a medium containing 500 μM Tempo, 55
mM tris-oxalato chromium (III), 1.4 mM ferricyanide and 8 mM
potassium phosphate pH 7.0. Correlation times (τ_B) were
calculated from the formula given by Marsh (1981).

Fig. 3 shows Arrhenius plots of the temperature dependence of the mobility of Tempo in chlorosomes from Chlorobium and Chloroflexus. Mobility is measured as the reciprocal of the apparent rotational correlation time. There is gradual decrease in fluidity with decreasing temperature, which is much more pronounced for Chloroflexus than for Chlorobium. In the case of Chloroflexus alone, there is also a clear discontinuity at around 25°C. This might result from the formation of a gel phase in the lipid. However, such apparent discontinuities in Arrhenius plots can also be the result of a number of other phenomena, and assignment to a phase transition requires confirmation by an independent technique such as calorimetry. In any case the temperature in question is well below the range where the phenomenon can have any physiological significance.

DISCUSSION

The results with the various spin labels can be readily reconciled with the idea that the chlorosome envelope is arranged as a monolayer. The dynamic properties of the envelope lipids are broadly similar to those of an artificial bilayer membrane, with some evidence for relatively less mobility at the ends of the hydrocarbon chains.

The results with both delocalised and localised probes demonstrate that the lipid phase of chlorosomes from Chloroflexus is more rigid than that of Chlorobium chlorosomes at around 30°C, while the two have similar dynamic properties at 55°C. There is thus a large difference between the fluidity of the two types at the respective temperatures of growth. However, it seems probable that the chlorosome envelope is only a structural element, acting as an interface between the aqueous cytoplasm and the hydrophobic chlorophyll-binding proteins, and plays no functional role for which lipid-phase fluidity is important.

ACKNOWLEDGEMENTS

M.M. is the recipient of a Research Fellowship from the Danish Council for Research Policy and Planning.

REFERENCES

Castenholz, R., and Pierson, B. K., 1981, Isolation of members of the family Chloroflexaceae, in: "The Prokaryotes, A Handbook on Habitats, Isolation and Identification of Bacteria," M. P. Starr, H. Stolp, H. G. Trüper, A. Balows, and H. G. Schlegel, eds., Springer-Verlag, Berlin.
Gerola, P. D., and Olson, J. M., 1986, A new bacteriochlorophyll a-protein complex associated with chlorosomes of green sulfur bacteria, Biochim. Biophys. Acta, 848:69.
Marsh, D., 1981, Electron spin resonance: spin labels, in: "Membrane Spectroscopy." E. Grell, ed., Springer-Verlag, Berlin.
Pfennig, N., and Trüper, H. G., 1981, Isolation of members of the families Chromatiaceae and Chlorobiaceae, in: "The Prokaryotes, A Handbook on Habitats, Isolation and Identification of Bacteria," M. P. Starr, H. Stolp, H. G. Trüper, A. Balows, and H. G. Schlegel, eds., Springer-Verlag, Berlin.
Quinn, P. J., 1981, The fluidity of cell membranes and its regulation, Progr. Biophys. Mol. Biol., 38:1.
Sprague, S. G., and Varga, A. R., 1986, Membrane architecture of anoxygenic photosynthetic bacteria, in: "Encyclopedia of Plant Physiology New Series," L. A. Staehelin, and C. J. Arntzen, eds., Springer-Verlag,

Berlin.

Staehelin, L. A., Golecki, J. R., Fuller, R. C., and Drews, G., 1978, Visualization of the supramolecular architecture of chlorosomes (*Chlorobium* type vesicles) in freeze-fractured cells of *Chloroflexus aurantiacus*, *Arch. Microbiol*., 119:269.

Staehelin, L. A., Golecki, J. R., and Drews, D., 1980, Supramolecular organisation of chlorosomes (*Chlorobium* vesicles) and of their membrane attachment sites in *Chlorobium limicola*, *Biochim. Biophys. Acta*, 589:30.

GROUND-STATE MOLECULAR INTERACTIONS OF BACTERIOCHLOROPHYLL C IN CHLOROSOMES

OF GREEN BACTERIA AND IN MODEL SYSTEMS: A RESONANCE RAMAN STUDY

M. Lutz[a] and G. van Brakel[b]

[a]Département de Biologie, CEN Saclay
 F-91191 Gif-sur-Yvette, France
[b]Biokemisk Institut, Odense Universitet
 DK-5230 Odense, Denmark

INTRODUCTION

Chlorosomes of green photosynthetic bacteria from the two families Chlorobiaceae and Chloroflexaceae contain large amounts of BChls c, d or e, which constitute peripheral antenna systems connected to the intrinsic membrane antenna (Olson, 1980). Chlorosomes of the two best studied species, Chlorobium limicola and Chloroflexus aurantiacus contain BChl c (Fig. 1). In both species these molecules appear highly organized. Their lowest singlet electronic transitions Q_y assume mutual coupling and the resulting transition moments exhibit non-random orientations, being predominantly parallel to the membrane (Olson, 1980; Betti et al., 1982; van Dorssen et al., 1986a, b; Gerola and Olson, 1986). BChl c is thought to be associated with small polypeptides (Schmitz, 1967; Wechsler et al., 1985; Gerola et al., 1988). These associations however must be somewhat weaker than in intrinsic membrane antenna complexes, considering the ease with which BChl c can be washed from chlorosomes (Feick and Fuller, 1984), and the fact that it has not yet been possible to extract any intact BChl c-protein from chlorosomes.

The ground-state interactions which result in the above-mentioned properties of BChl c in chlorosomes have been the subject of much interest, and several models have been proposed for the organization of BChl c in vivo. Most of these models were based on studies of in vitro systems thought to be relevant to chlorosome structure (Bystrova et al., 1979; Smith et al., 1983; Worcester at al., 1986). The model proposed by Wechsler et al. (1985) for Cf. aurantiacus was based on the primary sequence of the chlorosome protein. In all of these models, the BChl c molecules were proposed to form oligomers arranged in linear or bidimensional arrays (Fig. 2). In all of them, binding of adjacent molecules was assumed to involve the hydroxyethyl group (a substituent specific to BChls c, d, and e), through interactions with partner's keto group or magnesium atom, or both (Fig. 2). The magnesium atoms were generally proposed to assume 5-coordination, except in the model of Bystrova et al. (1979), which involved 6-coordinated magnesiums (see, however, Blankenship et al., 1988).

A possibility of directly observing ground-state molecular interactions of BChl c in intact chlorosomes, and hence of testing the validities of these models, was offered by resonance Raman (RR) spectroscopy. Indeed, this technique can be made selective enough so as to yield vibrational spectra of the photosynthetic pigments in their native environments (Lutz, 1984; Lutz and Robert, 1987). A RR study was thus conducted on chlorosomes of Cb. limicola and of Cf. aurantiacus, as well as on isolated BChl c and bacteriopheophytin c (BPheo c) in model systems. A conclusion of this study is that none of the yet published models agrees with the actual state of BChl c in chlorosomes, which consists of oligomers involving keto carbonyl-magnesium bonds. A preliminary account of this study was given at the 5th International Symposium of Photosynthetic Prokaryotes (Lutz and Robert, 1985).

MATERIALS AND METHODS

Chlorosomes were prepared from Cb. limicola f. thiosulfatophilum 6230 (Tassajara), according to the procedure of Gerola and Olson (1986). The major, "high density" fraction absorbing at 730-745 nm was studied. The method of Feick and Fuller (1984) was followed in preparing chlorosomes from Cf. aurantiacus OK-70-fl.

BChl c was extracted from wet cells of the above strain of Cb. limicola, using methanol/diethyl ether/petroleum ether (5:2:1) and was chromatographed on polyethylene powder by HPLC, using an acetone gradient (66-80% in water). BChl c was washed with methanol and dried, then washed with CCl$_4$ and dried 3 times. The preparation was a mixture of four homologs of farnesyl-BChl c (Fig. 1), in which 4-ethyl-5-ethyl BChl c and 4-n-propyl-5-ethyl BChl c should have predominated over 4-ethyl-5-methyl BChl c and 4-isobutyl-5-ethyl BChl c (Olson and Pedersen, 1988). Final dehydration of the samples used for RR spectroscopy was achieved by repeated dissolution in CCl$_4$ dried on CaH$_2$ followed by slow pumping down to 10^{-3} Pa. BPheo c was prepared by acidification of BChl c solutions.

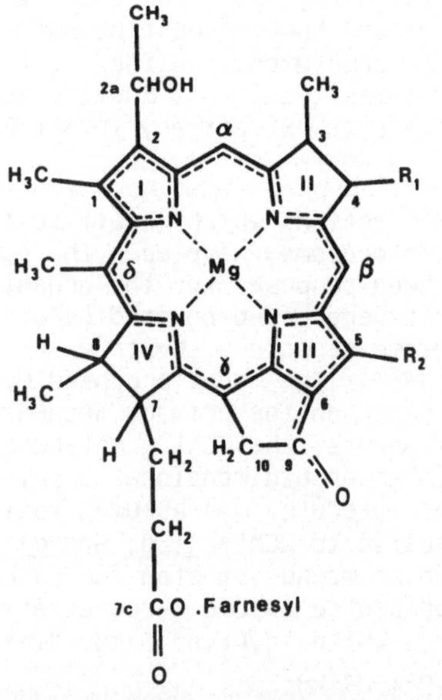

Fig. 1. Chemical structures of BChl c. R$_1$ = ethyl, n-propyl, or isobutyl. R$_2$ = methyl or ethyl. Intermolecular binding of BChl c may involve the magnesium atom, the C-2a hydroxyethyl group, the C-7c ester group and the C-9 keto carbonyl group.

Resonance Raman spectra were obtained on a Jobin-Yvon HG2S monochannel spectrometer, using an average 7 cm⁻¹ spectral resolution. The excitation lines were the 441.6 nm emission from a He-Cd laser (Liconix) and the 454.5 and 457.9 nm emissions from an Argon laser (Spectra-Physics 171.05). Signal to noise ratios were improved by summation of 2-12 scans in a multichannel analyzer (Tracor Nothern 1710). The samples were studied at ca. 20 K, using a He-circulating cryostat (SMC).

RESULTS AND DISCUSSION

Resonance Raman Spectra of BChl c

Raman spectra of BChl c were readily obtained, at Soret resonance conditions, from both model systems and from chlorosomes. Indeed RR spectra of the latter excited at 441.6 nm essentially arose from BChl c, with some contributions from carotenoids, but there were no sizable contributions for any other chlorosome constituents, including protein (Fig. 3).

As those of other chlorophylls, Raman spectra of BChl c obtained at resonance with the Soret and visible transitions should essentially arise from the π-conjugated parts of the phorbin macrocycle and from its conjugated substituents. As both are chlorins, and as both involve the C-9 keto carbonyl as their only conjugated substituent (Lutz, 1984), BChl c and

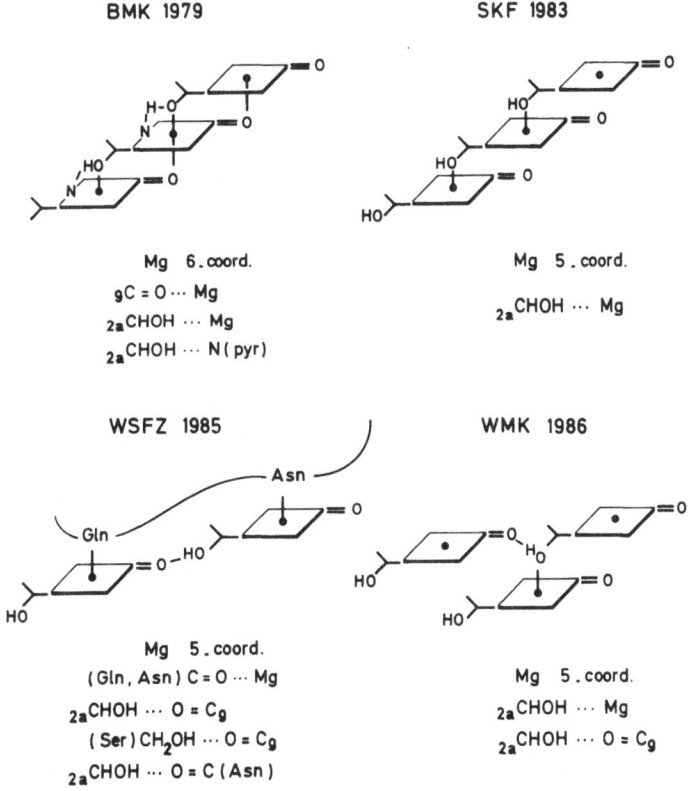

Fig. 2. Some models proposed for ground-state interactions of BChl c in chlorosomes. BMK: Bystrova et al., 1979. SKF: Smith et al., 1983. WSFZ: Wechsler et al., 1985. WMK: Worcester et al., 1986.

Chl <u>a</u> should thus yield very similar spectra, differing only to "second order", due to vibrational coupling of their different, non-conjugated substituents with the conjugated parts of the molecules. As a matter of fact, RR spectra of BChl <u>c</u> are very similar to those of Chl <u>a</u> and of pyrochlorophyll <u>a</u> (BChl <u>c</u>, lacking the carbomethoxy group at C-10 <u>a</u>, is a pyro derivative). In particular, most band frequencies match within 5 cm⁻¹ for these three derivatives indicating that the normal modes assignments performed for the <u>a</u>-type derivatives can largely be transposed to BChl <u>c</u> (Lutz, 1974, 1979, 1984).

These observations also explain why the chemical heterogeneity of BChl <u>c</u>, which concerns non-conjugated groups at C-4 and C-5, may only result in limited increases in band multiplicities. Indeed, only a very few features of RR spectra of BChl <u>c</u>, such as bands at 1483, 1465 and 1414 cm⁻¹, are lacking in those of the <u>a</u>-type derivatives and might originate in this heterogeneity.

For the same reasons again, the hydroxyethyl group at position C-2a may hardly contribute in RR spectra of BChl <u>c</u>. Comparisons between Raman spectra of phenyl-l-ethanol and of ethyl-benzene indicate that only BChl <u>c</u> bands at 1414, 1093 and 930-945 cm⁻¹ might involve sizable contributions from internal coordinates of the hydroxyethyl group.

A more detailed discussion of RR spectra of BChl <u>c</u> and BPheo <u>c</u> will be published elsewhere. For the present purpose of diagnosing molecular interaction states of BChl <u>c</u>, the following RR-active modes have been used: (i) the stretching mode of the C-9 keto carbonyl, which gives rise to a

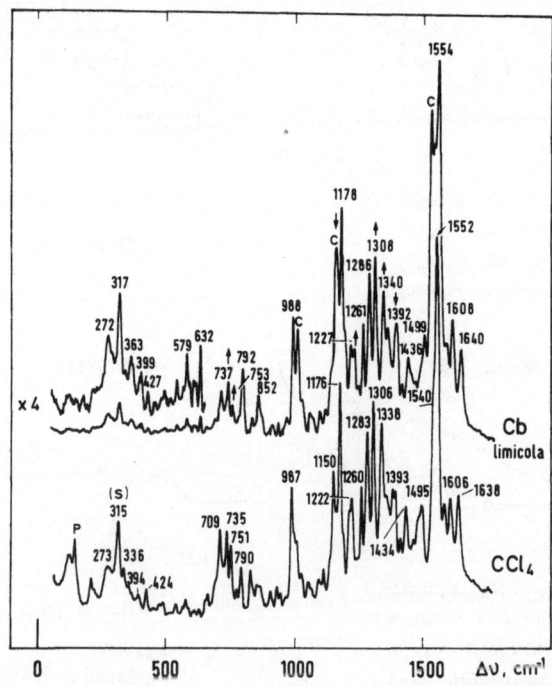

Fig. 3. Resonance Raman spectra of BChl <u>c</u> in chlorosomes of <u>Cb. limicola</u>, 441.6 nm excitation (upper trace) and in 1 mM solution in CaH₂-dried CCl₄, 457.9 nm excitation (lower trace). Sample temperatures 20 K, average spectral resolution 7 cm⁻¹. C: carotenoid bands. S: solvent contribution.

medium intensity band in the 1630-1690 cm⁻¹ range, (ii) a 1595-1615 cm⁻¹ mode, largely arising from stretching of methine bridges, and known to be sensitive to the coordination number of the Mg atom, and (iii) a 1540-1555 cm⁻¹ mode, similarly sensitive to Mg coordination, and involving stretching of C_bC_b and C_aC_m bonds.

Environmental Heterogeneity of BChl c in Chlorosomes

RR spectra of BChl c in Cb. limicola chlorosomes indicate that most if not all of the C-9 carbonyls of these molecules share identical interaction states, i.e., are most likely to each interact with chemically identical sites. Indeed, although the stretching modes of these groups are downshifted by about 50 cm⁻¹ from the wavenumber of the free vibrator, indicating a rather strong interaction, they give rise to a single, sharp band at 1639-1640 cm⁻¹ (Fig. 4). The halfwidth of this band is remarkably small (14 cm⁻¹), being not higher than those of many interaction-insensitive RR bands of the same spectrum. This situation is strikingly different from those previously observed in bacterial or plant antenna complexes, in which several environmental populations can be distinguished by their differing carbonyl stretching frequencies, which usually span a 50-60 cm⁻¹ range (Fig. 4). The numbers of these RR-distinguishable populations generally are close, if not identical, to the chlorophyll/protein stoichiometries in these complexes (Lutz et al., 1982; Robert and Lutz, 1985). The BChl c/polypeptide ratio being close to 7 in chlorosomes of Cb. limicola, the above RR observations suggest that the state of BChl c may differ qualitatively from that of other chlorophylls in other antenna complexes (see below).

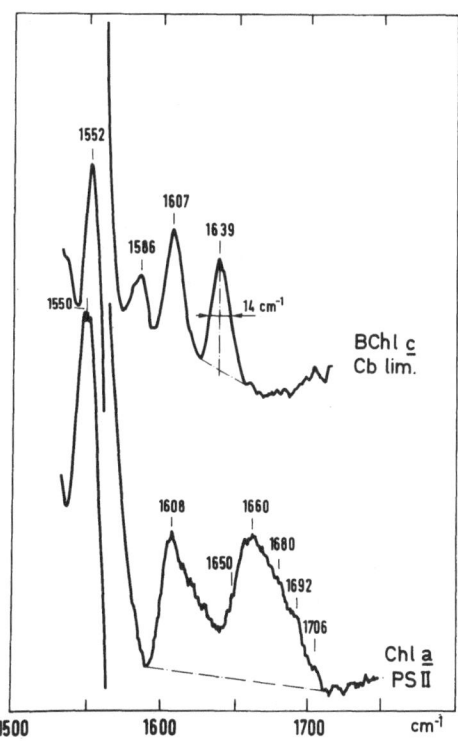

Fig. 4. Higher frequency regions (1500-1750 cm⁻¹) of RR spectra of BChl c in chlorosomes of Cb. limicola (upper trace) and Chl a in PSII particles from Chlamydomonas (lower trace). Samples at 20 K, excitation 441.6 nm, spectral resolution 6 cm⁻¹.

The C-9 keto stretching band of BChl c in Cb. limicola involves a very weak shoulder at ca. 1650 cm⁻¹ (Fig. 4). This suggests that, in these chlorosomes, a small proportion of BChl c molecules (probably less than 1:10) might assume a different interaction state.

A similar situation is found in chlorosomes of Cf. aurantiacus, in which a majority of BChl c keto carbonyls vibrate at 1641-1642 cm⁻¹, hence probably assuming the same liganding as in Cb. limicola (Fig. 5). A more pronounced shoulder at 1656 cm⁻¹ indicates that a higher proportion of BChl c molecules might assume another interaction state.

The RR characterisation of the interaction state common to most of the BChl c in Cb. limicola and in Cf. aurantiacus has been carried out with RR spectra of chlorosomes of the former.

Coordination States of BChl c Magnesium

The frequencies of the ca. 1600 cm⁻¹, C_aC_m stretching mode of BChl c in various solvents occur within two discrete, narrow frequency ranges, 1590-1595 and 1607-1611 cm⁻¹ (Fig. 6). Solutions of BChl c at 20K in pyridine and tetrahydrofuran yield bands in the former range, while solutions in CCl₄, CH₂Cl₂, benzyl alcohol, phenyl-l-ethanol, isopropanol and acetone yield bands in the latter range. RR spectra of solutions in the two latter solvents may contain additional, minor components at 1592 cm⁻¹ (Fig. 6).

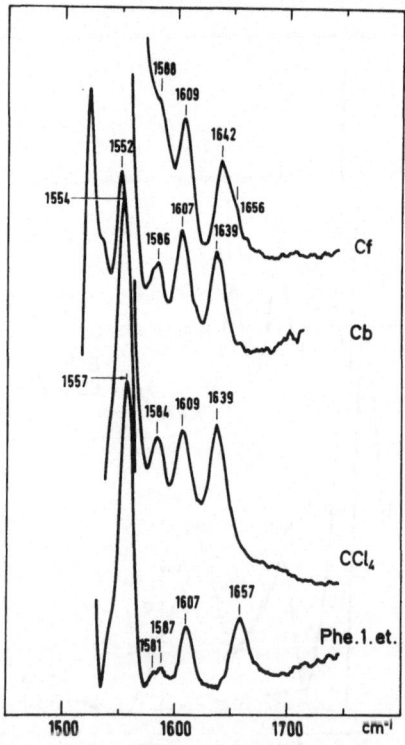

Fig. 5. Higher frequency regions (1500-1750 cm⁻¹) of RR spectra of BChl c at 20 K in, from top to bottom, chlorosomes of Cf. aurantiacus, chlorosomes of Cb. limicola, 1 mM solution in dry CCl₄, solution in phenyl-l-ethanol. Excitations at 441.6 nm, except CCl₄ solution, 457.9 nm.

This behavior of the ca. 1600 cm⁻¹ RR band of BChl c is the same as
that of the homologous bands of various chlorophylls and bacteriochlorophylls. It has been shown to originate in the sensitivity of this ν C$_a$C$_m$
mode to the coordination number of the magnesium atom, the lower and higher
frequency ranges corresponding to 6- and 5-coordinated states, respectively
(Cotton and Van Duyne, 1981; Robert and Lutz, 1985; Fujiwara and Tasumi,
1986; Lutz and Robert, 1987; Tasumi and Fujiwara, 1987). The solvent-
dependence of the ν C$_a$C$_m$ mode being qualitatively the same for BChl c as for
other chlorophylls, it can be safely concluded that for this molecule also a
1590-1595 cm⁻¹ frequency indicates a 6-coordinated state, and a 1607-1611
cm⁻¹ frequency a 5-coordinated state. RR spectra of BChl c in Cb. limicola
yield a single, sharp (12 cm⁻¹ halfwidth) band at 1607 cm⁻¹, which
indicates a 5-coordinated state. A deep trough occurs between this band and
the 1580-1585 cm⁻¹ skeletal band, leaving no possibility for any sizable
6-coordinated population of BChl c in the chlorosome (Fig. 6).

Fig. 5 shows that the ca. 1540-1555 cm⁻¹ band is also sensitive to
magnesium coordination, as observed for Chl a by Tasumi and Fujiwara (1986).
The 1552 cm⁻¹ frequency of BChl c in Cb. limicola confirms 5-coordination of
the magnesiums.

Both the 1552 and 1607 cm⁻¹ values observed for BChl c in chlorosomes
fall at the lower frequency limits of the "5-coordination ranges" for these
two bands. This fact is not to be taken as indicative of any intermediate
state of the magnesiums, e.g., strongly asymmetric 6-coordination. Indeed,

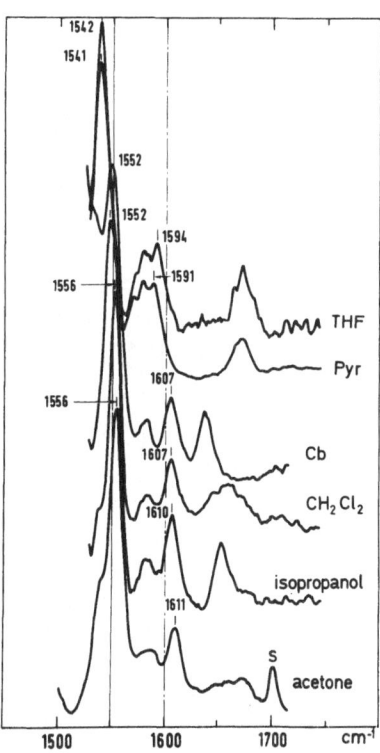

Fig. 6. Higher frequency regions (1500-1750 cm⁻¹) of RR spectra of BChl c
in (from top to bottom), tetrahydrofuran, pyridine, chlorosomes of
Cb. limicola, CH₂Cl₂, isopropanol and acetone. Samples at 20 K,
excitation 441.6 nm, spectral resolution 6 cm⁻¹.

the frequencies of these modes are also sensitive to the dielectric permittivity of the chlorophyll environment (Lutz et al., 1982), being lower for lower permittivity media (Fig. 6). Hence, in so far as the frequencies observed for BChl $\underline{c}$ in chlorosomes are also those observed in CH_2Cl_2 and CCl_4 (Fig. 6, Fig. 7), their relatively lower values indicate that the chlorosome environment of BChl $\underline{c}$ has a low dielectric constant, as expected.

The ca. 1550 and 1610 cm^{-1} bands of chlorophylls are most probably sensitive to magnesium coordination through differences in constraints on the tetrapyrrolic ring resulting from the in-plane or out-of-plane positions of 6- or 5-coordinated magnesium, respectively (Cotton and Van Duyne, 1982; Lutz and Robert, 1985). As shown further by Tasumi and Fujiwara (1987), monotonic, almost linear laws relate the frequencies of the coordination-sensitive RR bands to the nitrogen-macroring center (C_t) distance. Hence, asymmetric 6-coordination of BChl $\underline{c}$ in chlorosomes as proposed by Krasnovskii's group (Bystrove et al., 1979) and by Blankenship's group (Blankenship et al., 1988), should result in intermediate C_t-N distances, which should result in intermediate frequencies of the ca. 1550 and 1610 cm^{-1} bands. This is actually not observed in chlorosome spectra (Fig. 6).

We thus conclude, on the basis of the present Raman spectra, that in chlorosomes of $\underline{Cb.\ limicola}$, each magnesium of the BChl $\underline{c}$ molecules is involved in a single intermolecular bond. Probably this conclusion also holds for chlorosomes of $\underline{Cf.\ aurantiacus}$ (Fig. 5).

Liganding of the Keto Carbonyls of BChl c in Chlorosomes

When free from intermolecular binding, the keto carbonyl group of BChl

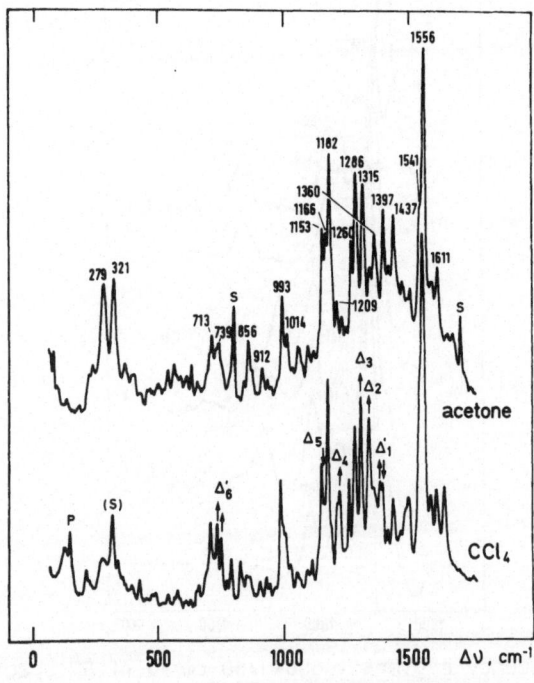

Fig. 7. RR spectra of BChl $\underline{c}$ at 20 K in acetone solution, 441.6 nm excitation (upper trace) and in dry CCl_4 solution, 457.9 excitation (lower trace). $\Delta_1,\ \ldots\ \Delta_6$: see text. S: solvent contributions.

$\underline{c}$ vibrates at 1685-1690 cm^{-1}. Such frequencies are indeed observed for monomeric BChl $\underline{c}$ in pyridine and tetrahydrofuran (Fig. 6), as well as for BPheo $\underline{c}$ (not shown).

These values are somewhat smaller than those observed for the C-9 carbonyls of other chlorophylls, including Chl $\underline{a}$, which occur around 1700 cm^{-1}. Since pyrochlorophyll $\underline{a}$ also exhibits a rather low ν C=O frequency at 1690 cm^{-1} (Lutz, 1979, 1984), we ascribe this fact to the absence of the carbomethoxy group at C-10a in both BChl $\underline{c}$ and pyroChl $\underline{a}$.

The 1639 cm^{-1} frequency observed for BChl $\underline{c}$ in chlorosomes of $\underline{Cb.}$ $\underline{limicola}$ thus corresponds to a ca. 50 cm^{-1} downshift, indicating a rather strong intermolecular interaction. This downshift may be compared with those induced by interaction of the C-9 carbonyl of BChl $\underline{c}$ with alcohols and with the magnesiums of other BChl $\underline{c}$, inasmuch as these two types of ligands have been involved in previous models of BChl $\underline{c}$ state in chlorosomes (Fig. 2).

Figs. 5 and 6 show that when H-bonded to the hydroxyl groups of the secondary alcohols isopropanol and phenyl-l-ethanol, the C-9 carbonyl of BChl $\underline{c}$ vibrates at 1655 and 1657 cm^{-1}, respectively. These frequencies are thus 16-18 cm^{-1} higher than that observed in chlorosomes. This constitutes strong evidence against involvement of the hydroxyethyl at C-2a in molecular interactions of BChl $\underline{c}$ $\underline{in\ vivo}$, particularly if one considers that phenyl-l-ethanol constitutes a reasonable model for this function.

On the other hand, concentrated (> 1 mM), very dry solutions of BChl $\underline{c}$ in CCl$_4$ yielded a C-9 keto stretching mode occurring at almost exactly the same frequency as in chlorosomes (1638-1640 cm^{-1}, Fig. 5).

The BChl $\underline{c}$ population corresponding to this ν C=O frequency most probably has a Soret band slightly more to the red than those of two other, minor populations also present, the keto carbonyls of which vibrated at 1665 and 1685 cm^{-1}. Indeed, excitation of such preparations at 454.5 and 457.9 nm rather than at 441-6 nm selectively enhanced the contribution of the 1639 cm^{-1} vibrating species (Figs. 3, 5, 7). As the Mg atoms of this predominant species clearly are 5-coordinated (bands at 1554 and 1607 cm^{-1}, Fig. 5), and as the 1639 cm^{-1} frequency is too low to correspond to C-9 carbonyls H-bonded to C-2a hydroxyethyl groups of adjacent BChl $\underline{c}$, we identify this species with (BChl $\underline{c}$)$_n$ oligomers involving $_9$C=O$\cdots$Mg bonds only.

This assignment is strongly supported by comparisons of relative intensity changes induced in RR spectra by formation of these (BChl $\underline{c}$)$_n$ oligomers and by formation of (Chl $\underline{a}$)$_n$ and of (pyroChl $\underline{a}$)$_n$ oligomers from the monomers. RR spectra of BChl $\underline{c}$ solutions in CCl$_4$ excited at 457.9 nm differ from those of monomeric BChl $\underline{c}$ in acetone or in phenyl-l-ethanol by their relative band intensities in the skeletal "finger-print", 700-1500 cm^{-1} region. The most conspicuous of these have been labelled as Δ_1 ... Δ_6 in Fig. 7. This set of variations concerns the same bands as those previously shown to be sensitive to oligomer formation in RR spectra of Chl $\underline{a}$ and of pyroChl $\underline{a}$ (Lutz, 1979, 1984). In particular, for these three molecules, oligomer formation results in decreases of the 1340/1355 and 1160/1150 cm^{-1} band intensity ratios (Δ_2 and Δ_5), and in increases of the 1225/1210 and 753/705 cm^{-1} band intensity ratios (Δ_4 and Δ_6), respectively (Fig. 14 in Lutz, 1984) (Fig. 7). As such variations are very characteristic of oligomer structure (Lutz, 1984), and as (Chl $\underline{a}$)$_n$ and (pyroChl $\underline{a}$)$_n$ oligomers are known to involve $_9$C=O$\cdots$Mg intermolecular bands only (Katz et al., 1978; Lutz, 1984), the same structure also must be involved in the present (BChl $\underline{c}$)$_n$ oligomers.

This predominant involvement of $_9$C=O$\cdots$Mg bonds in (BChl $\underline{c}$)$_n$ oligomers formed in very dry nonpolar solvents is not unexpected. A large body of

evidence has indeed shown that, due to its conjugation with the phorbin
ring, the C-9 carbonyl of chlorophylls constitutes a better electron donor
to their magnesiums than many other Lewis bases. Moreover, a larger free
energy of $_9C=0\cdots Mg$ bond formation is observed for the pyro-derivatives,
which, like BChl c, lack the carbomethoxy group at C-10a (Katz et al.,
1978).

Fig. 3 shows that RR spectra of BChl c in Cb. limicola differ from
those of monomeric BChl c by Δ_1 ... Δ_6 features which are very similar (Δ_1)
or identical (Δ_2 - Δ_6) to those characterizing the formation of chlorophyll
oligomers via $_9C=0\cdots Mg$ bonds.

From this observation, and from the above discussion of C-9 carbonyl
stretching frequencies, we conclude that in chlorosomes of Cb. limicola,
BChl c occurs as oligomers involving a single, C-9 keto carbonyl ligand on
each magnesium atom (Fig. 8). Inasmuch as symmetrical chlorophyll dimers
involving two $_9C=0\cdots Mg$ bonds are unlikely to occur (Katz et al., 1978), the
necessarily small proportions of C-9 carbonyls vibrating, in the chlorosome,
at frequencies other than 1639 cm^{-1} indicate that BChl c should be arranged
as rather long (n > 10 ?), monodimensional oligomers. Shorter oligomers
might be involved in Cf. aurantiacus chlorosomes.

Conclusion: State of BChl c in Chlorosomes

The present RR data show that, in chlorosomes of Cb. limicola, most of
the BChl c molecules assume the same ground-state interactions, involving
5-coordinated magnesiums and keto C=0$\cdots$Mg bonding between adjacent
molecules, thus forming linear oligomers (Fig. 7). Chlorosomes thus
constitute the first known example of a photosynthetic system in which
relative orientations and distances between pigments should be largely
determined by pigment-pigment bonding, rather than by bonding of individual
pigments to a protein core (Lutz, 1977, 1984; Robert and Lutz, 1985, 1986).

Along those linear oligomers of BChl c, both the C-2a hydroxyethyl and
C-7c ester groups remain available for additional molecular interactions,
e.g., H-bonding with the chlorosome protein. In particular, chlorosome
proteins of Chlorobiaceae as well as of Chloroflexaceae appear to contain
Glu, Gln, Asp and Asn residues, which could interact with these BChl c
groupings (Wechsler et al., 1985; Zuber, 1988). Although possibly not
involving all of the groups available along BChl c oligomers, such
interactions might result in preferential orientation of (BChl c)$_n$ chains
along the axes of α-helices which are likely to be formed by chlorosome
proteins (Wechsler et al., 1985), thus explaining the observed preferential
orientations of the Q$_y$ transitions (van Dorssen et al., 1987a). Current
computer modelling should help in understanding the relationships between

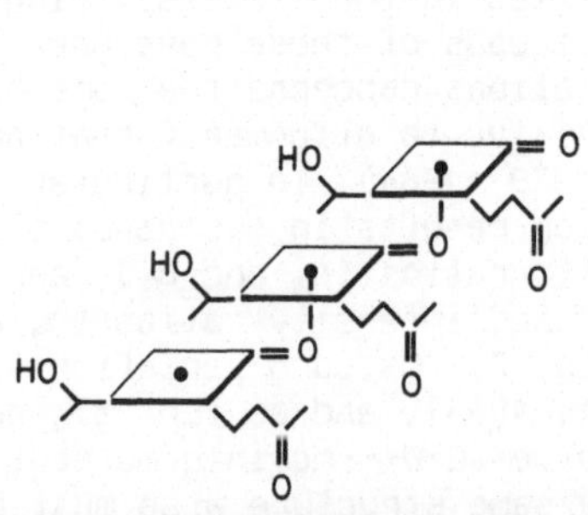

Fig. 8. Schematic model for the ground-state interactions assumed by BChl c
in chlorosomes of Cb. limicola and, likely, of Cf. aurantiacus. The
C-2a hydroxyethyl and C-7c ester groups remain available for further
interactions with the protein (see text).

chlorosome proteins and BChl c oligomers (Bréhamet, L., Gilquin, B. and Lutz, M., 1987, unpublished).

ACKNOWLEDGEMENTS

    Multiple and fruitful interactions with J.M. Olson are gratefully acknowledged. We are indebted to A. Vermeglio for preparations of chlorosomes of Chloroflexus, as well as to J. Kléo and J.P. Pedersen for preparation and purification of BChl c and BPheo c. We wish to thank R. E. Blankenship and M. Tasumi for preprints of work in press. This study was supported in part by an EMBO grant to G.v.B.

REFERENCES

Betti, J. A., Blankenship, R. E., Natarajan, L. V., Dickinson, L. C., and Fuller, R. C., 1982, Antenna organization and evidence for the function of a new antenna pigment species in the green photosynthetic bacterium Chloroflexus aurantiacus, Biochim. Biophys. Acta, 680:194.

Blankenship, R. E., Brune, D. C., Freeman, J. M., King, G. H., McManus, J. H., Nozawa, T., Trost, J. T., and Wittmershaus, B. P., 1988, Energy trapping and electron transfer in Chloroflexus aurantiacus, This volume.

Bystrova, M. I., Malgosheva, I. N., and Krasnovskii, A. A., 1979, Study of molecular mechanism of self-assembly of aggregated forms of bacteriochlorophyll c, Mol. Biol. USSR, 13:582 (Engl. trans.: 440).

Cotton, T. M., and Van Duyne, R. P., 1981, Characterization of bacteriochlorophyll interactions in vitro by resonance Raman spectroscopy, J. Am. Chem. Soc., 103:6020.

Feick, R. G., and Fuller, R. C., 1984, Topography of the photosynthetic apparatus of Chloroflexus aurantiacus, Biochem., 23:3693.

Fujiwara, M., and Tasumi, M., 1986, Resonance Raman and infrared studies on axial coordination to chlorophylls a and b in vitro, J. Phys. Chem., 90:250.

Gerola, P. D., and Olson, J. M., 1986, A new bacteriochlorophyll a-protein complex associated with chlorosomes of green sulfur bacteria, Biochim. Biophys. Acta, 848:69.

Gerola, P. D., Højrup, P., Knudsen, J., Roepstorff, P., and Olson, J. M., 1988, The bacteriochlorophyll c-binding protein from chlorosomes of Chlorobium limicola f. thiosulfatophilum, This volume.

Katz, J. J., Shipman, L. L., Cotton, T. M., and Janson, T. R., 1978, Chlorophyll aggregation: coordination interactions in chlorophyll monomers, dimers, and oligomers, in: "The Porphyrins, Vol. 5," D. Dolphin, ed., Academic Press, New York.

Lutz, M., 1974, Resonance Raman spectra of chlorophyll in solution, J. Raman Spectrosc., 2:497.

Lutz, M., 1977, Antenna chlorophyll in photosynthetic membranes: a study by resonance Raman spectroscopy, Biochim. Biophys. Acta, 460:408.

Lutz, M., 1979, Diffusion Raman de resonance des chlorophylles: application a l'étude de l'organisation de la membrane photosynthetique, Thèse de Doctorat d'Etat, Université Pierre et Marie Curie, Paris, France.

Lutz, M., 1984, Resonance Raman studies in photosynthesis, Adv. Infrared and Raman Spectrosc., 11:211.

Lutz, M., and Robert, B., 1985, Local environments of bacteriochlorophylls in antenna structures of Rhodospirillales and of Chlorobiales, Abstracts, Fifth Int. Symp. on Photosynthetic Prokaryotes, Grindelwald, Switzerland, 27-28 Sept. 1985.

Lutz, M., and Robert, B., 1987, Chlorophylls and the photosynthetic membrane, in: "Biological Applications of Raman Spectroscopy, Vol. 3," T. G. Spiro, ed., Wiley-Heyden, New York (in press).

Lutz, M., Hoff, A. J., and Bréhamet, L., 1982, Bacteriochlorophyll a-protein interactions in a complex from <u>Prosthecochloris aestuarii</u>: a resonance Raman study, <u>Biochim. Biophys. Acta</u>, 679:631.

Olson, J. M., 1980, Chlorophyll organization in green photosynthetic bacteria, <u>Biochim. Biophys. Acta</u>, 594:33.

Olson, J. M., and Pedersen, J. P., 1988, Bacteriochlorophyll <u>c</u> aggregates in carbon tetrachloride as models for chlorophyll organization in green photosynthetic bacteria, <u>in</u>: "Organization and Function of Photosynthetic Antennas," H. Scheer and S. Scheider, eds., Walter de Gruyter, Berlin (in press).

Robert, B., and Lutz, M., 1985, Structures of antenna complexes of several <u>Rhodospirillales</u> from their resonance Raman spectra, <u>Biochim. Biophys. Acta</u>, 807:10.

Robert, B., and Lutz, M., 1986, Structure of the primary donor of <u>Rhodopseudomnas sphaeroides</u>: difference resonance Raman spectroscopy of reaction centers, <u>Biochemistry</u>, 25:2303.

Schmitz, R., 1967, Über die Zusammensetzung der pigmenthaltigen Strukturen aus Prokaryoten. II. Undersuchungen an Chromatophoren von <u>Chlorobium thiosulfatophilum</u>, Stamm Tassajara, <u>Arch. Microbiol.</u>, 56:238.

Smith, K. M., Kehres, L. A., and Fajer, J., 1983, Aggregation of the bacteriochlorophylls <u>c</u>, <u>d</u>, and <u>e</u>. Models for the antenna chlorophylls of green and brown photosynthetic bacteria, <u>J. Am. Chem. Soc.</u>, 105:1387.

Tasumi, M., and Fujiwara, M., 1987, Vibrational spectra of chlorophylls, <u>Adv. Infrared and Raman Spectrosc.</u>, 14 (in press).

van Dorssen, R. J., Vasmel, H., and Amesz, J., 1986a, Pigment organization and energy transfer in the green photosynthetic bacterium <u>Chloroflexus aurantiacus</u>. II. The chlorosomes, <u>Photosynth. Res.</u>, 9:33.

van Dorssen, R. J., Gerola, P. D., Olson, J. M., and Amesz, J., 1986b, Optical and structural properties of chlorosomes of the photosynthetic green sulfur bacterium <u>Chlorobium limicola</u>, <u>Biochim. Biophys. Acta</u>, 848:77.

Wechsler, T., Suter, F., Fuller, R. C., and Zuber, H., 1985, The complete amino acid sequence of the bacteriochlorophyll <u>c</u> binding polypeptide from chlorosomes of the green photosynthetic bacterium <u>Chloroflexus aurantiacus</u>, <u>FEBS Lett.</u>, 181:173.

Worcester, D. L., Michalski, T. J., and Katz, J. J., 1986, Small-angle neutron scattering studies of chlorophyll micelles: models for bacterial antenna chlorophyll, <u>Proc. Natl. Acad. Sci. U.S.A.</u>, 83:3791.

Zuber, H., 1988, Structural studies on the antenna complexes and polypeptides of <u>Chloroflexus aurantiacus</u> and other green photosynthetic bacteria, This volume.

GROWTH RATE AND THE CONTROL OF DEVELOPMENT OF THE PHOTOSYNTHETIC APPARATUS

IN CHLOROFLEXUS AURANTIACUS AS STUDIED ON THE BASIS OF CYTOPLASMIC MEMBRANE

STRUCTURE AND CHLOROSOME SIZE

J. Oelze and J. R. Golecki

Institut für Biologie II (Mikrobiologie)
Universität Freiburg, Schänzlestr. 1
D-7800 Freiburg, F.R.G.

INTRODUCTION

Chloroflexus aurantiacus synthesizes two different kinds of bacteriochlorophyll, BChl a and BChl c (Pierson and Castenholz, 1974). Both pigments are, however, not only allotted to cytologically different structures, but also fulfil different functions in the photosynthetic apparatus of Cf. aurantiacus (Fuller and Redlinger, 1985). While BChl a-protein complexes constitute light-harvesting and photochemical reaction-center units located in the cytoplasmic membrane, BChl c-protein complexes harvest light in bag-like structures, the chlorosomes, which are attached to the cytoplasmic membrane. Chlorosomes are linked to the cytoplasmic membrane by a baseplate which contains a third BChl a-protein complex isolated at a constant low proportion together with chlorosomes (Feick et al., 1982).

Since the early investigations on the formation of BChl a and BChl c complexes it has been known that light and oxygen are involved in the control of both holochromes and, consequently, in the control of the photosynthetic apparatus in Cf. aurantiacus (Pierson and Castenholz, 1974; Schmidt et al., 1980; Fuller and Redlinger, 1985). But an exact definition of developmental stages of the photosynthetic apparatus on the basis of BChl a and BChl c determinations has been hampered by two facts: First, it is almost impossible to grow Cf. aurantiacus reproducibly in batch culture (Schmidt, 1980) and, second, the development of the photosynthetic apparatus is highly asynchronous within individual cells of batch cultures (Sprague et al., 1981). This may be the reason for the as yet unresolved question as to whether a change in the BChl c content of cells represent a change in the cellular chlorosome number or a change of chlorosome size or both (Pierson and Castenholz, 1974; Schmidt et al., 1980; Belti et al., 1982; Feick et al., 1982).

Thus, in order to study different developmental stages of the photosynthetic apparatus on a structural and functional basis, it is necessary to grow Cf. aurantiacus not only reproducibly but also under conditions which facilitate the formation of homogeneous cell populations. It will be shown in this paper that these criteria can be met when Cf. aurantiacus is grown in a chemostat. Continuous cultivation in a chemostat is based on the limitation of growth by, for example, the supply of the substrate. This opens the possibility to study whether, as suggested by Pierson and Castenholz (1974), growth rate is a third factor involved in the control of development of the photosynthetic apparatus in Cf. aurantiacus after oxygen and light.

MATERIALS AND METHODS

Organism and Growth Conditions

Chloroflexus aurantiacus strain J-10-Fl was grown in medium D with 0.1% (w/v) yeast extract, 0.25% casamino acids, and 0.08% glycylglycine at pH 8.3 (Pierson and Castenholz, 1974). Cultures for the inoculation of chemostats were grown anaerobically in screw-capped bottles at 55°C in the light. For continuous cultivation jacketed culture vessels (1 l, Wheaton-Celstir) were used (for further details, see Oelze and Fuller, 1987). Unless stated otherwise, the cultures were stirred at 100 rpm and irradiated with 240 W m^{-2} of white light measured at the surface of the culture vessel. All measurements were performed with steady-state cultures.

Chemical Analyses

BChl a and BChl c were determined after extraction of sedimented cell samples with acetone/methanol (7:2 v/v). Molar absorption coefficients were ε_{767} = 76 $mM^{-1}cm^{-1}$ for BChl a (Clayton, 1966) and ε_{666} = 74 $mM^{-1}cm^{-1}$ for BChl c (Feick et al., 1982). Protein was determined according to Lowry et al. (1951).

Electron Microscopic Analyses

Samples for freeze-fracture experiments were frozen without glycerol-impregnation in liquid propane after glutaraldehyde fixation at the growth temperature (Staehelin et al., 1978).

Freeze-fracture replicas were prepared in a Balzers UHV-freeze-etch unit BAF 500 under a vacuum of between 2 x 10^{-4} and 2 x 10^{-7} Pa and temperatures between -110°C and -260°C. Electron microscopic examinations and morphometric measurements were performed as described (Golecki et al., 1980). Preparation of ultra-thin sections of samples was carried out with standard methods (Staehelin et al., 1978).

Total membrane areas, as well as number and sizes of chlorosome baseplates, were measured on freeze-fractured cells. Ultra-thin-sectioned and freeze-fractured cells were used to determine the height, width and length of chlorosomes. On the basis of these parameters the volumes of chlorosomes were calculated (Golecki and Oelze, 1987). Comparable measurements were performed in order to determine sizes and total membrane areas of cells.

The number of BChl molecules and photochemical reaction centers per chlorosome were calculated on BChl content either per culture volume or per cell protein, as well as on the number of cells per culture volume (Golecki and Oelze, 1987). In order to calculate numbers of BChl a and BChl c molecules, as well as reaction centers per chlorosome, a constant molar ratio of BChl c/BChl a = 25:1 and 20 molecules of BChl a per reaction center were assumed (Feick et al., 1982).

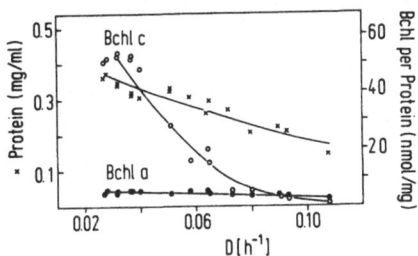

Fig. 1. Steady-state protein concentrations (x), as well as specific
       contents of BChl a (●) and BChl c (o) in continuous chemostat
       cultures of Cf. aurantiacus growing phototrophically at different
       dilution rates (D). The temperature was 52°C and the incident light
       intensity was 240 Wm⁻².

RESULTS AND DISCUSSION

       Chemostat cultures are operated by adding fresh growth medium to a
constant culture volume at the same rate at which culture suspension is
withdrawn (Herbert et al., 1956). Steady states, which become established
under these conditions, are characterized by the fact that the growth rate,
$\mu(h^{-1})$, of a culture is equal to the diluation rate, $D(h^{-1})$. This is because
fresh medium contains the substrate, limiting both the growth rate and the
amount of biomass produced. In the case of Cf. aurantiacus the complex
medium of Pierson and Castenholz (1974), containing yeast extract and
casamino acids, is required in order to obtain a reasonable biomass level
and BChl content in chemostat culture.

       When the dilution rate was increased, the steady-state protein level,
and the specific BChl c and BChl a contents decreased although the
incubation temperature (52°C) and irradiance (240 Wm⁻²) were kept constant
(Fig. 1). Since the decrease of BChl c was much more pronounced than the
decrease of BChl a, the molar ratio BChl c/BChl a decreased as well. These
results suggest that the formation of BChl complexes can be controlled by
the rate at which the limiting substrate is supplied. Since, however, cells
formed the highest BChl levels at the highest densities, one might argue

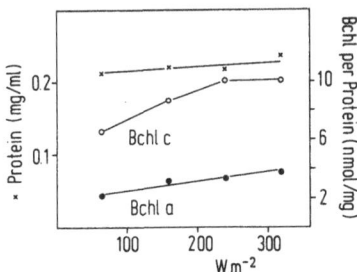

Fig. 2. Steady-state protein levels (x) and specific contents of BChl a (●)
       and BChl c (o) in chemostat cultures of Cf. aurantiacus growing
       phototrophically at a dilution rate of 0.074 h⁻¹, a temperature of
       48°C and different light irradiances as indicated.

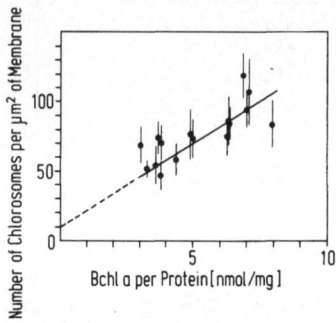

Fig. 3. Chlorosome numbers per μm² of cytoplasmic membrane and specific BChl
a content of Cf. aurantiacus grown in continuous culture as
described in Fig. 1. The connecting line is based on a 99%
confidence level. (From Golecki and Oelze, 1987.)

that changes in the internal light conditions, because of the different
degrees of self-shading, rather than growth rate were responsible for the
observed changes in BChl content.

Previously, we provided evidence supporting the involvement of growth
rate rather than of light in the control of BChl complexes (Oelze and
Fuller, 1987). This was corroborated in the present investigation after
measuring protein, as well as BChl levels of chemostat cultures grown at
constant D (0.074 h⁻¹) and temperature (48°C) but at different irradiances
(Fig. 2). While protein levels stayed largely constant, BChl levels slightly
increased when irradiance was increased. Nevertheless, the molar ratio BChl
c/BChl a remained constant. Thus on the basis of previous results and the
data reported herein it may be concluded that formation of BChl complexes
in Cf. aurantiacus is controlled by a factor represented by the growth
rate.

In order to relate the BChl content of a culture to structural
parameters of the photosynthetic apparatus of individual cells, it is
necessary that all cells of a population are homogeneously equipped with
chlorosomes, which indicate the presence of the photosynthetic apparatus. It
was mentioned before that these conditions cannot be met with batch cultures

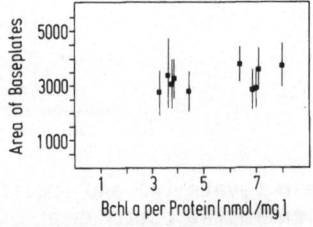

Fig. 4. Specific BChl a content of Cf. aurantiacus and average area (nm²) of
baseplates of chlorosomes. The organisms were grown in continuous
culture as described in Fig. 1. The mean baseplate area was 3.2 ±
0.8 nm². (From Golecki and Oelze, 1987.)

Table 1. Morphometric Measurements of 9 Chloroflexus aurantiacus Cultures with Different BChl Contents. The organisms were grown in continuous culture at constant light conditions (240 W m⁻² of incandescent light) but at different growth rates (Oelze and Fuller, 1987). (From Golecki and Oelze, 1987).

Cul-ture	BChl a (nmol/mg) cell protein	BChl c (nmol/mg) cell protein	No of chlorosomes per 1 μm² cytoplasmic membrane	No of chlorosomes per bacterium	% of covered cytoplasmic membrane	area of baseplate (nm²)	baseplate length (nm)	baseplate width (nm)	chlorosome height (nm)	chlorosome vol. (nm³)	cell length (nm) *1)	number of molecules per chlorosome BChl a *2)	number of molecules per chlorosome BChl c *2)	number of molecules per chlorosome P865 *3)
1	3.2	7.7	51	276	13.9	2,770	96	34	12	29.900	4.2	1,436	3,389	65
2	3.6	3.3	54	309	15.7	3,400	109	40	-	-	4.5	1,513	1,392	72
3	3.8	24.5	70	312	22.0	3,320	113	36	-	-	3.5	1,222	7,850	45
4	4.4	7.8	57	284	14.0	2,813	-	-	-	-	3.8	1,699	3,046	78
5	6.5	50.4	84	343	27.8	3,810	111	41	-	-	2.9	1,546	12,277	52
6	6.8	51.6	119	482	31.5	2,900	103	37	-	-	3.2	1,295	9,736	45
7	7.0	66.1	94	327	25.0	2,940	-	-	-	-	2.7	1,667	15,772	51
8	7.0	56.5	106	412	34.4	3,650	108	38	19	55.500	3.0	1,496	12,011	50
9	7.9	48.5	83	-	27.0	3,765	118	37	18	57,500	-	-	-	-

*1) The diameter was in all cultures 0.4 μm; *2) localized in the chlorosome; *3) localized in the reaction center in the cytoplasmic membrane

(Sprague et al., 1981). However, in the present case of continuous cultivation, inspection of more than one hundred freeze-fractured cells revealed that between 98 and 100% of all cells housed chlorosomes. On the basis of this high degree of homogeneity the following parameters were determined: First, number of chlorosomes per cytoplasmic membrane area and per bacterium; second, size of baseplates; third, volume of chlorosomes.

The data compiled in Table 1 and depicted in Fig. 3 show that an increase of cellular BChl $\underline{a}$ content resulted in a linear increase of chlorosomes per area of cytoplasmic membrane, as well as per cell. The average area of the baseplates, however, was rather constant, irrespective of considerable variations in BChl $\underline{a}$ and BChl $\underline{c}$ content (Fig. 4 and Table 1).

It should be noted that the number of chlorosomes was determined from the number of baseplates, i.e., chlorosome attachment sites at the cytoplasmic membrane. These sites are assumed to represent the BChl $\underline{a}$-containing moiety of the membrane-bound photosynthetic apparatus (Feick and Fuller, 1984; Fuller and Redlinger, 1985). Thus the direct proportionality between the number of baseplates and cellular BChl $\underline{a}$ content suggests that BChl $\underline{a}$ molecules are confined to baseplates, as well as to the membrane area underlying baseplates. This is important information, because it has not been known as yet whether BChl $\underline{a}$ molecules are evenly distributed over the entire cytoplasmic membrane area or whether they are confined to defined areas within the membrane.

Constancy of baseplate areas was interpreted by Feick et al. (1982) to indicate that $\underline{Cf.\ aurantiacus}$ adapted the number rather than the volume of chlorosomes to different conditions. Detailed histograms (Fig. 5) show, however, that lengths of chlorosomes ranged from about 60 to 205 nm while widths were largely constant at 30-40 nm.

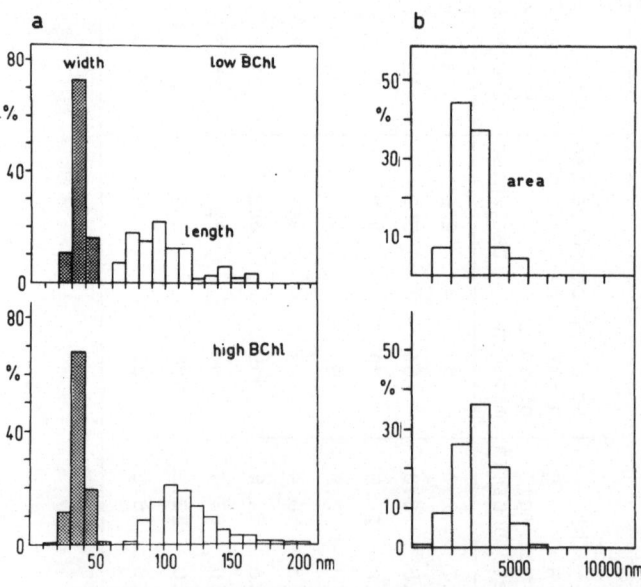

Fig. 5. Size histograms of baseplates from cultures of $\underline{Cf.\ aurantiacus}$ with low (3.8 nmol per mg of protein) and high (7.9 nmol per mg of protein) BChl $\underline{a}$ contents. Baseplate width (dotted columns) and length (open columns) (a); baseplate area (b). The organisms were grown as described in Fig. 1. From Golecki and Oelze, 1987).

Interestingly, cells with low BChl a content exhibited higher numbers of shorter chlorosomes than cells with high BChl a content. Consequently, the baseplate areas were smaller in cells of low BChl a content than those in cells of higher BChl a content. This suggests that not only the numbers but also the areas of baseplates increased when the BChl a content increased. The results support the above conclusion on the localization of BChl a molecules within the areas of the cytoplasmic membrane forming the baseplates.

The data of Fig. 1 demonstrate that specific BChl c content increased almost exponentially and BChl a content linearly when the dilution rate was decreased. Since a relationship between BChl a content and baseplate numbers and areas was suggested by the data discussed above, the question arises whether BChl c is related to the volume of chlorosomes, which exclusively house BChl c. Indeed, an increase in specific BChl c content corresponded to an increase not only in the number but also in the height and thus the volume of chlorosomes (Table 1). However, the number of BChl c molecules per chlorosome showed a higher variation than volume (Table 1). This suggests that the number of BChl c molecules per unit of chlorosome volume, i.e., the packing of BChl c molecules, increased when cellular BChl c content increased.

With the results presented so far the number of BChl a molecules per chlorosome could be calculated. The data of Table 1 demonstrate that 1200-1700 molecules of BChl a (mean value 1484 ± 165) were present per chlorosome. Assuming 20 molecules of BChl a per photochemical reaction center implies that there were 45-78 reaction centers (mean value 58 ± 12) per chlorosome.

SUMMARY

When growing in substrate-limited chemostats, Cf. aurantiacus populations become homogeneously equipped with chlorosomes, which may be taken to represent the photosynthetic apparatus. In such cultures, however, growth rate (i.e., the supply of the limiting substrate) rather than light controls the development of the photosynthetic apparatus. Morphometric measurements on cells of different BChl a and BChl c content led to the suggestion that baseplates represent cytoplasmic membrane areas housing BChl a-protein complexes while BChl c contents correlate with chlorosome volume. In the latter case, however, steep increases of cellular BChl c content can only be explained on the assumption of increased packing of BChl c-protein complexes. Taken together, the results suggest that variations of cellular BChl a are accompanied by variation of number and length of baseplates while variation of BChl c content correlates not only with number and volume of chlorosomes but also with the density of packing of BChl c complexes within chlorosomes.

ACKNOWLEDGEMENT

This investigation was financially supported by the Deutsche Forschungsgemeinschaft (Dr29/31-1F).

REFERENCES

Betti, J. A., Blankenship, R. E., Natarajan, L. V., Dickinson, L. C., and Fuller, R. C., 1982, Antenna organization and evidence for the function of a new antenna pigment species in the green photosynthetic bacterium Chloroflexus aurantiacus, Biochim. Biophys. Acta, 680:194.

Clayton, R. K., 1966, Spectroscopic analysis of bacteriochlorophyll in vitro and in vivo, Photochem. Photobiol., 5:669.

Feick, R. G., and Fuller, R. C., 1984, Topography of the photosynthetic apparatus of Chloroflexus aurantiacus, Biochemistry, 23:3693.

Feick, R. G., Fitzpatrick, M., and Fuller, R. C., 1982, Isolation and characterization of cytoplasmic membranes and chlorosomes from the green bacterium Chloroflexus aurantiacus, J. Bacteriol., 150:905.

Fuller, R. C., and Redlinger, T. E., 1985, Light and oxygen regulation of the development of the photosynthetic apparatus in Chloroflexus, in: "Molecular Biology of the Photosynthetic Apparatus", K. E. Steinbeck, S. Bonitz, C. J. Arntzen, and L. Bogorad, eds., Cold Spring Harbor Laboratory, Cold Spring Harbor, New York.

Golecki, J. R., and Oelze, J., 1987, Quantitative relationship between bacteriochlorophyll content, cytoplasmic membrane structure and chlorosome size in Chloroflexus aurantiacus, Arch. Microbiol., in press.

Golecki, J. R., Schumacher, A., and Drews, G., 1980, The differentiation of the photosynthetic apparatus and the intracytoplasmic membrane in cells of Rhodopseudomonas capsulata upon variation of light intensity, Eur. J. Cell Biol., 23:1.

Herbert, D., Elsworth, R., and Telling, R. C., 1956, The continuous culture of bacteria: a theoretical and experimental study, J. Gen. Microbiol., 14:661.

Lowry, O. H., Rosebrough, N. J., Farr, A. L., and Randall, R. F., 1951, Protein measurement with the Folin phenol reagent, J. Biol. Chem., 193:254.

Oelze, J., and Fuller, R. C., 1987, Growth rate and control of development of the photosynthetic apparatus in Chloroflexus aurantiacus, Arch. Microbiol., 148:132.

Pierson, B. K., and Castenholz, R. W., 1974, Studies of pigments and growth in Chloroflexus aurantiacus, a phototrophic filamentous bacterium, Arch. Microbiol., 100:283.

Schmidt, K., 1980, A comparative study on the composition of chlorosomes (chlorobium vesicles) and cytoplasmic membranes from Chloroflexus aurantiacus strain OK-70-fl and Chlorobium limicola f. thiosulfatophilum strain 6230, Arch. Microbiol., 124:21.

Schmidt, K., Maarzahl, M., and Mayer, F., 1980, Development and pigmentation of chlorosomes in Chloroflexus aurantiacus strain OK-70-fl, Arch. Microbiol., 127:87.

Sprague, S. G., Staehelin, L. A., DiBartolomeis, M. J., and Fuller, R. C., 1981, Isolation and development of chlorosomes in the green bacterium Chloroflexus aurantiacus, J. Bacteriol., 147:1021.

Staehelin, L. A., Golecki, J. R., Fuller, R. C., and Drews, G., 1978, Visualization of the supramolecular architecture of chlorosomes (chlorobium type vesicles) in freeze-fractured cells of Chloroflexus aurantiacus, Arch. Microbiol., 119:269.

THE BACTERIOCHLOROPHYLL C--BINDING PROTEIN FROM CHLOROSOMES OF <u>CHLOROBIUM</u>

<u>LIMICOLA</u> F. <u>THIOSULFATOPHILUM</u>

P.D. Gerola[a], P. Højrup[b], J. Knudsen[c], P. Roepstorff[b], and
J.M. Olson[b]

[a]Department of Ecology, University of Calabria
 I-87030 Arcavacata di Rende (Cosenza), Italy
[b]Institute of Molecular Biology
 Odense University, DK-5230 Odense M, Denmark
[c]Institute of Biochemistry
 Odense University, DK-5230 Odense M, Denmark

INTRODUCTION

In green bacteria the main light-harvesting pigment (BChl c, d or, e)
is located inside oblong bodies (chlorosomes) which lie in the cytoplasm and
adhere to the cytoplasmic membrane, where the reaction centers are
localized.

In green sulfur bacteria the BChl c molecules appear to be associated
with 15-kDa proteinaceous subunits organized hexagonally in 10-nm diameter
rod elements running the length of each chlorosome (Olson, 1980). Excitation
energy transfer is from BChl c to BChl a inside the chlorosome (van Dorssen
et al., 1986), then to BChl a in the two-dimensional crystal of
water-soluble BChl a-protein between the chlorosome and the cytoplasmic
membrane and finally to the BChl a antenna and/or the reaction centers
embedded in the membrane.

When chlorosomes were characterized by polyacrylamide gel
electrophoresis (PAGE), there was only one band observed at approx. 4-5 kDa
(M_r) after Coomassie brilliant blue staining (Gerola and Olson, 1986).
Silver staining revealed the presence of 6 minor bands at about 9, 20, 26,
27, 32 and 38 kDa (M_r). Since the BChl c-binding protein is undoubtedly the
major protein of the chlorosome, it was assumed to be a constituent of the
4-5 kDa band seen after PAGE.

Recently Wechsler et al. (1985) determined the amino acid (aa) sequence
of a BChl c-binding protein from chlorosomes of the green filamentous
bacterium <u>Chloroflexus aurantiacus</u>, which is only distantly related to <u>Cb.
limicola</u> (S_{AB} = 0.2) (Gibson et al., 1985; Woese, 1987). In <u>Chloroflexus</u> the
BChl c molecules have stearyl tails, whereas in <u>Chlorobium</u> the tails are
farnesyl. Furthermore the rod elements in <u>Chloroflexus</u> are only 5.2 nm in
diameter compared to 10 nm in <u>Chlorobium</u>. Nevertheless both kinds of
chlorosomes, in addition to BChl c, contain small amounts of a special BChl
a (792 nm in <u>Chloroflexus</u> and 794 nm in <u>Chlorobium</u>) possibly localized in
the lipid monolayer thought to cover the chlorosome. The BChl c absorption
spectra in the two kinds of chlorosomes are quite similar, but important

differences are observed in the circular dichroism spectra (Betti et al., 1982; Olson et al., 1985). We wanted to compare the primary and secondary structures of the two BChl c-binding proteins in view of the large evolutionary distance between green sulfur bacteria and green filamentous bacteria (Gibson et al., 1985; Woese, 1987), and the apparent similarity between the BChl c antenna systems in these two organisms.

MATERIALS AND METHODS

Chlorosome Preparation

Cb. limicola f. thiosulfatophilum 6230 (Tassajara) was grown, and cells were broken as described previously (Gerola and Olson, 1986), except that 2.5 M NaBr was used in place of 2 M NaSCN during cell breakage. The broken bacteria were centrifuged 30 min at 18,000 x g. The supernatant and floating pellet were combined, loaded onto 50% sucrose and centrifuged for 17 h at 5°C in a SB 110 rotor (International ultracentrifuge B-60) at 100,000 x g. The chlorosomes banded at the top of NaBr.

When pure chlorosomes were needed, a purification gradient was run. The chlorosome fraction was loaded onto a small amount of 50% sucrose and a continuous sucrose gradient (20-35%, w/w) was loaded over the sample. Finally a small amount of buffer was added at the top. The gradient was centrifuged for 18 h at 5°C in a SB 283 swinging bucket rotor at 240,000 x g.

BChl c-Binding Protein Purification

The chlorosomes were extracted with 80% acetone at a concentration corresponding to A = 15 cm^{-1} at the BChl c Q$_y$-peak and centrifuged for 4 min at 17,000 x g. The supernatant was mixed with diethyl ether and water in the volume ratio 80% acetone: diethyl ether: water = 3:5:2. This mixture was centrifuged for 10 min at 3,000 x g, and most of the ether phase was removed from the top. The acetone water phase was repeatedly washed with fresh ether until the ether wash was colorless. The acetone-water phase was then dried and the residue dissolved in 0.1% trifluoroacetic acid in isopropanol:water (1:1). The entire extraction procedure was carried out in the dark. The dissolved residue was loaded on a Lichrosorp RP8 column (25 x 0.4 cm). The column was equilibrated with 0.1% (v/v) trifluoroacetic acid and 50% (v/v) isopropanol in water. The protein was eluted at 20-25°C by increasing the percentage of isopropanol from 50% to 82% over 35 min at a flow rate of 1-2 ml/min. The BChl c-binding protein was eluted from the column at about 77% isopropanol. The protein in the eluent was monitored by its absorbance at 220 nm.

PAGE was carried out as described previously (Gerola and Olson, 1986).

Digestion with Trypsin and Chymotrypsin

The protein (5-7 μmole) was dissolved in 100 μl of 0.1 M NH$_4$HCO$_3$ and 1 mM CaCl$_2$, pH 8.0. Trypsin (0.5 μg) or chymotrypsin (0.5 μg) was added in 5 μl of the same buffer, and the digestion was allowed to proceed for two hours at 37°C until the digestion was terminated by injection of the sample on HPLC. Separation of the peptides was carried out on a reverse phase ODS column (25 x 0.4 cm). The column was equilibrated with 98% buffer A (0.1% trifluoroacetic acid in water) and 2% buffer B (0.1% trifluoroacetic acid in 96% ethanol). The peptides were eluted with a gradient of 1.5% buffer B/min at a flow rate of 1.0 ml/min at 50°C.

<u>Protein Sequence Determination</u> (Højrup et al., in preparation)

For sequencing the intact protein, 30 nmole was applied to a Beckman 890C spinning cup sequencer. This determined the first 54 residues with a few ambiguities. Tryptic and chymotryptic peptides were sequenced on an Applied Biosystems gas-phase sequencer using the 2NVAC program supplied by the manufacturer. For all samples the molecular weight was determined by plasma desorption mass spectrometry.

RESULTS

Pure chlorosomes ($A=1800$ cm^{-1} at the BChl <u>c</u> Q$_y$-peak) were incubated in 10 mM Tris-HCl, pH 8, with chymotrypsin (10 µg/ml) and elastase (10 µg/ml) for 0.75, 4 and 16 min. The reaction was stopped by the addition of an acetone-HCl mixture to final concentrations of 80% and 0.1 M respectively. The absorption spectrum (BChl <u>c</u>) was unaffected by the proteolytic treatment even after the longest treatment. PAGE of the samples showed that, while the 4-5 kDa (M_r) protein band was apparently unaffected by the treatment, the other proteins present as minor components were gradually destroyed and gave rise to new bands (data not shown). These results suggested that the 4-5-kDa (M_r) protein (or the major component of the 4-5 kDa band) is inside the chlorosome, as is the BChl <u>c</u>, protected from proteolytic digestion, while the other proteins are on the surface of the chlorosome, exposed to the proteolytic enzyme action. When either purified or partially purified chlorosomes were extracted with 80% acetone, the extract contained mainly the 4-5-kDa (M_r) proteins with almost no trace of the others, because the proteins of higher M_r were precipitated by the acetone treatment. When the BChl-free 80% acetone extract was analyzed by HPLC, a main double peak (B, the BChl <u>c</u>-binding protein) and a minor peak (A) were eluted (Fig. 1).

Although the purified BChl <u>c</u>-binding protein was eluted from the HPLC column as a double peak, the amino acid compositions of the two peaks were very similar, but quite different from that of the minor peak (Højrup et al., in preparation).

After PAGE both the BChl <u>c</u>-binding protein and the protein associated with the HPLC minor peak gave rise to bands at about 4-5 kDa (M_r) which could not be separated (data not shown). The BChl <u>c</u>-binding protein molecular weight was calculated from the aa sequence to be 7488 (verified by mass spectrometry, Højrup et al., in preparation) compared to 5592 for the <u>Chloroflexus</u> protein (Wechsler et al., 1985). The amino acid compositions of the BChl <u>c</u>-binding proteins from <u>Chlorobium</u> and <u>Chloroflexus</u> are compared in

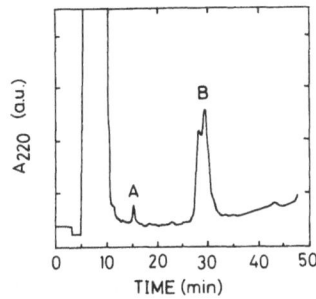

Fig. 1. Elution pattern of 4-5 kDa (M_r) proteins by reverse-phase HPLC. See MATERIALS AND METHODS for details. The BChl <u>c</u>-binding protein is marked by the letter B.

45

Table 1. Amino Acid Composition of the BChl c-Binding Proteins from Chloroflexus (Cf.) and Chlorobium (Cb.)

Residue	Number of Residues Cf.	Cb.	Residue	Number of Residues Cf.	Cb.
Ala	9	14	Leu	1	8
Arg	3	-	Lys	-	5
Asn	4	5	Met	1	1
Asp	2	4	Phe	2	3
Cys	-	-	Pro	1	2
Gly	4	6	Ser	4	3
Gln	5	4	Thr	2	8
Glu	1	2	Trp	3	-
His	1	-	Tyr	1	-
Ile	3	4	Val	4	5

Table 1. In Fig. 2 the sequences of the Chlorobium and Chloroflexus BChl c-binding proteins are aligned for maximum homology, and in Fig. 3 the hydropathy plots (Kyte and Doolittle, 1982) of the two proteins are shown.

The secondary structures for the Chlorobium and Chloroflexus BChl c-binding proteins have been tentatively predicted according to the method of Chou and Fasman (1978; 1979). The α-helix potential values of 6-aa running segments are shown in Fig. 4 and the difference between α-helix and β-sheet potential values of 6-aa running segments are shown in Fig. 5.

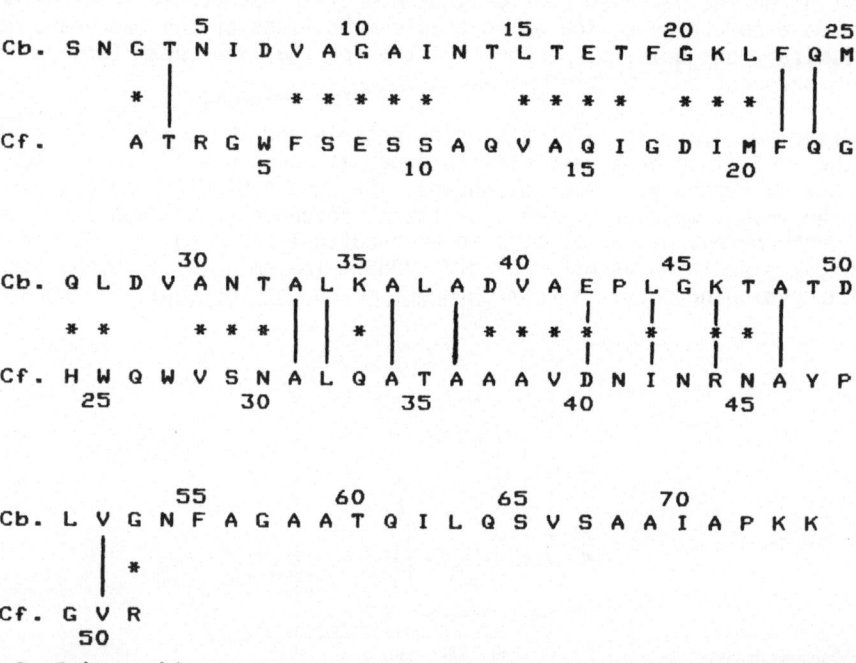

Fig. 2. Amino acid sequences of BChl c-binding proteins from Chlorobium (Cb.) and Chloroflexus (Cf.). The homologies are indicated by solid vertical lines. Functional similarities are indicated by broken lines. Amino acid differences which can be explained by single mutation events are indicated by asterisks.

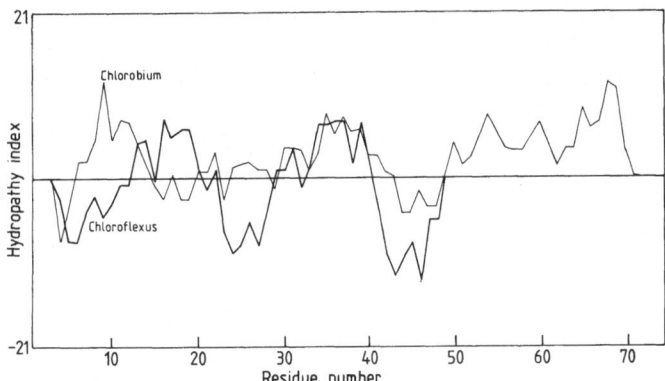

Fig. 3. Hydropathy plots (Kyte and Doolittle, 1982) for the BChl c-binding
proteins from Chlorobium (light line) and Chloroflexus (heavy line).
The average hydropathy was determined within a 7-aa running segment.
The residue number for the Chloroflexus protein has not been
adjusted to give maximum homology as in Fig. 2.

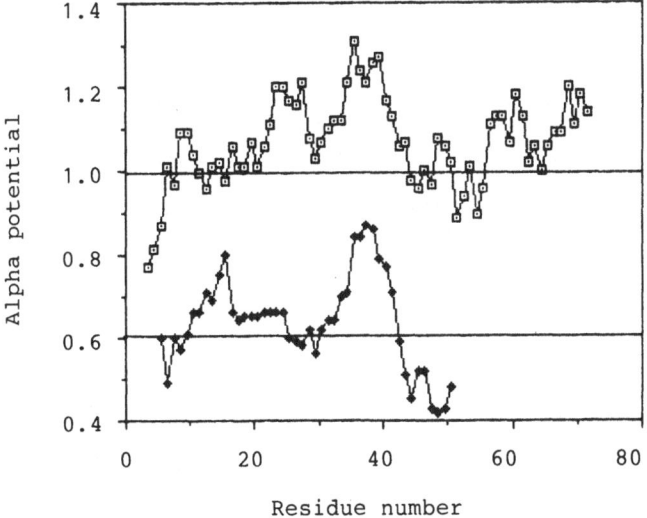

Fig. 4. α-helix potential values of 6-aa running segments for the BChl
c-binding proteins from Chlorobium (upper curve) and Chloroflexus
(lower curve). The α-helix potential values for Chloroflexus are
shifted by -0.4 to enhance clarity. The residue numbers for
Chloroflexus are shifted by 2 to align the two proteins according to
homology as in Fig. 2.

DISCUSSION

The BChl c-binding proteins from Chlorobium and Chloroflexus are quite different in length (74 and 51 aa respectively), and that might explain the difference in the rod element diameters in the chlorosomes of the two organisms. Both proteins are characterized by a large number of amide-containing residues and both lack Cys; however in the Chlorobium protein 4 other amino acids (Tyr, Trp, His and Arg) are missing, while Lys is missing in the Chloroflexus protein. The number of aromatic residues is higher in the Chloroflexus than in the Chlorobium protein. The differences in protein length and between aa sequences are to be expected because of the large evolutionary distance between the two organisms. Most of the aa sequence differences can be explained by single mutation events (asterisks in Fig. 2). In light of so little overall homology between the two proteins, the high homology (30%) in the region between Chlorobium residues 31 and 43 is particularly interesting from a functional point of view. This aspect is reinforced by the hydropathy plots (7-aa running segment) for the two proteins shown in Fig. 3. The two profiles are very similar from Chlorobium running segments 31 through 48, but are completely different from Chlorobium running segments 3 through 30. (If the hydropathy indices for individual aa residues are considered, the two proteins are quite similar from Chlorobium residue 30 to residue 50 (Chloroflexus residues 28-48).) Wechsler et al. (1985) have proposed a model for the secondary structure of the BChl c-binding protein from Chloroflexus which was stated to be based on the method of Chou and Fasman. The model consists of an α-helix between Trp 5 and Ile 42 with 7 amide-containing residues proposed as BChl c-binding sites: Gln 12, 15, 22, 26, 33 and Asn 30 and 41. However the method of Chou and Fasman (1978, 1979) for predicting secondary structures from aa sequences does not support the model proposed by Wechsler et al. (1985). α-potentials of 4-aa running segments (data not shown) indicate the presence 11 tetrapeptides which are α-helix breakers (Pα < 1) in the Trp 5 - Ile 42 region. Moreover the 6-aa segments 21-26, 22-27, 23-28, 25-30, have Pα < 1

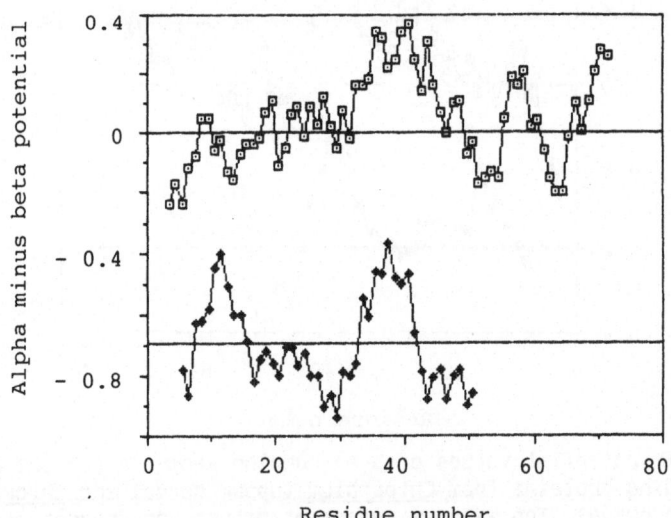

Fig. 5. Differences between α-helix and β-sheet potential values of 6-aa running segments for the BChl c-binding proteins from Chlorobium (upper curve) and Chloroflexus (lower curve). The values for Chloroflexus are shifted by -0.3 to enhance clarity. The residue numbers for Chloroflexus are shifted by 2 to align the two proteins according to homology as in Fig. 2.

(Fig. 4), and all the 6-aa segments from 12-17 up to 27-32 have $P_\alpha < P_\beta$
(Fig. 5). It is thus not possible according to the method of Chou and Fasman
(1978, 1979) to predict an unbroken α-helix between Trp 5 and Ile 42. A
similar conclusion is reached when the method of Garnier et al. (1978) is
used to predict the secondary structure.

In our opinion methods to predict secondary structure based only on the
primary structure are not very reliable when applied to proteins like the
BChl c-binding proteins. The predictive methods are based on known secondary
structures of water-soluble globular proteins, while the BChl c binding
proteins are in hydrophobic environments, and moreover the BChl c molecules
bound to the proteins might influence their secondary structures. However if
one wants to try to predict the secondary structure for the BChl c-binding
protein of *Chloroflexus* according to the method of Chou and Fasman, an
α-helix region is clearly predicted from Ala 31 to Val 39 and possibly from
Trp 5 or Ala 11 to Ile 16. β-sheet should be present from Ile 19 to Val 28,
with a possible turn in the region 22-25 (Fig. 6).

More difficult is the prediction for the *Chlorobium* protein, as aa
sequences with high P_α and P_β values overlap in several regions. An α-helix
region is clearly predicted from Thr 32 to Ala 41, while other α-helix
regions might be present between Ile 6 and Ile 12, Lys 21 and Ala 30, Lys 46
and Val 52 and between Phe 55 and Ala 71. A β-sheet region might be
predicted from Asn 13 to Thr 16 and from Gln 61 to Ser 65 (not shown in
Fig. 6), and a turn between Glu 42 and Gly 45 (Fig. 6). In Fig. 7 helical
wheels are drawn according to Schiffer and Edmundson (1967) representing the
α-helix regions from Ala 31 to Val 39 of the *Chloroflexus* protein and from
Lys 21 to Ala 30 and from Thr 32 to Ala 41 for the *Chlorobium* protein. On
each wheel all the hydrophobic residues lie on one side, while almost all
polar and charged residues are localized on the other side. There are 2
exceptions: Gln 33 in the *Chloroflexus* protein and Gln 26 in the *Chlorobium*

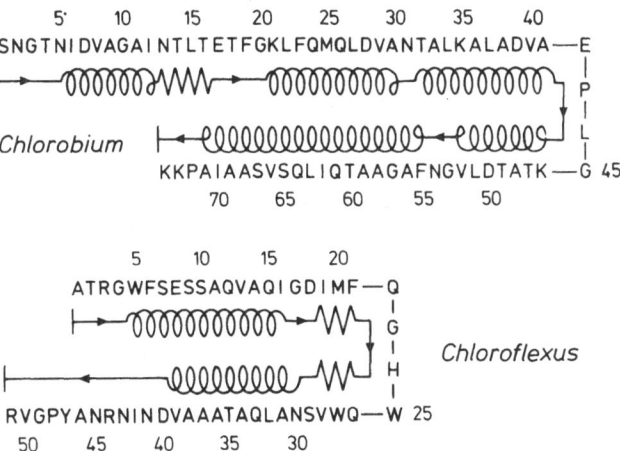

Fig. 6. Secondary structures predicted according to Chou and Fasman (1978,
1979) for the BChl c-binding proteins from *Chlorobium* and
Chloroflexus. The symbol ⓜⓜ indicates α-helix, and the symbol
ⱮⱮ indicates β-sheet. There is one turn (Glu 42 to Gly 45) in
the *Chlorobium* protein and another (Gln 22 to Try 25) in the
Chloroflexus protein. Gln 22 and Try 25 may also be considered to
be part of the β-sheet structures in the *Chloroflexus* protein. One
α-helix in the *Chloroflexus* protein should have been shown starting
at Ala 31 instead of at Asp 30.

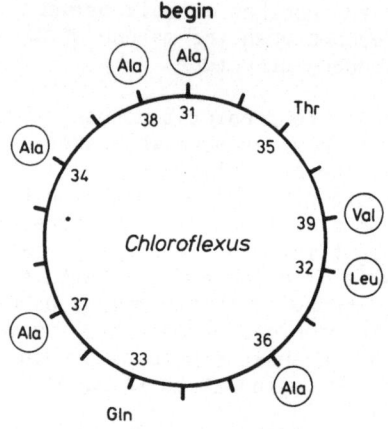

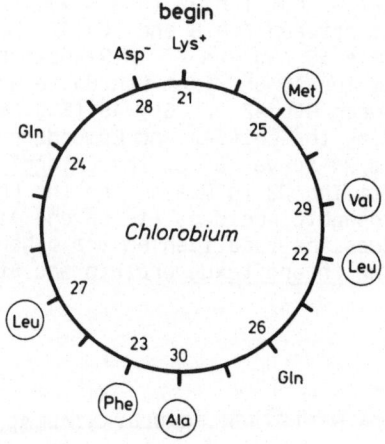

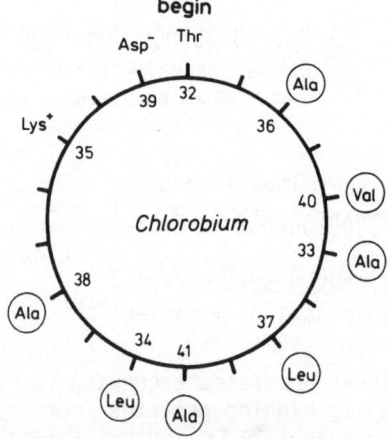

Fig. 7. Helical wheels drawn according to Schiffer and Edmundson (1967)
representing the α-helix regions from Ala 31 to Val 39 in the
protein from <u>Chloroflexus</u> and frum Lys 21 to Ala 30 and from Thr 32
to Ala 41 in the protein from <u>Chlorobium</u>.

protein. Each of these charged residues is on the hydrophobic side of its wheel. Because of their locations these two amide-containing residues are likely candidates for BChl c binding sites. Also the other α-helix regions predicted by the Chou-and-Fasman method, when represented as helical wheels, show distinct hydrophobic arcs, and the hydrophobic arcs of the two α-helix regions (Lys 21-Ala 30 and Thr 32-Ala 41) in Chlorobium lie on the same side of the protein, if we construct an unbroken α-helix wheel from Lys 21 to Ala 41. According to the Chou-and-Fasman method α-helix is predicted for both proteins from Chlorobium residues 33 to 41 (Chloroflexus 31 to 39). This is the only region in the two proteins with similar profiles of α-helix potential values of 6-aa running segments (Fig. 4). It is also the protein region showing the highest homology (Fig. 2) and similar hydropathy profiles (Fig. 3). These facts suggest a functional role for that protein region, i.e., involvement in binding BChl c molecules. In their model for the Chloroflexus protein Wechsler et al. (1985) proposed a rod-like protein with 75% of the residues forming a long α-helix (57Å) in which 7 Gln/Asn residues were proposed as binding sites for BChl c molecules. However, taking into account the P_α values and the presence of tetrapeptide α-helix breakers, a long α-helix does not agree with our analysis according to the method of Chou and Fasman (1978,1979). The presence of turns and bends in both the Chloroflexus and Chlorobium proteins suggests that both proteins may have somewhat globular structures.

Wechsler et al. (1985) proposed a chlorosome subunit composed of 6 dimers of BChl c-binding proteins (12 polypeptide chains) binding 84 BChl c molecules. The molecular mass of the BChl c-binding polypeptide (7.5 kDa) in Chlorobium might permit a dimeric form of that protein also. According to the model proposed by Olson (1980) the six subunits forming each ring in the rod element are made up of the BChl c-binding protein. Unfortunately the size estimation (ca. 15 kDa, Olson, 1980) of these subunits based on electron microscopy is not precise enough to distinguish between 1 or 2 polypeptides in each subunit. If each subunit consists of 1 polypeptide, it should bind 6 ± 1 BChl c molecules, but if there are 2 polypeptides in each subunit, it should bind 12 ± 2 BChl c molecules. It is questionable whether there is enough room for 12 ± 2 BChl c molecules per subunit in the rod element of Chlorobium. Experiments with crosslinking agents are needed to resolve this matter.

ACKNOWLEDGEMENTS

This work has been supported in part by grant GM 3210 to JMO from the US National Institute of General Medical Sciences. PDG has been partially supported by Ministero P.I. (Italy).

REFERENCES

Betti, J. A., Blankenship, R. E., Natarajan, L. V., Dickinson, L. C., and Fuller, R. C., 1982, Antenna organization and evidence for the function of a new antenna pigment species in the green bacterium Chloroflexus aurantiacus, Biochim. Biophys. Acta, 680:194.

Chou, P. Y., and Fasman, G. D., 1978, Prediction of the secondary structure of proteins from their amino acid sequence, Adv. Enzymol. Relat. Areas Mol. Biol., 47:45.

Chou P. Y., and Fasman, G. D., 1979, Prediction of β-turns, Biophys. J., 26:367.

Garnier, J., Osgathorpe, D. V., and Robson, B., 1978, Analysis of the accuracy and implications of simple methods for predicting the secondary structure of globular proteins, J. Mol. Biol., 120:97.

Gerola, P. D., and Olson, J. M., 1986, A new bacteriochlorophyll a-protein complex associated with chlorosomes of green sulfur bacteria, *Biochim. Biophys. Acta*, 848:69.

Gibson, J., Ludwig, W., Stackebrandt, E., and Woese,C. R., 1985, The phylogeny of the green photosynthetic bacteria: absence of a close relationship between *Chlorobium* and *Chloroflexus*, *System. Appl. Microbiol.*, 6:152.

Kyte, J., and Doolittle, R. F., 1982, A simple method for displaying the hydropathic character of a protein, *J. Mol. Biol.*, 157:105.

Olson, J. M., 1980, Chlorophyll organization in green photosynthetic bacteria, *Biochim. Biophys. Acta*, 594:33.

Olson, J. M., Gerola, P. D., van Brakel, G. H., Meiburg, R. F., and Vasmel, H., 1985, Bacteriochlorophyll a- and c-protein complexes from chlorosomes of green sulfur bacteria compared with bacteriochlorophyll c aggregates in CH_2Cl_2-hexane, *in*: "Antennas and Reaction Centers of Photosynthetic Bacteria," M. E. Michel-Beyerle, ed., Springer-Verlag, Berlin.

Schiffer, M., and Edmundson, A. B., 1967, Use of helical wheels to represent the structures of proteins and to identify segments with helical potential, *Biophys. J.*, 7:121.

van Dorssen, R. J., Gerola, P. D., Olson, J. M., and Amesz, J., 1986, Optical and structural properties of chlorosomes of the photosynthetic green sulfur bacterium *Chlorobium limicola*, *Biochim. Biophys. Acta*, 848:77.

Wechsler, T., Suter, F., Futter, R. C., and Zuber, H., 1985, The complete amino acid sequence of the bacteriochlorophyll c binding polypeptide from chlorosomes of the green photosynthetic bacterium *Chloroflexus aurantiacus*, *FEBS Lett.*, 181:173.

Woese, C. R., 1987, Bacterial evolution, *Microbiol. Rev.*, 51:211.

STRUCTURAL STUDIES ON THE ANTENNA COMPLEXES AND POLYPEPTIDES OF CHLOROFLEXUS

AURANTIACUS AND OTHER GREEN PHOTOSYNTHETIC BACTERIA

H. Zuber

Institut für Molekularbiologie und Biophysik
Eidg. Technische Hochschule, ETH-Hönggerberg
CH-8093 Zürich, Switzerland

SUMMARY

Biochemical and biophysical studies revealed that for functional
reasons the light-harvesting antenna of photosynthetic bacteria represent a
highly organized pigment-polypeptide system. It is comprised of an aggregate
of antenna complexes which in turn consist of specific arrays of
polypeptides or pigments. This particular organization is the basis for the
very effective heterogeneous energy transfer within and between these
complexes (Zuber, 1985). Green photosynthetic bacteria which are able to
survive under extremely low light intensity contain an extended
heterogeneous energy-transfer system to the reaction center including the
intramembrane antenna complexes close to the reaction center, the
extramembrane antenna complexes within the chlorosomes at the surface of the
membrane and a BChl a-protein complex (B790) located between the chlorosome
and the intramembrane complexes. The most detailed biochemical analysis and
characterization of antenna complexes was performed for Chloroflexus
aurantiacus, particularly by Feick and Fuller (1984). They also identified 5
polypeptides, one antenna polypeptide of the intramembrane antenna complex,
and 2 antenna polypeptides, the BChl c-binding polypeptide, the 5.8-kDa
"base-plate" protein (B790) and two 11- and 18-kDa polypeptides.

During the last 2 years we isolated and characterized 3 antenna
polypeptides of the antenna complexes of Chloroflexus and determined their
primary structures (Wechsler et al., 1985a, b; 1987). From the intramembrane
antenna complex B808-866 2 antenna polypeptides could be isolated after
extraction of the membrane with chloroform/methanol/ammonium acetate and
fractionation of the polypeptides on LH-60, DE-32, Biogel P-10
(α-polypeptide) or Pro RPC-FPLC (β-polypeptide). Their primary structures
were established by automated and manual Edman degradation of the whole
polypeptides and of polypeptide fragments (BNPS-Skatol, propionic acid). One
antenna polypeptide consists of 44 amino acid residues (Wechsler et al.,
1985a). Its amino acid sequence is most homologous (27-30%) to the
α-polypeptides of the B870/890 (B1015) antenna complexes of purple bacteria.
This implies that it is the α-antenna polypeptide of the core complex
B808-866, located close to the reaction center. The same is true for the
other antenna polypeptide consisting of 51 amino acid residues and
representing the β-polypeptide of the B 808-866 core antenna complex
(Wechsler et al., 1987). Thus the α- and β-polypeptide pair forms, as in
purple bacteria, the basic unit (heterodimer) of the B808-866 complex. Its

sequence homology to the α- and β-polypeptides of purple bacteria shows the phylogenetic relationship between the antenna systems of purple and green bacteria. Both polypeptides have the 3-domain structure, which is the basis for the formation of the transmembrane helix, and show the hypical His residues as possible BChl a binding sites. Interestingly, they possess the functional important clusters of aromatic residues (including Arg), typical for the α- or β-polypeptides of the core complexes of purple bacteria.

The BChl c-binding (antenna) polypeptide of the B740 antenna complex from the chlorosomes of Cf. aurantiacus was extracted from whole cells and from chlorosomes by chloroform/methanol/ammonium acetate/acetic acid and extraction and purification by LH-60 chromatography (Wechsler et al., 1985b). Its amino acid sequence was determined by automated and manual Edman degradation and carboxypeptidase A and B digestion (Wechsler et al., 1985b). It also consists of 51 amino acid residues but has no 3-domain structure. It possibly forms a helix between Trp 5 and Ile 42 (length 57 A). Most interestingly, all 5 Gln and 2 Asn residues are located asymmetrically at one side of the helix. They represent most probably the binding sites for the 7 BChl c molecules found per polypeptide. In this arrangement the BChl c molecules can specifically interact via the carbonyl function of ring V and the hydroxyethyl group of ring I of the tetrapyrrole. A similar type of BChl c-BChl c interaction was shown with BChl model compounds, without the polypeptide matrix (Smith et al., 1983). In this basic α-helix unit clusters of BChl c molecules are formed, which are most probably exciton-coupled via the Q_y-transition dipoles. It is further hypothesized that in the rod-shaped elements revealed by electron microscopy (Staehelin et al., 1978) the helices are associated in aggregates of 12 polypeptides (dimensions 52 x 60 A). On the basis of the specific BChl c clusters all BChl c molecules within and between the rod-shaped elements are coupled for energy transfer. Since the BChl c molecules have their transition dipoles oriented about parallel to the helix axis and the helix axes in the rod elements are about parallel to the long axis of the chlorosome, the overall transition dipoles of the BChl c molecules are about parallel to the long axis of the chlorosome.

From the chlorosomes of the green phototrophic sulfur bacterium Pelodictyon luteolum (Pfennig and Trüper, 1971) a polypeptide was isolated and its primary structure determined (Wagner et al., in preparation). This polypeptide is most probably the BChl c-binding polypeptide. It consists also of 51 amino acid residues and is sequence homologous, particularly in two regions, to the BChl c-binding polypeptide of Cf. aurantiacus. The hydropathy plots of the Pelodictyon and Chloroflexus antenna polypeptides are very similar. Most interestingly, 3 Glu residues and 1 Asp residue of the Pelodictyon polypeptide were found in identical positions as 3 Gln residues and 1 Asp residue of the Chloroflexus polypeptide. It is hypothetically assumed that these acid amino acid residues also form BChl c binding sites. Practically no sequence homology can be detected between the BChl c-binding polypeptides of Cb. limicola (Gerola et al., 1988) and P. luteolum.

REFERENCES

Feick, R. G., and Fuller, R. C., 1984, Topography of the photosynthetic apparatus of Chloroflexus aurantiacus, Biochem., 23:3693.
Gerola, P., Højrup, P., Knudsen, J., Roepstorff, P., and Olson, J. M., 1988, The bacteriochlorophyll c-binding protein from chlorosomes of Chlorobium limicola f. thiosulfatophilum, This volume.
Pfennig, N., and Trüper, H. G., 1971, New nomenclatural combinations in the phototrophic sulfur bacteria, Int. J. Systematic Bact., 21:11.
Smith, K. M., Kehrer, L. A., and Fajer, J., 1983, Aggregation of the bacteriochlorophylls c, d, and e. Models for the antenna chlorophylls

of green and brown photosynthetic bacteria, J. Am. Chem. Soc.,
105:1378.

Staehelin, L. A., Golecki, J. R., Fuller, J. R., and Drews, G., 1978,
Visualization of the supramolecular architecture of chlorosomes
(chlorobium type vesicles) in freeze-fractured cells of Chloroflexus
aurantiacus, Arch. Microbiol., 119:269.

Wechsler, T., Brunisholz, R., Suter, F., Fuller, R. C., and Zuber, H.,
1985a, The complete amino acid sequence of a bacteriochlorophyll a
binding polypeptide isolated from the cytoplasmic membrane of the
green photosynthetic bacteria Chloroflexus aurantiacus, FEBS Lett.,
191:34.

Wechsler, T., Suter, F., Fuller, R. C., and Zuber, H., 1985b, The complete
amino acid sequence of the bacteriochlorophyll c binding polypeptide
from chlorosome of the green photosynthetic bacterium Chloroflexus
aurantiacus, FEBS Lett., 181:173.

Wechsler, T., Brunisholz, R., Frank, G., Suter, F., and Zuber, H., 1987, The
complete amino acid sequence of the antenna polypeptide B806-866β
from the cytoplasmic membrane of the green bacterium Chloroflexus
aurantiacus, FEBS Lett., 210:189.

Zuber, H., 1985, Structure and function of light-harvesting complexes and
their polypeptides, Photochem. Photobiol., 42:821.

ENERGY TRAPPING AND ELECTRON TRANSFER IN CHLOROFLEXUS AURANTIACUS

R. E. Blankenship, D. C. Brune, J. M. Freeman, G. H. King,
J. D. McManus, T. Nozawa[a], T. Trost and B. P. Wittmershaus

Department of Chemistry, Arizona State University
Tempe, AZ 85287-1604, U.S.A.
[a] Permanent Address: Chemical Research Institute of Non-
Aqueous Solutions, Tohoku University, Sendai 980, Japan

INTRODUCTION

Green photosynthetic bacteria are characterized by the presence of chlorosomes, antenna structures attached to the cytoplasmic side of the inner cell membrane. These complexes contain BChl c, d, or e and function by absorbing light energy and transferring it to the photochemical reaction center which is embedded in the cytoplasmic membrane (Olson, 1980). In the reaction center a portion of the photon energy is conserved by a series of electron transfer reactions, and the products of these reactions are used to drive cellular processes.

Two families of green photosynthetic bacteria have been described, the Chlorobiaceae, or green sulfur bacteria, and the Chloroflexaceae, or green gliding bacteria (Pfennig and Trüper, 1983). According to analysis of 16S rRNA, the two families are only distantly related (Gibson et al., 1985). The chlorosomes found in the two types of green bacteria are rather similar, while the membrane-bound antenna complexes and reaction centers are very different (Olson, 1980; Blankenship, 1985). Two different models for the organization of BChl c in the chlorosome have been proposed. Bystrova et al. (1979) suggested on the basis of spectral properties of in vitro aggregates of BChl c that the pigment in vivo was present in an oligomeric form, and several other groups have presented evidence in support of this view (Smith et al., 1983; Olson et al., 1985; Brune et al., 1987a,b). A different view was given by Wechsler et al. (1985), who proposed on the basis of the primary sequence of the protein thought to be associated with BChl c in the chlorosome that 7 BChl c molecules are coordinated to glutamine and asparagine residues, forming a pigment-protein complex. The proposed geometries of the two models are significantly different, with the oligomer model predicting substantial overlap of the BChl c macrocycles along rings I and III, and the central Mg probably coordinated to the C-2 hydroxyethyl on one side and the C-9 keto on the other side. The Wechsler-et-al. (1985) model predicts that the Mg is coordinated to amino acid groups on one side; no group is specifically assigned as a ligand to the sixth position, although adjacent molecules may provide ligands. Fig. 1 shows a general model of the chlorosome and membrane, emphasizing the probable pathway of excitation transfer.

Fig. 2 shows electron transfer pathways of the two families of green bacteria along with pathways for the purple bacteria and O_2-evolving photosynthetic organisms. The Chloroflexaceae contain a reaction center incorporating a 2-quinone acceptor complex and cyclic electron transfer system that is very similar to that found in the purple photosynthetic bacteria and somewhat similar to the acceptor region of Photosystem 2. These reaction centers are referred to as "quinone type".

The reaction centers of the Chlorobiaceae (and of Heliobacterium chlorum) contain low potential Fe-S centers as early electron acceptors and bear a considerable similarity to the acceptor region of Photosystem 1. These reaction centers are called "Fe-S type". Although definitive evidence is lacking, it is tempting to suggest that the reaction centers of PS 1 and 2 of O_2-evolving organisms bear a direct evolutionary link to the "quinone type" and "Fe-S type" reaction centers found in various anoxygenic photosynthetic bacteria.

Our laboratory has been examining the properties of the antenna complexes, the photochemical reaction center and the secondary electron transfer components of the thermophilic green gliding bacterium Chloroflexus aurantiacus. Our results presented in this communication and in other publications (Brune and Blankenship, 1987; Brune et al., 1987a,b) tend to support the oligomer model for chlorosome structure. Work from our laboratory as well as from others has given rise to the view that the reaction center of Cf. aurantiacus is a "quinone type" complex (Bruce et al., 1982; Blankenship et al., 1983; Pierson and Thornber, 1983; Vasmel and Amesz, 1983; Vasmel et al., 1983; Kirmaier et al., 1983, 1984, 1986; Parot et al., 1985; Shuvalov et al., 1986).

MATERIALS AND METHODS

Chloroflexus aurantiacus J-10-fl was obtained from the American Type Culture Collection and grown at 55°C in the medium described by Pierson and Castenholz (1974). Chlorosomes were isolated by the method of Feick and Fuller (1984), and carotenoid-depleted chlorosomes as described by Brune et al. (1987a). Reaction centers were isolated by a modification of the method of Pierson and Thornber (1983). Cyt c-554 was isolated from the supernatant resulting from a 1.5% lauryl dimethylamineoxide (LDAO) extraction of membranes (A865 = 16), followed by ion-exchange chromatography on DEAE Sephacel. Auracyanin was solubilized by salt treatment (200 mM NaCl, 20 mM $MgCl_2$, 20 mM Tris, pH 8.0) of membranes, and purified by ammonium sulfate precipitation and DEAE and gel filtration (G-100 or S-300) chromatography.

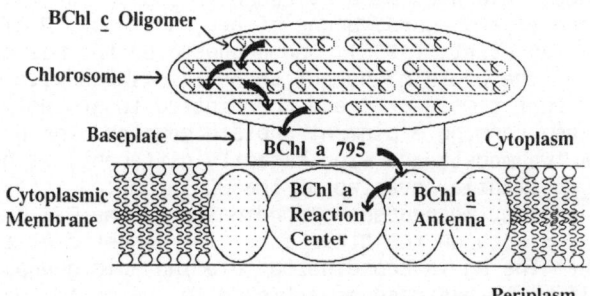

Fig. 1. Schematic picture of the energy transfer pathway in the Chloroflexaceae. The chlorosome baseplate is defined differently than in the Chlorobiaceae.

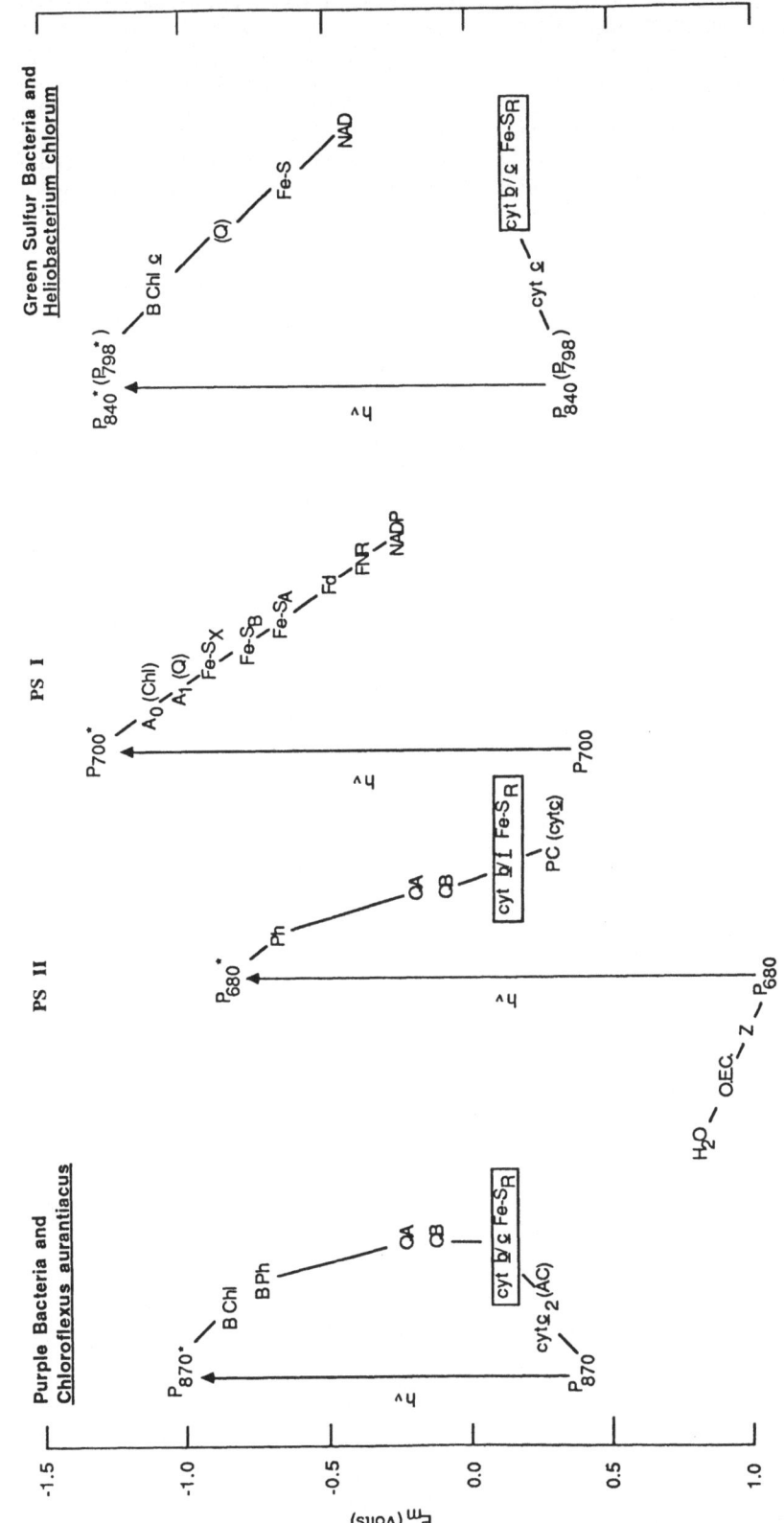

Fig. 2. Electron transfer pathways of photosynthetic organisms. The "quinone type" of reaction center includes that found in the purple bacteria, the Chloroflexaceae and Photosystem 2 of O₂-evolving organisms. The "Fe-S type" of reaction center is found in the Chlorobiaceae, *Heliobacterium chlorum* and Photosystem 1.

In both cases final purification was accomplished using a Sota chromatography DEAE analog HPLC column on an ISCO HPLC system. Redox titrations on Cyt c-554 (50 mM Tris, pH 8.0, 0.05% Triton X-100) were performed on a Cary 219 UV-VIS spectrophotometer using a stirred cuvette bubbled with Ar. Redox potentials were measured using Radiometer K401 calomel and P101 Pt electrodes and an Orion 701A pH meter. Quoted values are expressed relative to the normal hydrogen electrode. Redox mediators (10 μM each) were 2,3,5,6-tetramethyl-p-phenylenediamine, 1,4-napthoquinone, phenazinemethosulfate, phenazineethosulfate, duroquinone and pyocyanin. Ferricyanide and dithionite served as titrants.

BChl c was purified and 740-nm pigment oligomers prepared as described by Brune et al. (1987a). Fourier Transfer Infrared (FTIR) spectra were obtained on dried films of chlorosomes and BChl c oligomers using a Nicolet MX-1 spectrometer. Pigments were converted to monomeric form by injecting 5-10 μl of pyridine into the bottom of the sample cell and allowing the vapor to bathe the sample. Streak camera fluorescence experiments were carried out at the University of Rochester in collaboration with R. S. Knox and S. Lin. Samples of isolated chlorosomes (A532<0.1; A740 0.2 in 10 mM Tris, pH 8.0) were illuminated with single 30-ps, 532-nm pulses from a Nd³⁺:YAG laser system. Time-resolved fluorescence was measured using a low-jitter streak camera and an optical multichannel analyzer and analyzed as described by Wittmershaus et al. (1985, 1987). Circular dichroism (CD) spectra were measured on the instrument described by Brune et al. (1987b) incorporating a Hinds Instruments PEM-80 photoelastic modulator and polarizing prism in place of the chopper.

RESULTS AND DISCUSSION

Chlorosome Structure

Absorption spectra of carotenoid-depleted chlorosomes, the 740-nm

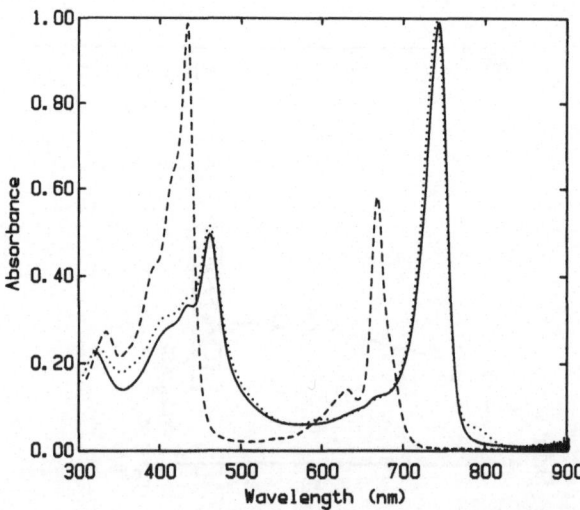

Fig. 3. UV/VIS spectra of carotenoid extracted chlorosomes (dotted line), synthetic BChl c oligomer in hexane (solid line) and BChl c in methylene chloride (dashed line). The methylene chloride solution is mostly monomeric, but contains some dimer as indicated by the shoulder at 680 nm.

absorbing oligomer of BChl c and monomeric Bchl c are shown in Fig. 3. The absorbance maximum of BChl c is shifted from the 665-670 nm wavelength characteristic of the monomer to 740 nm when the pigment is diluted into dry hexane. The absorption spectrum of this self-aggregated oligomeric form of BChl c is remarkably similar to the absorption spectrum of BChl c in chlorosomes. The wavelength maxima in both the Q_y- and Soret bands (740 nm and 462 nm, respectively), as well as the relative intensity of the bands are similar in the chlorosome and the oligomer. Formation of the oligomer leads to a considerable shift of oscillator strength from the Soret band to the Q_y-band. The small amount of BChl a in the chlorosome sample (peaks at 795 nm and 600 nm) is not present in the oligomer. Chlorosomes that have not been depleted of carotenoid have similar spectra in the near IR, but exhibit increased absorption in the 400-500 nm region and a shoulder at 520 nm (not shown).

Room temperature CD spectra of chlorosomes and BChl c oligomers are shown in Fig. 4. The CD spectra exhibit some similar features, but are not identical. Both spectra show a derivative-shaped feature centered at 747 nm; the chlorosome sample has additional negative features at 714 nm and 806 nm. The latter feature is probably due to the BChl a present in chlorosomes. This spectrum is similar to a 77 K spectrum shown by van Dorssen et al. (1986). The major feature of both spectra is significantly red-shifted from the absorbance maximum (747 nm vs 740 nm). Olson et al. (1985) measured CD spectra of chlorosomes isolated from Chlorobium limicola and found them to be sign reversed from the spectrum shown in Fig. 4. They also measured the CD spectrum of oligomeric farnesyl BChl c, which is similar to the spectrum reported here, except for a negative feature at 720 nm similar to that observed in the CD of Chloroflexus chlorosomes. They also found that the rotational strengths (724 and 753 nm) of the farnesyl BChl c aggregates in Chlorobium chlorosomes were about 7-8 times the rotational strengths (736 and 766 nm) of the farnesyl BChl c oligomers in CH_2Cl_2-hexane for equivalent concentrations of BChl c. We, on the other hand, found that the CD intensity of the synthetic stearyl BChl c oligomer is approx. 6 times stronger than that of Chloroflexus chlorosomes.

FTIR spectra of dried films of chlorosomes and oligomers (not shown)

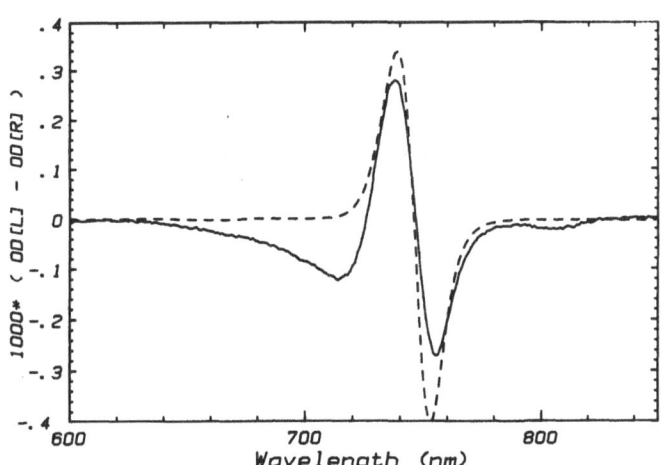

Fig. 4. Circular dichroism spectra of chlorosomes from Cf. aurantiacus (solid line) and 740-nm absorbing BChl c oligomer in hexane (dashed line). The chlorosome spectrum has been multiplied by a factor of five.

exhibit features typical of BChl $\underline{c}$ (Katz et al., 1966). In addition the chlorosome sample has contributions from protein and carotenoids, in particular a strong peptide carbonyl vibration at 1653 cm⁻¹ characteristic of α-helix (Breton and Nabedryk, 1984). FTIR difference spectra of chlorosomes and oligomers are shown in Fig. 5. These spectra reflect the differences induced in the vibrations by treating with pyridine vapor, which causes the pigment Q_y-absorption to shift from 740 nm to 670 nm in both the chlorosome and the oligomer. This treatment did not affect the 795-nm absorbance of the BChl $\underline{a}$ in the chlorosome, and presumably did not affect the protein conformation, although definitive evidence on this point is lacking. The most conspicuous change in the FTIR spectra induced by the pyridine treatment is the change of frequency of the C-9 keto group from 1653 cm⁻¹ in the intact chlorosome and oligomer to 1680 cm⁻¹ in the pyridine-treated samples. These shifts are characteristic of the conversion of the keto from an "associated" to a "free" state and implicate the C-9 keto group in the pigment organization of the chlorosome (Cotton et al., 1978; Lutz, 1984). Similar results for oligomers were obtained by Bystrova et al. (1979). The probable involvement of the C-2 hydroxyethyl group is indicated by the finding of Smith et al. (1983) that oligomers do not form if the hydroxyethyl group is converted to a vinyl group, and by the fact that this is a unique, conserved structural feature of the BChl $\underline{c}$, $\underline{d}$, or $\underline{e}$ found in chlorosomes.

Recent resonance Raman results of Nozawa and Tasumi et al. (unpublished observations) indicate that the BChl $\underline{c}$ in chlorosomes has C = C vibrational frequencies in the 1540-1560 cm⁻¹ region similar to those observed in 5-coordinate BChl $\underline{c}$ and Chl $\underline{a}$. In 5-coordinate chlorophyll the Mg is out of the plane of the chlorin ring (Chow et al., 1975). These C = C vibrations are characteristic of the out of plane nature of the Mg rather than the coordination number per se (Fujiwara and Tasumi, 1986a,b). We propose that the BChl $\underline{c}$ in chlorosomes and 740-nm oligomers is asymmetrically 6-coordinate, with one axial ligand being the C-9 keto carbonyl and the other the oxygen of the C-2 hydroxyethyl. Under these conditions the Mg may still be out of the plane of the chlorin because the interactions from the axial ligands are nonequivalent. Fig. 6 shows a schematic model of this view of

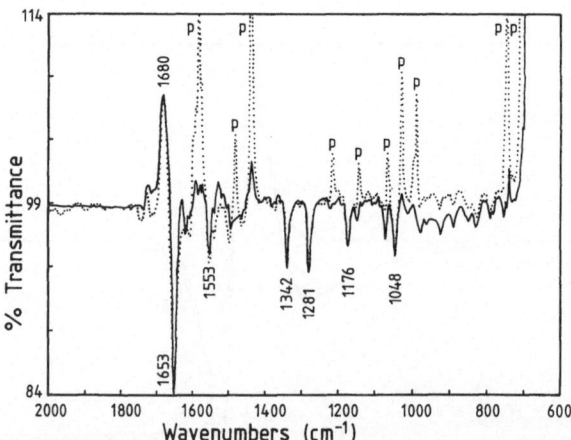

Fig. 5. Fourier Transform Infrared difference spectra of chlorosomes (solid line) and 740-nm absorbing BChl $\underline{c}$ oligomer (dashed line) obtained by subtracting the FTIR spectrum of chlorosomes (or oligomer) exposed to pyridine vapor from that of intact chlorosomes (or oligomer). Peaks due to infrared absorbance of pyridine are marked "P". Numbers adjacent to other spectral features indicate wavenumbers in cm⁻¹.

the BChl c organization.

The fluorescence lifetime of the BChl c in both chlorosomes and 740-nm
oligomers is < 100 ps (Brune et al., 1987a,b). Fig. 7 shows a time-resolved
fluorescence trace from chlorosomes taken with a streak camera system. The
lifetime of the 750-nm emission is 14 ± 5 ps. According to the results of
Vos et al. (1987), the excitation fluence in our experiment was sufficiently
low to avoid significant excitation annihilation effects. The slower decay
process is 805-nm emission from the 795-nm absorbing BChl a in the
baseplate, as shown by experiments using interference filters (not shown).
The lifetime of the BChl c in oligomers has not yet been measured with this
time resolution. The fluorescence from the 740-nm absorbing oligomer is
emitted at 750 nm and is extremely weak, indicating that nonradiative
processes dominate the decay of the excited state (Brune et al., 1987a).
This very short lifetime of the oligomer, which has no energy acceptor such
as the 795-nm BChl a, is difficult to reconcile with highly efficient
collection of energy by the membrane-bound pigments in chlorosomes unless
additional quenching processes (such as quenching by impurities) are present
in oligomers and not in chlorosomes. The energy-transfer efficiency is
essentially 100% in freshly isolated whole cells, indicating that the system
is capable of highly efficient energy capture (Brune et al., 1987b).
However, the efficiency drops to about 50-60% with even gentle manipulations
of the samples. More work is needed to understand the nature of the coupling
of the chlorosome to the membrane and the exact pathway of energy transfer
from BChl c to the reaction center. One possibility currently being tested
in our laboratory is that the BChl a 795 can transfer energy not only to the
membrane-bound antenna complex (B808-865) but also directly to the reaction
center, depending on conditions.

Electron Transfer Components

Bruce et al. (1982) provided evidence for a membrane-bound c-type
cytochrome, Cyt c-554, which donates electrons to the oxidized reaction
center in Chloroflexus. Additional experiments (Blankenship et al., 1985;
Pierson, 1985; Zannoni and Ingledew, 1986; Foster et al., 1986; Wynn et al.,

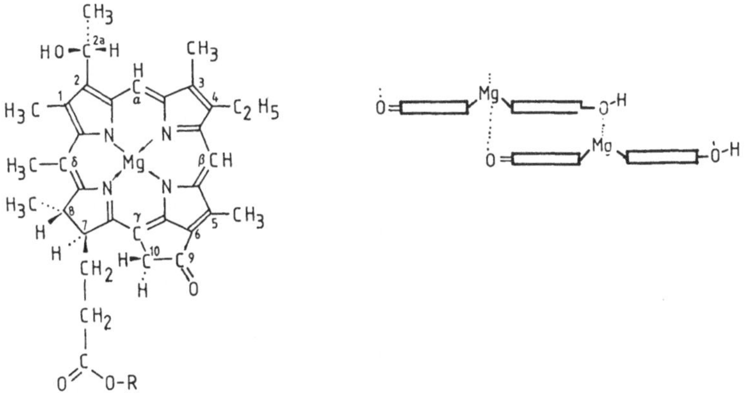

Fig. 6. Chemical structure of BChl c and proposed asymmetric 6-coordinate
model of the molecular environment of Bchl c in chlorosomes and
oligomers. The Mg is coordinated to the oxygen of the C-2
hydroxyethyl group and the C-9 keto carbonyl. The Mg is out of the
plane, as indicated by resonance Raman data. The Mg is shown toward
the hydroxyethyl side of the macrocycle, although this is not
certain.

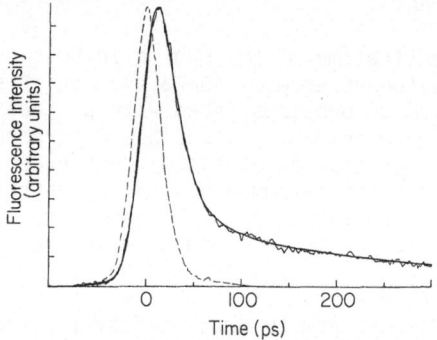

Fig. 7. Time-resolved fluorescence of isolated chlorosomes from
         Cf. aurantiacus at 295 K. Fluorescence was observed
         through a 715-nm long-pass filter (solid line) and is the average
         of 100 laser pulses. A sum of two exponential-decay components
         (smooth solid line) of lifetimes τ_1 = 14 ps and τ_2 = 258 ps, and
         relative amlitudes A_1 = 92.3 and A_2 = 7.7 was used to fit the
         curve. The excitation pulse (dashed line) was 30 ps at full-width
         half-maximum with a fluence of 5 x 10^{13} photons cm^{-2} at 532 nm.

1987) indicated that multiple hemes are present in both the membrane and the
isolated cytochrome and that Cyt c-554 is only present in phototrophically
grown cells. Fig. 8 shows a redox titration of isolated Cyt c-554. Complex
behavior involving apparently at least three distinct redox centers is
apparent. These titrations are similar to those reported by Zannoni and

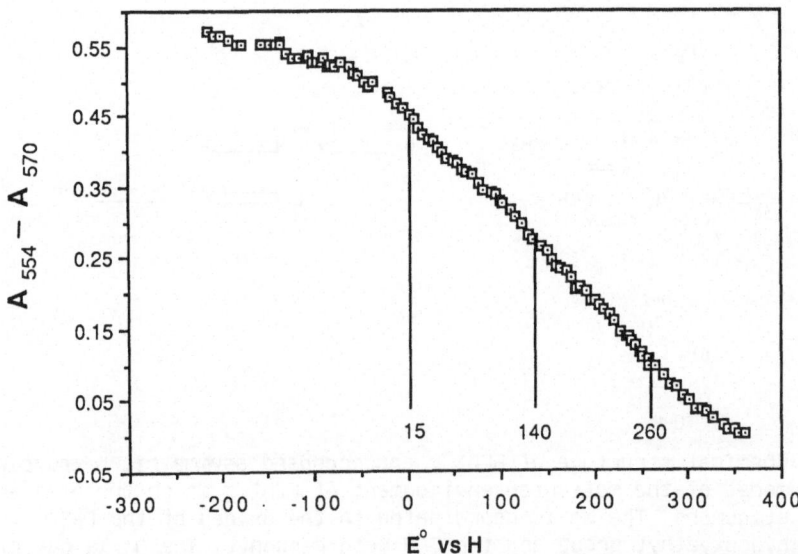

Fig. 8. Redox titration of purified Cyt c-554 from Cf. aurantiacus. Midpoint
        potentials of +15 mV, +140 mV and +260 mV (vs NHE) are indicated.

Ingledew (1986) for membranes, and we find that Cyt c-554 is the most abundant heme-staining peptide in the membranes (Foster et al., 1986; Wynn et al., 1987; Freeman and Blankenship, unpublished). It is not yet possible to equate the low potential wave observed in titrations of membranes with that observed in the isolated cytochrome, because the isolation procedure may alter the redox properties of the hemes. Work is in progress to determine the number of heme groups in isolated Cyt c-554 (an estimate of 2 hemes per peptide by Blankenship et al. (1985) may be low due to the presence of contaminating protein). In contrast to the results reported by Zannoni and Ingledew (1986), our phototrophically grown membranes contain only very small amounts of protoheme and no observable b-type cytochrome absorption bands in the 560-nm region (Bruce et al., 1982; Wynn et al., 1987; Freeman and Blankenship, unpublished).

Chloroflexus aurantiacus apparently does not contain any soluble c-type cytochromes, in contrast to essentially all other photosynthetic bacteria. We have recently isolated a small water-soluble blue copper protein (auracyanin) by salt-washing isolated membranes (Trost et al., 1987). Fig. 9 shows a UV/VIS absorption spectrum of auracyanin, which shows a pattern of bands (280 nm, 460 nm, 596 nm and 715 nm) that is typical of type I or blue copper proteins such as plastocyanin. The monomer molecular mass of this protein determined by SDS-PAGE is 12.8 kDa, and its redox potential is +240 mV (vs NHE). This redox potential is consistent with a role in the cyclic electron transport chain, although kinetic evidence on this point is not yet available. Possible sites of donation, based on redox properties of auracyanin and other carriers, are directly to the oxidized reaction center (+360 mV) or to the high potential (+260 mV) heme group of Cyt c-554.

ACKNOWLEDGEMENTS

This research was supported by grants from the United States Department of Energy Biological Energy Storage Program (DE-FG02-85ER13388) and the Competitive Grants Research Program of the U.S. Department of Agriculture (84-CRCR-1-1523). T.N. was supported by a grant from the Japanese Ministry of Education (Japan-U.S. Cooperative Photoconversion and Photosynthesis

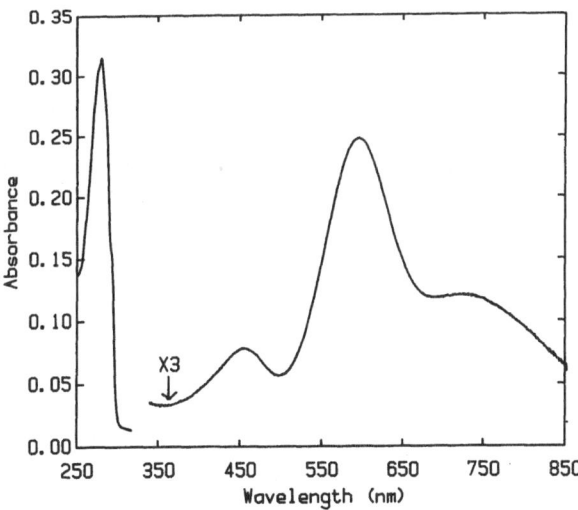

Fig. 9. UV/VIS absorption spectrum of auracyanin, a blue copper protein isolated from Cf. aurantiacus, suspended in 20 mM Tris pH 8.0.

Program). Time-resolved fluorescence measurements were supported in part by the sponsors of the Laser Fusion Feasibility Project of the University of Rochester.

Note added in proof: In the paper by Lutz and van Brakel (1988) resonance Raman data are used to propose a 5-coordinate model for BChl c oligomers and pigment organization in chlorosomes. In their model, the hydroxyethyl group is left unassociated, while the C-9 keto group is a ligand to the Mg. The involvement of the hydroxyethyl group in oligomer formation is implicated by considerable evidence, including data of Bystrova et al. (1979), Smith et al. (1983) and Brune et al. (1988). These groups showed that analogous chlorophylls lacking the hydroxyethyl group at the 2a position do not form extremely red-shifted oligomers. The hydroxyl group is expected to be a stronger nucleophile than the keto group, so a structure like that proposed by Lutz and van Brakel where the OH group is positioned near the Mg but does not interact seems chemically unreasonable. No model compounds with asymmetric 6 coordination have been examined using resonance Raman, so we feel it is premature to rule out this structure. Other structures that include a direct role for both the hydroxyethyl and C-9 keto groups in oligomer formation and yet maintain the Mg as 5-coordinate are also possible. Two such models are presented by Brune et al. (1988).

REFERENCES

Blankenship, R., 1985, Electron transport in green photosynthetic bacteria, Photosynth. Res., 6:317.

Blankenship, R. E., Feick, R., Bruce, B. D., Kirmaier, C., Holten, D., and Fuller, R. C., 1983, Primary photochemistry in the facultative green photosynthetic bacterium Chloroflexus aurantiacus, J. Cell. Biochem., 22:251.

Blankenship, R. E., Huynh, P., Gabrielson, H., and Mancino, L. J., 1986, Purification, physical properties and kinetic behavior of cytochrome c-554 from Chloroflexus aurantiacus, Biophys. J., 47:2a.

Bruce, B. D., Fuller, R. C., and Blankenship, R.E., 1982, Primary photochemistry in the facultatively aerobic green photosynthetic bacterium Chloroflexus aurantiacus, Proc. Natl. Acad. Sci. U.S.A., 79:6532.

Brune, D. C., and Blankenship, R. E., 1987, Light absorption and fluorescence of BChl c in chlorosomes from Chloroflexus aurantiacus and in an in vitro model, in: "Progress in Photosynthesis Research Vol. I," J. Biggins, ed., Martinus Nijhoff, Dordrecht.

Brune, D. C., Nozawa, T., and Blankenship, R. E., 1987a, Antenna organization in green photosynthetic bacteria I. Aggregated bacteriochlorophyll c as a model for the 740 nm absorbing bacteriochlorophyll c in Chloroflexus aurantiacus chlorosomes, Biochemistry, in press.

Brune, D. C., King, G. H., Infosino, A., Steiner, T., Thewalt, M. L. W., and Blankenship, R. E., 1987b, Antenna organization in green photosynthetic bacteria II. Excitation transfer in detached and membrane-bound chlorosomes from Chloroflexus aurantiacus, Biochemistry, in press.

Brune, D. C., King, G. H., and Blankenship, R. E., 1988, Intermolecular interactions between bacteriochlorophyll c in in vitro oligomers and in chlorosomes, in: "Photosynthetic Light-Harvesting Systems," H. Scheer and S. Schneider, eds., Walter de Gruyter, Berlin, in press.

Bystrova, M. I., Mal'gosheva, I. N., Krasnovskii, A. A., 1979, Study of molecular mechanism of self-assembly of aggregated forms, Mol. Biol., 13:582.

Cotton, T. M., Loach, P. A., Katz, J. J., and Ballschmiter, K., 1978,

Studies of chlorophyll-chlorophyll and chlorophyll-ligand interactions by visible absorption and infrared spectroscopy at low temperature, Photochem. Photobiol., 27:735.

Feick, R. G., and Fuller, R. C., 1984, Topography of the photosynthetic apparatus of Chloroflexus aurantiacus, Biochemistry, 23:3693.

Foster, J. M., Redlinger, T.E., Blankenship, R. E., and Fuller, R. C., 1986, Oxygen regulation of development of the photosynthetic membrane system in Chloroflexus aurantiacus, J. Bacteriol., 167:655.

Fujiwara, M., and Tasumi, M., 1986a, Resonance raman and infrared studies on axial coordination to chlorophylls a and b in vitro, J. Phys. Chem., 90:250.

Fujiwara, M., and Tasumi, M., 1986b, Metal-sensitive bands in the ramen and infrared spectra of intact and metal-substituted chlorophyll a, J. Phys. Chem., 90:5646.

Gibson, J., Ludwig, W., Stackebrandt, E., Woese, C. R., 1985, The phylogeny of the green photosynthetic bacteria: absence of a close relationship between Chlorobium and Chloroflexus, System. Appl. Microbiol., 6:152.

Katz, J. J., Dougherty, R. C., and Boucher, L. J., 1966, Infrared and nuclear magnetic resonance spectroscopy of chlorophyll, in: "The Chlorophylls," L. P. Vernon and G. R. Seely, eds., Academic Press, New York.

Kirmaier, C., Holten, D., Feick, R., and Blankenship, R. E., 1983, Picosecond measurements of the primary photochemical events in reaction centers isolated from the facultative green photosynthetic bacterium Chloroflexus aurantiacus, FEBS Lett., 158:73.

Kirmaier, C., Holten, D., Mancino, L. J., Blankenship, R. E., 1984, Picosecond photodichroism studies on reaction centers from the green photosynthetic bacterium Chloroflexus aurantiacus, Biochim. Biophys. Acta, 765:138.

Kirmaier, C., Blankenship, R. E., Holten, D., 1986, Formation and decay of radical-pair state P+I- in Chloroflexus aurantiacus, Biochim. Biophys. Acta, 850:275.

Lutz, M., 1984, Resonance raman studies in photosynthesis, Adv. Infrared and Raman Spectrosc., 11:211.

Lutz, M., and van Brakel, G., 1988, Ground-state molecular interactions of bacteriochlorophyll c in chlorosomes of green bacteria and in model systems: a resonance Raman study, This volume.

Olson, J. M., 1980, Chlorophyll organization in green photosynthetic bacteria, Biochim. Biophys. Acta, 594:33.

Olson, J. M., Gerola, P. D., van Brakel, G. H., Meiburg, R. F., and Vasmel, H., 1985, Bacteriochlorophyll a- and c-protein complexes from chlorosomes of green sulfur bacteria compared with bacteriochlorophyll c aggregates in CH_2Cl_2-hexane, in "Antennas and Reaction Centers of Photosynthetic Bacteria," M. E. Michel-Beyerle, ed., Springer-Verlag, Berlin.

Parot, P., Delmas, N., Garcia, D., and Vermeglio, A., 1985, Structure of Chloroflexus aurantiacus reaction center: Photoselection at low temperature, Biochim. Biophys. Acta, 809:137.

Pfennig, N., and Trüper, H. G., 1983, Taxonomy of phototrophic green and purple bacteria: A review, Ann. Microbiol., 134B:9.

Pierson, B. K., and Castenholz, R. W., 1974, Studies of pigments and growth in Chloroflexus aurantiacus, a phototrophic filamentous bacterium, Arch. Microbiol., 100:283.

Pierson, B. K., and Thornber, J. P., 1983, Isolation and spectral characterization of photochemical reaction centers from the thermophilic green bacterium Chloroflexus aurantiacus strain J-10-fl, Proc. Natl. Acad. Sci. U.S.A., 80:80.

Pierson, B. K., 1985, Cytochromes in Chloroflexus aurantiacus grown with and without oxygen, Arch. Microbiol., 143:260.

Shuvalov, V. A., Shkuropatov, A. Ya., Kulakova, S. M., Ismailov, M. A.,

Shkuropatova, V. A., 1986, Photoreactions of bacteriopheophytins and bacteriochlorophylls in reaction centers of <u>Rhodopseudomonas sphaeroides</u> and <u>Chloroflexus aurantiacus</u>, <u>Biochim. Biophys. Acta</u>, 849:337.

Smith, K. M., Kehres, L. A., and Fajer, J., 1983, Aggregation of the bacteriochlorophylls <u>c</u>, <u>d</u>, and <u>e</u>. Models for the antenna chlorophylls of green and brown photosynthetic bacteria, <u>J. Am. Chem. Soc.</u>, 105:1387.

Trost, J. T., Freeman, J. M., and Blankenship, R. E., 1987, Purification and properties of auracyanin, a blue copper protein from <u>Chloroflexus aurantiacus</u>, <u>Biophys. J.</u>, 51:309a.

van Dorssen, R. J., Vasmel, H., and Amesz, J., 1986, Pigment organization and energy transfer in the green photosynthetic bacterium <u>Chloroflexus aurantiatic</u>, <u>Photosynth. Res.</u>, 9:33.

Vasmel, H., Meiburg, R. F., Kramer, H. J. M., De Vos, L. J., and Amesz, J., 1983, Optical properties of the photosynthetic reaction center of <u>Chloroflexus aurantiacus</u> at low temperature, <u>Biochim. Biophys. Acta</u>, 724:333.

Vasmel, H., and Amesz, J., 1983, Photoreduction of menaquinone in the reaction center of the green photosynthetic bacterium <u>Chloroflexus aurantiacus</u>, <u>Biochim. Biophys. Acta</u>, 724:118.

Vos, M., Nuijs, A. M., van Grondelle, R., van Dorssen, R. J., Gerola, P. D., and Amesz, J., 1987, Excitation transfer in chlorosomes of green photosynthetic bacteria, <u>Biochim. Biophys. Acta</u>, 891:275.

Wechsler, T., Suter, F., Fuller, R. C., Zuber, H., 1985, The complete amino acid sequence of the bacteriochlorophyll c binding polypeptide from chlorosomes of the green photosynthetic bacterium <u>Chloroflexus aurantiacus</u>, <u>FEBS Lett.</u>, 181:173.

Wittmershaus, B. D., Nordlund, T. M., Knox, W. H., Knox, R. S., Geacintov, N. E., and Breton, J., 1985, Picosecond studies at 77 K of energy transfer in chloroplasts at low and high excitation intensity, <u>Biochim. Biophys. Acta</u>, 806:93.

Wittmershaus, B. D., Berns, D. S., and Huang, C., 1987, Picosecond time-resolved fluorescence from detergent-free photosystem I particles, <u>Biophys. J.</u>, in press.

Wynn, R. M., Redlinger, T. E., Foster, J. M., Blankenship, R. E., Fuller, R. C., Shaw, R. W., Knaff, D. B., 1987, Electron-transport chains of phototrophically and chemotrophically grown <u>Chloroflexus aurantiacus</u>, <u>Biochim. Biophys. Acta</u>, 891:216.

Zannoni, D., and Ingledew, W. J., 1985, A thermodynamic analysis of the plasma membrane electron transport components in photoheterotrophically grown cells of <u>Chloroflexus aurantiacus</u>, <u>FEBS Lett.</u>, 193:93.

TRANSFER OF EXCITATION ENERGY IN CHLOROFLEXUS AURANTIACUS

R. J. van Dorssen, M. Vos and J. Amesz

Department of Biophysics
Huygens Laboratory of the State University
P.O. Box 9504, NL-2300 RA Leiden, The Netherlands

INTRODUCTION

Green bacteria are characterized by the presence of chlorosomes, oblong bodies attached to the inner surface of the cytoplasmic membrane. These chlorosomes contain BChl c, d, or e as the major light-harvesting pigment and a small amount of BChl a (1-5% of the BChl c content) (Schmidt, 1980; Feick et al., 1982; Gerola and Olson, 1986; van Dorssen et al., 1986a,b). In addition to the chlorosome the light-harvesting system of the green photosynthetic bacterium Chloroflexus aurantiacus consists of a BChl a-containing complex, B808-866, which is contained in the membrane. Both the chlorosome and the B808-866 complex have been isolated in a pure form (Feick et al., 1982) and studied by spectroscopic means (van Dorssen et al., 1986b; Vasmel et al., 1986). The absorption and fluorescence characteristics of the B808-866 complex resemble those of the B800-850 complex present in Rhodobacter sphaeroides and Rb. capsulatus (van Grondelle et al., 1982). Excitation is transferred from BChl a 808 to BChl a 866 with an efficiency close to 100% (Vasmel et al., 1986). Emission spectra of isolated chlorosomes on excitation of BChl c show bands of both BChl c and BChl a, indicating the transfer of excitation energy from the former to the latter pigment.

It has been postulated that the minor BChl a component in the chlorosome (BChl a 798) is involved in the transfer of excitation from the chlorosome to the membrane (Betti et al., 1982; Amesz and Vasmel, 1986; van Dorssen et al., 1986b), but this transfer has not yet been directly demonstrated. For this reason it was of interest to study the energy transfer in whole cells of Cf. aurantiacus and the possible role of the chlorosomal BChl a in this process, both at low temperature and under more physiological conditions. The results are reported in this paper, together with measurements on isolated chlorosomes.

MATERIALS AND METHODS

Cells of Cf. aurantiacus strain J-10-fl were grown at 52°C in medium D as described by Pierson and Castenholz (1974). For measurements between 50 and 10°C the cells were diluted with glycerol to reduce scattering and transferred directly to the cuvette. When experiments at low temperature were performed the cells were harvested by centrifugation and resuspended in

10 mM Tris, pH = 8.0. Prior to measurements glycerol was added (66% v/v), again to reduce scattering and to prevent crystallization at low temperature. Isolated chlorosomes were prepared as described by Feick et al. (1982) and resuspended in a glycerol-containing medium.

The apparatus to measure absorption, fluorescence emission and excitation spectra has been described by Kramer and Amesz (1982). Linear dichroism (LD) spectra of cells were measured after orientation by biaxial pressing of a polyacrylamide gel (Vasmel et al., 1986). Excitation annihilation was measured using 35-ps flashes of 532 nm obtained from a frequency-doubled mode-locked Nd-YAG laser by means of the apparatus described by Vos et al. (1986).

RESULTS

Fig. 1 shows the excitation spectrum of the long-wave emission of isolated chlorosomes from Cf. aurantiacus measured at 4 K. The band at 800 nm is attributed to the Q_y transition of BChl a 798; the corresponding Q_x transition is located at 613 nm. In addition to the bands of BChl a, contributions of BChl c at 742 and 462 nm and of carotenoid at 521 nm are observed. This indicates that transfer of excitation energy occurs from the major light-harvesting pigments BChl c and carotenoid to the minor fraction of BChl a molecules present in the chlorosome. From the absorption (1-T, where T is the transmittance) spectrum the efficiency of this energy transfer to BChl a 798 can be calculated to be 55% and 35% for BChl c and carotenoid, respectively.

Information of the rate of energy transfer between BChl c molecules at 4 K was obtained by measurements of excitation annihilation, by plotting the relative yield of BChl c fluorescence as a function of the intensity of 35-ps flashes of 532 nm. From the excitation spectrum for BChl c fluorescence it followed that light energy absorbed at this wavelength is transferred with an efficiency of 50% to BChl c. From the absorption at 532 nm the number of excited BChl c molecules in the sample was then calculated for each flash energy. With high flash energies a strong quenching of the BChl c fluorescence was observed due to singlet-singlet annihilation. The extent of this annihilation is dependent on the rate of energy transfer between the BChl c molecules and on the size of the domain, i.e. the ensemble of BChl c

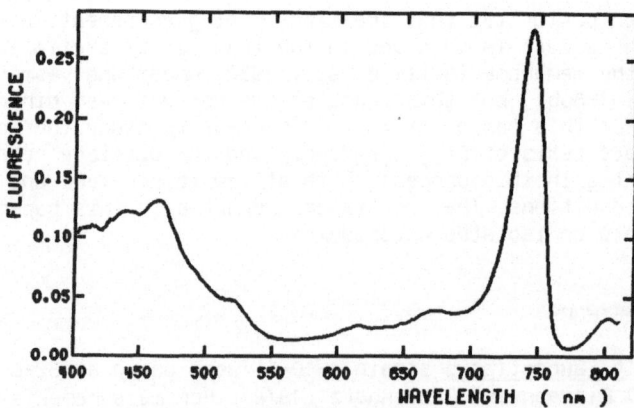

Fig. 1. Fluorescence excitation spectrum of isolated chlorosomes of Cf. aurantiacus at 4 K detected at 835 nm. The absorbance at 742 nm was 0.3 (4 K).

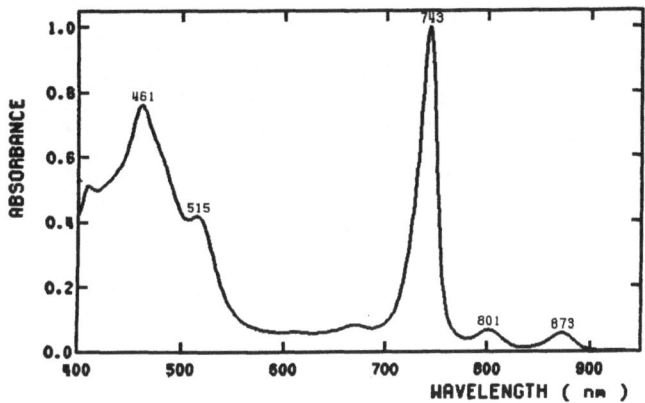

Fig. 2. Absorption spectrum of whole cells measured at 4 K.

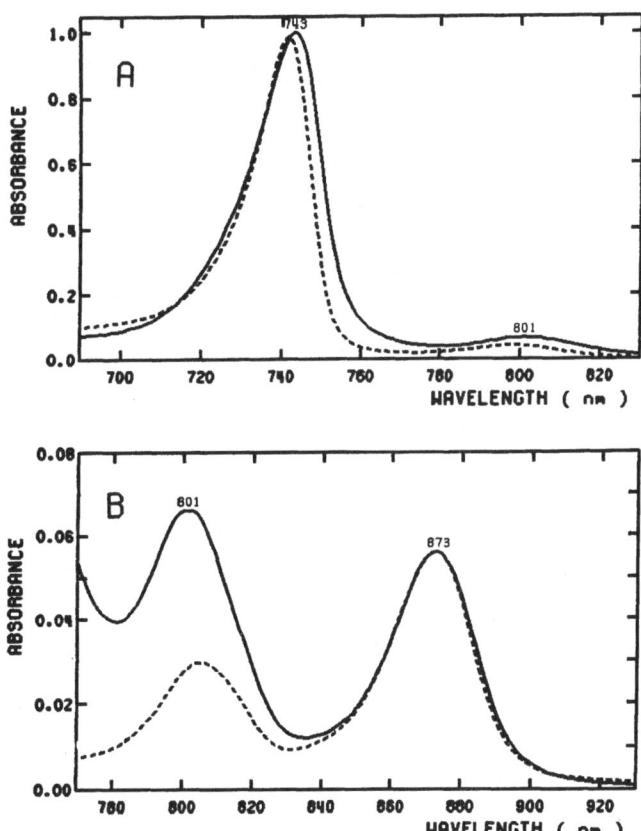

Fig. 3. Q_y region of the 4 K absorption spectrum of whole cells. The dashed lines give the spectra of the isolated chlorosomes (A) and of isolated membrane fragments (taken from Vasmel et al., 1986) (B) see text).

molecules among which energy transfer occurs. The data were analyzed using the theory described by Paillotin et al. (1979). In this way a domain size at 4 K of ≥ 100 BChl c molecules was obtained.

By use of the random walk model of Den Hollander et al. (1983), the rate constant for energy transfer between BChl c molecules was calculated. The rate of mono-excitation decay, needed for this calculation was obtained

by comparing the fluorescence yield of the chlorosomes with those of chromatophores of Rhodospirillum rubrum. The fluorescence yield at 4 K was found to be 1.2%, corresponding to a life-time of the excited state of 25 ps. Together with a domain size of ≥ 100 BChls, this yielded an average rate constant for energy transfer of ≤ 4.3 x 10^{11} s^{-1}. This rate is faster than observed in the antenna of purple bacteria (Vos et al., 1986), in agreement with the high pigment density in chlorosomes. However, structural data indicate that BChl c occurs in clusters of perhaps about 14 molecules (Zuber et al., 1987), among which strong interaction and consequently very fast energy transfer may be assumed to occur. This consideration does not affect the calculated domain size, but the rate of energy transfer between clusters reduces to ≤ 2.2 x 10^{10} s^{-1}, if it is assumed that energy transfer within a cluster is very fast.

To obtain information on the efficiency of energy transfer from the chlorosome to the membrane and on the possible role of BChl a 798 in this energy transfer, the spectral characteristics of whole cells of Cf. aurantiacus were studied. Fig. 2 shows the low temperature absorption spectrum of whole cells of Cf. aurantiacus. The important light-harvesting function of the chlorosome is reflected by the dominance of the BChl c Q$_y$ band at 743 nm in the absorption spectrum. In the blue region bands of carotenoid (515 nm) and of the Soret transitions of BChl c (461 nm) are observed. BChl a bands are seen at 801 and 873 nm. Part of the spectrum is replotted on an extended scale for a comparison with the spectra of isolated chlorosomes (Fig. 3A) and the membrane fragments (Fig. 3B). The Q$_y$ bands of BChl c in the spectrum of whole cells and of isolated chlorosomes (Fig. 3A) do not coincide. The latter band is located at somewhat shorter wavelength and is much more asymmetric resulting in a considerably lower intensity beyond 745 nm.

In the near-infrared region (Fig. 3B) the shape and position of the maximum of the band of BChl 866 is very similar for the spectra of whole cells and the isolated membrane fragments, but below 840 nm the two spectra deviate strongly and it is clear that the additional absorption must arise from the chlorosomal BChl a 798. Subtraction of the two spectra indeed yielded a band at 798 nm, which accounted for 61% of the absorption at this wavelength. A ratio of 34 for the BChl c: BChl a 798 content was calculated based on in vivo extinction coefficients of 102 and 137 mM$^-$ cm^{-1} for BChl c and BChl a, respectively (van Dorssen et al., 1986b; Vasmel et al., 1986). This ratio is somewhat higher than those reported for isolated chlorosomes (van Dorssen et al., 1986b; Feick et al., 1982).

The bands of BChl a 798 and BChl a 808 were not resolved in second and fourth derivative spectra. However, well-separated contributions of these two pigments can be observed in the LD spectrum of whole cells. From the LD of isolated chlorosomes (van Dorssen et al., 1986b) it is known that the Q$_y$ transition of BChl c is oriented nearly parallel to the long axis of the chlorosome. Similar experiments with isolated membrane fragments have shown that the Q$_y$ transitions of BChl a 866 are lying in a plane parallel to the plane of the cytoplasmic membrane. In agreement with this, the low-temperature LD of whole cells (Fig. 4) shows positive signals of BChl c at 744 nm and of BChl a at 866 at 872 nm. The latter observation indicates that the cells are squeezed to align their membranes preferentially parallel

to the axis of orientation. From the positive LD signal at 744 it can be
concluded that the chlorosomes are not randomly oriented, but that they are
aligned with their long axis more or less parallel to the cytoplasmic
membrane, in agreement with the observations made by electron microscopy
(Staehelin et al., 1978). Upon closer examination of the region around 800
nm (see inset of Fig. 4) negative signals of BChl a 798 at 796 nm and of
BChl a 808 at 810 nm are observed. Both Q_y transitions thus make a fairly
large angle with the plane of the membrane in agreement with earlier
observations on the isolated systems (van Dorssen et al., 1986b; Vasmel et
al., 1986).

Upon excitation of the cells at 690 nm in the tail of the Q_y band of
BChl c the low temperature emission spectrum (Fig. 5) shows maxima at 755,
817 and 891 nm of BChl c, BChl a 798 and BChl a 866, respectively. The
positions of the latter two bands agree well with those reported for the
isolated chlorosomes and membranes, but the BChl c emission maximum is
located at somewhat shorter wavelength than in the isolated chlorosomes (van
Dorssen et al., 1986b). It should be noted, however, that the maximum of the
BChl c emission varied by as much as 6 nm for different batches of cells.
Smaller variations of up to 2 nm were observed for the positions of the
emission bands of BChl a 798 and BChl a 866.

Fig. 6 shows the low temperature excitation spectrum of the BChl a 866
emission. The most significant feature of this spectrum is the low
contribution of the BChl c bands relative to those of BChl a. Comparison
with the absorption spectrum indicated that BChl c transfers its excitation
energy to BChl a 866 with an overall efficiency of only 15%. The height of
the band at 803 nm is also relatively low. A high rate of energy transfer is
known to occur within the B808-866 complex (Vasmel et al., 1986). Therefore,
the reduced height at 803 nm must reflect a reduced transfer efficiency
between BChl a 798 and BChl a 808 of the B808-866 complex; this transfer
efficiency was calculated to be 30%. By measuring the excitation spectrum of
the BChl a 798 emission at 825 nm (not shown) the transfer efficiency from
BChl c to BChl a 798 can be calculated to be 49%, in agreement with results
obtained with isolated chlorosomes (Fig. 1). These results strongly support
the hypothesis that BChl a 798 functions as an intermediate in energy
transfer from the chlorosome to the membrane. At low temperature the
efficiency of energy transfer from BChl c to BChl a 798 is about 50%, from
BChl a 798 to BChl a 808 and hence to BChl a 866 it is 30%, and together
these efficiencies result in an overall efficiency for energy transfer from
BChl c to BChl a 866 of only 15% at low temperature.

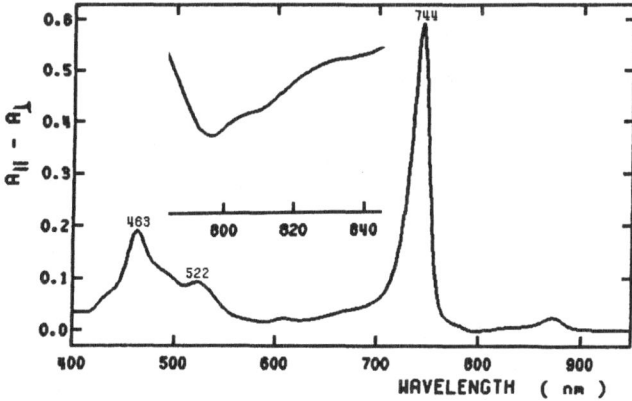

Fig. 4. Linear dichroism of whole cells oriented by biaxial pressing and
measured at 4 K. Absorbance at 742 nm was 0.74 (293 K).

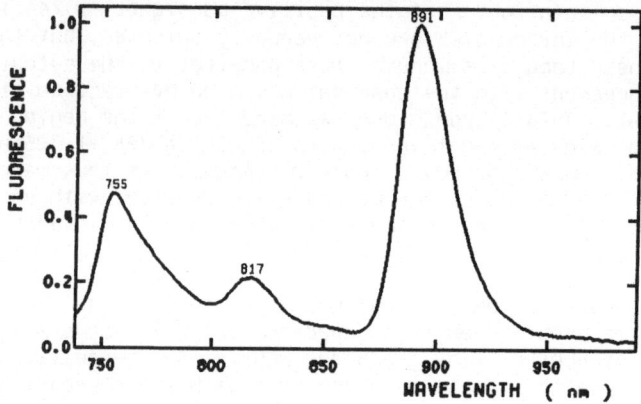

Fig. 5. Emission spectrum of whole cells at 4 K excited at 690 nm. A.U.,
    arbitrary units.

The above efficiencies were derived by comparing the amplitudes in the
maxima of the corresponding bands. When the transfer efficiency from BChl c
to BChl a 866 was calculated across the BChl c band the value was constant
until 745 nm and then dropped sharply on the long-wavelength side to a value
of only half of that observed between 700 and 745 nm. Similar observations
were made for the calculated efficiencies across the BChl c bands in the
excitation spectra of the BChl a 798 and BChl c fluorescence.

Energy transfer from the chlorosome to the membrane was much more
efficient at physiological temperatures. Nevertheless, the rate of energy
transfer from BChl c to BChl a 866 varied strongly over the temperature
range of 10-50°C. This is illustrated by the excitation spectra of the
long-wave emission, plotted together with the corresponding absorption
spectra in Fig. 7 for two different temperatures. At 52°C the excitation
spectrum closely follows the absorption spectrum (Fig. 7A) indicating a 100%

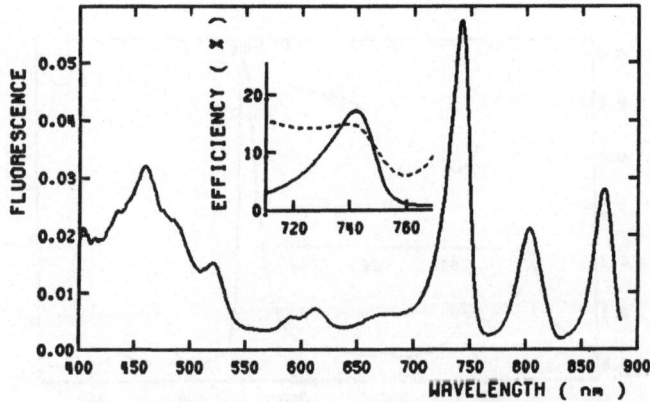

Fig. 6. Fluorescence excitation spectrum of the long-wave emission at 4 K
    detected at 900 nm. The inset shows the region around 740 nm and the
    calculated transfer efficiency (dashed line).

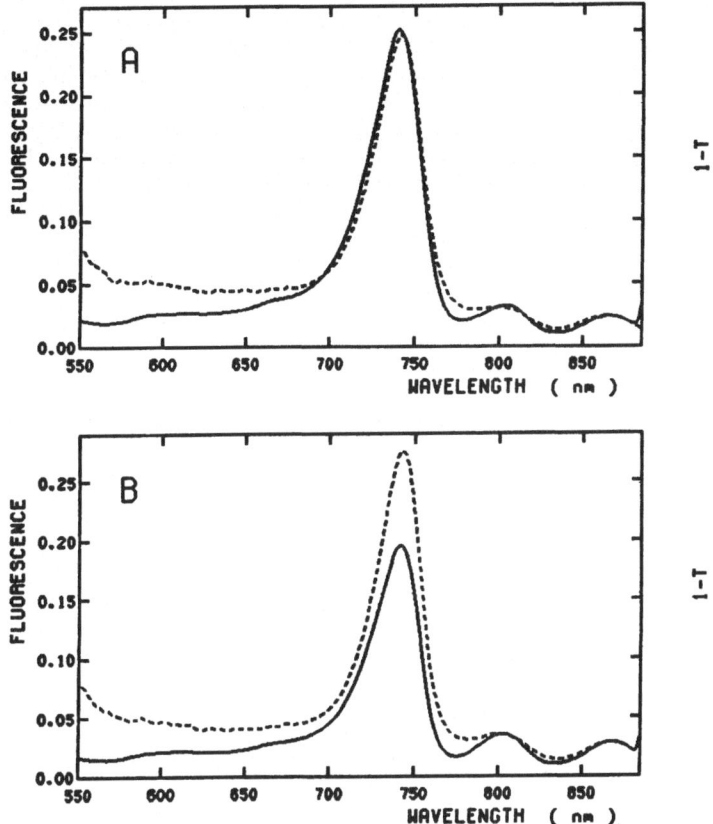

Fig. 7. Fluorescence excitation spectra of whole cells measured at 52°C (A) and 10°C (B). The corresponding absorption spectra are given by the dashed lines.

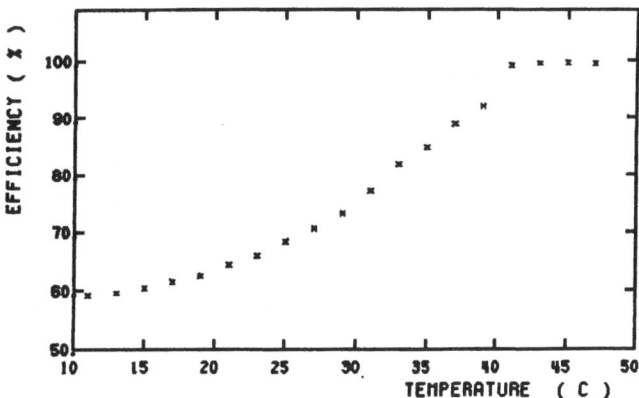

Fig. 8. Calculated transfer efficiency between BChl c and BChl a 866 as a function of temperature.

efficiency of energy transfer. Upon lowering the temperature to 10°C (Fig. 7B) this value dropped to 66%. The height of the band near 800 nm was not changed, indicating that the effect is due to a lowered efficiency of energy transfer from BChl c to BChl a 798 in the chlorosome. Fig. 8 shows the calculated transfer efficiency from BChl c to BChl a 866 as a function of temperature. Above 42°C a constant value of 100% is observed. Over the region of 40 to 20°C a sudden decrease occurs and upon further cooling the efficiency versus temperature curve flattens out. The shape of this curve seems to suggest that the observed decrease of the efficiency is related to a lipid phase transition. The full reversibility of this process provides additional evidence for this explanation.

The fluorescence emission also changed strongly over the same temperature range. Fig. 9 shows the emission spectra measured at 10 and 52°C upon BChl c excitation. At 52°C maxima were observed at 755, 805 and 882 nm. Upon cooling to 10°C small increases are observed in the BChl c and BChl a 798 emissions and a strong increase in the BChl a 866 fluorescence. At room temperature (20°C) the intensity of the exciting light was sufficient to bring the cells into a condition of maximum fluorescence within a few hundred ms. The ratio F_{max}/F_o was 2.5 for the BChl a 866 fluorescence. No induction was observed in the fluorescence of BChl c and BChl a 798. This shows that in contrast to green sulfur bacteria (Clayton, 1965) no "back transfer" of energy to the chlorosome occurs, presumably because of the larger energy difference in Cf. aurantiacus. Comparison with the fluorescence yield of R. rubrum chromatophores gave an absolute yield of fluorescence (F_{max}) of 8% for BChl a 866 at 20°C. The yield of BChl c fluorescence in whole cells at 20°C was 1.3%. In isolated chlorosomes the yield was about 8 times lower. This indicates that strong quenching processes occur in the isolated chlorosomes. Such a strong quenching was also observed in chlorosomes of Chlorobium limicola, where it could partially be relieved by the addition of dithionite (van Dorssen et al., 1986a).

As it is known that in green plants and algae the Chl a fluorescence can be used as a native probe of the physical state of the thylakoid membrane (Murata and Fork, 1975) we also investigated the possibility of a similar function of the BChl c and BChl a fluorescences in whole cells of Cf. aurantiacus. Fig. 10 shows the intensity of the BChl a 866 emission excited at 860 nm as a function of temperature. The intensity of the

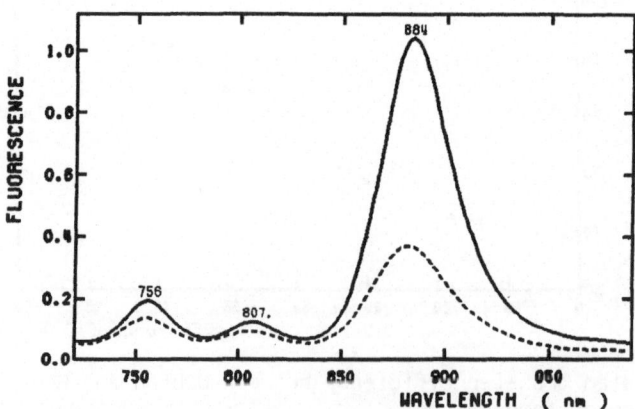

Fig. 9. Emission spectra of whole cells measured at 10°C (___) and 52°C (---) excited at 690 nm. The two spectra are drawn to the same scale.

emission increases monotonically with decreasing temperature, but two plateaus are observed at 37 and 28°C, indicating the occurrence of lipid phase transitions at these temperatures. In the plot of the BChl c fluorescence versus temperature excited at 690 nm (not shown) similar phase transitions were seen at 37 and 28°C.

DISCUSSION

Chlorosomes of green bacteria are eminently suited for their important light-harvesting function as they contain large quantities of antenna pigments (about 10,000 BChl c molecules) at a very high concentration. The chlorosome serves as a common antenna for several reaction centers. Rapid energy transfer occurs among the highly ordered BChl c chromophores. Previous reports have dealt with the spectral characterization of chlorosomes and membranes of Cf. aurantiacus in isolated form (van Dorssen et al., 1986b; Vasmel et al., 1986). In this study we have investigated the transfer of excitation energy within the intact system of whole cells, both at low temperature and under more physiological conditions.

It has been speculated that the minor fraction of BChl a 798 observed in isolated chlorosomes is involved as an intermediate in energy transfer between the chlorosome and the membrane (Betti et al., 1982; van Dorssen et al., 1986b). Evidence for this notion was found in the favorable overlap and orientation of BChl a 798 with respect to BChl a 808 in the membrane. Measurements of the LD of oriented cells confirm our earlier conclusions (van Dorssen et al., 1986a,b) that BChl a 798 and BChl a 808 have a very similar orientation with respect to the plane of the membrane, favoring energy transfer between them. The transfer efficiencies that have been calculated from the low temperature excitation spectra of the BChl a 798 and BChl a 866 emissions provide strong evidence that energy transfer between the chlorosome and the membrane indeed proceeds by means of BChl a 798.

The light-harvesting system of photosynthetic bacteria, algae and higher plants is in general characterized by a very high efficiency (> 90%) of energy transfer within antenna complexes and between different antenna complexes, both a room and low temperature (van Grondelle, 1985). In this context the low efficiencies that are observed at low temperature for Cf. aurantiacus are remarkable. At room temperature the transfer efficiency between BChl c and BChl a 798 is already rather low (60-70%), whereas the

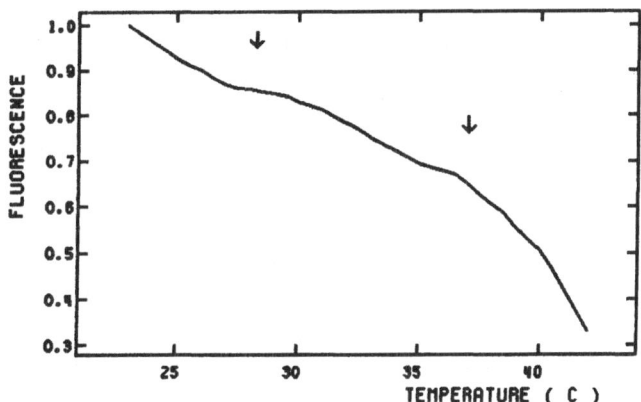

Fig. 10. Fluorescence of BChl a 866 detected at 900 nm and excited at 860 nm as a function of temperature.

77

transfer from BChl $\underline{a}$ 798 to the membrane still seems to proceed in a highly efficient manner. When the different transfer efficiencies are monitored as a function of temperature two effects are observed. Raising the temperature to 50°C, near the temperature where the cultures are grown, results in a fairly sudden and strong increase to a value of close to 100% for the transfer efficiency between BChl $\underline{c}$ and BChl $\underline{a}$ 798. Upon lowering the temperature to 4 K the transfer efficiency from BChl $\underline{c}$ to BChl $\underline{a}$ 798 remains more or less constant, but a decrease of the transfer efficiency between BChl $\underline{a}$ 798 and BChl $\underline{a}$ 808 to a value of 30% occurs.

The rate of energy transfer between two molecules can be described by the well-known Förster equation (van Grondelle, 1985). Several parameters can influence this transfer and a decrease in this rate can occur by an increase in the donor-acceptor distance or an increase in the rate of other non-radiative deactivation processes of the donor. The shape of the efficiency vs. temperature curve in the region 10-50°C (Fig. 8) and the observed yield changes in the BChl $\underline{a}$ 866 fluorescence in the same temperature range (Fig. 10) suggest the involvement of a lipid phase transition. Phase transitions in this temperature region have also been observed by microcalorimetry in membranes of $\underline{Cf.\ aurantiacus}$ (Oelze and Fuller, 1983). In model systems containing only a single lipid species the transition from the solid to the liquid crystalline state occurs over a small temperature range (T 1°C), but in natural bilayers a broader temperature range of 10-30°C is observed, due to a heterogeneity in the lipid content (Overath et al., 1976). Important aspects of lowering the temperature of a membrane system below the transition temperature are the disappearance of the rapid lateral diffusion of the lipid molecules, a decrease in their spacing and an increase in the thickness of the bilayer.

The observed drop in the transfer efficiency between BChl $\underline{c}$ and BChl $\underline{a}$ 798 from close to 100% at 52°C to approximately 66% at 10°C could result from an increase in the spacing between both chromophores related to the lipid phase transition. If this were the sole explanation a much stronger increase of the yield of the BChl $\underline{c}$ fluorescence should be observed. This indicates that a change in the rate of the non-radiative de-excitation of BChl $\underline{c}$ may result in a more effective competition of this process with energy transfer to BChl $\underline{a}$ 798. The analysis is further complicated by the presence of a BChl $\underline{c}$ fraction which at 4 K has a low fluorescence yield and a low transfer efficiency (Figs. 5 and 6), and which seems to manifest itself also at higher temperatures, as indicated by the drop in efficiency in the excitation spectrum at 760-780 nm (Fig. 7).

Cooling the cells from 10°C to 4 K appears to affect mainly the efficiency of energy transfer between BChl $\underline{a}$ 798 and BChl $\underline{a}$ 808. This effect may be largely due to a decrease in the Förster overlap, since the emission band of BChl $\underline{a}$ 798 sharpens strongly and shifts upon cooling from 805 to 819 nm (van Dorssen et al., 1986b), whereas the absorption band of BChl $\underline{a}$ 808 shows a blue shift of 3 nm (Vasmel et al., 1986).

REFERENCES

Amesz, J., and Vasmel, H., 1986, Fluorescence properties of photosynthetic bacteria, in: "Light Emission by Plants and Bacteria," Govindje, J. Amesz, and D. C. Fork, eds., Academic Press, New York.
Betti, J. A., Blankenship, R. E., Natarajan, L. V., Dickinson, L. C., and Fuller, R. C., 1982, Antenna organization and evidence for the function of a new antenna pigment species in the green photosynthetic bacterium Chloroflexus aurantiacus, Biochim. Biophys. Acta, 680:194.
Clayton, R. K., 1965, Characteristics of fluorescence and delayed light from green photosynthetic bacteria and algae, J. Gen. Microbiol., 48:633.

Den Hollander, W. T. F., Bakker, J. G. C., and van Grondelle, R., 1983, Trapping, loss and annihilation of excitations in a photosynthetic system, <u>Biochim. Biophys. Acta</u>, 725:492.

Feick, R. G., Fitzpatrick, M., and Fuller, R. C., 1982, Isolation and characterization of cytoplasmic membranes and chlorosomes from the green bacterium <u>Chloroflexus aurantiacus</u>, <u>J. Bacteriol.</u>, 150:905.

Gerola, P. D., and Olson, J. M., 1986, A new bacteriochlorophyll <u>a</u>-protein complex associated with chlorosomes of green sulfur bacteria, <u>Biochim. Biophys. Acta</u>, 848:69.

Kramer, H. J. M., and Amesz, J., 1982, Anisotropy of the emission and absorption bands of spinach chloroplasts by fluorescence polarization and polarized excitation spectra at low temperature, <u>Biochim. Biophys. Acta</u>, 682:201.

Murata, N., and Fork, D. C., 1975, Temperature dependence of chlorophyll <u>a</u> fluorescence in relation to the physical phase of membrane lipids in algae and higher plants, <u>Plant Physiol.</u>, 56:791.

Oelze, J., and Fuller, R. C., 1983, Temperature dependence of growth and membrane-bound activities of <u>Chloroflexus aurantiacus</u> energy metabolism, <u>J. Bacteriol.</u>, 155:90.

Overath, P., Thilo, L., and Träuble, H., 1976, Lipid phase transitions and membrane function, <u>Trends Biochem. Sci.</u>, 1:186.

Paillotin, G., Swenberg, C. E., Breton, J., and Geacintov, N. E., 1979, Analysis of picosecond laser-induced fluorescence phenomena in photosyntehtic membranes utilizing a master equation, <u>Biophys. J.</u>, 25:513.

Pierson, B. K., and Castenholz, R. W., 1974, A phototrophic gliding filamentous bacterium of hot springs. <u>Chloroflexus aurantiacus</u>, gen. and sp. nov., <u>Arch. Microbiol.</u>, 100:5.

Schmidt, K., 1980, A comparative study on the composition of chlorosomes and cytoplasmic membranes from <u>Chloroflexus aurantiacus</u> strain OK-70-fl and <u>Chlorobium limicola</u> f. <u>thiosulfatophilum</u> strain 6230, <u>Arch. Microbiol.</u>, 124:21.

Staehelin, L. A., Golecki, J. R., Fuller, R. C., and Drews, G., 1978, Visualization of the supermolecular architecture of chlorosomes (chlorobium type vesicles) in freeze-fractured cells of <u>Chloroflexus aurantiacus</u>, <u>Arch. Microbiol.</u>, 119:269.

van Dorssen, R. J., Gerola, P. D., Olson, J. M., and Amesz, J., 1986a, Optical and structural properties of chlorosomes in the photosynthetic green sulfur bacterium <u>Chlorobium limicola</u>, <u>Biochim. Biophys. Acta</u>, 848:77.

van Dorssen, R.J., Vasmel, H., and Amesz, J., 1986b, Pigment organization and energy transfer in the green photosynthetic bacterium <u>Chloroflexus aurantiacus</u>. II. The chlorosome, <u>Photosynth. Res.</u>, 9:33.

van Grondelle, R., 1985, Excitation energy transfer, trapping and annihilation in photosynthetic systems, <u>Biochim. Biophys. Acta</u>, 811:147.

van Grondelle, R., Kramer, H. J. M., and Rijgersberg, C. P., 1982, Energy transfer in the B800-850 carotenoid light-harvesting complex of various mutants of <u>Rhodopseudomonas sphaeroides</u> and <u>Rhodopseudomonas capsulata</u>, <u>Biochim. Biophys. Acta</u>, 682:208.

Vasmel, H., van Dorssen, R. J., de Vos, G.J., and Amesz, J., 1986, Pigment organization and energy transfer in the green photosynthetic bacterium <u>Chloroflexus aurantiacus</u>. I. The cytoplasmic membrane, <u>Photosynth. Res.</u>, 7:281.

Vos, M., van Grondelle, R., van der Kooij, F. W., van de Poll, D., Amesz, J., and Duysens, L. N. M., 1986, Singlet-singlet annihilation at low temperatures in the antenna of purple bacteria, <u>Biochim. Biophys. Acta</u>, 850:501.

Zuber, H., Brunisholz, R., and Sidler,W., 1987, Structure and function of light-harvesting pigment-protein complexes, <u>in</u>: "Photosynthesis," J. Amesz, ed., Elsevier, Amsterdam.

EXCITATION ENERGY TRANSFER IN LIVING CELLS OF THE GREEN BACTERIUM CHLOROBIUM

LIMICOLA STUDIED BY PICOSECOND FLUORESCENCE SPECTROSCOPY

A.M. Freiberg[a], K.E. Timpmann[a] and Z.G. Fetisova[b]

[a]Institute of Physics, Estonian SSR Academy of Sciences
 202400 Tartu, U.S.S.R.
[b]A.N. Belozersky Laboratory of Molecular Biology and
 Bio-organic Chemistry, Moscow State University
 119899 Moscow, U.S.S.R.

INTRODUCTION

During the past few years advances have been made in understanding the structure and the function of the photosynthetic unit (PSU) of green bacteria (Olson, 1980; Fetisova and Borisov, 1980; Betti et al., 1982; Wechsler et al., 1984; Amesz, 1985; Zuber, 1985; Amesz and Vasmel, 1986; van Dorssen et al., 1986a,b; Fetisova et al., 1986; Fetisova et al., 1987). In these studies the progress of preparative biochemistry proved to be decisive, since a major reason for the paucity of data on the primary events in the photosynthesis of green bacteria was undoubtedly the large antenna. The available data provide information about the structure and function of various isolated subcellular pigmented fractions from green bacteria, whereas hardly anything is known about the functioning of the PSU in intact cells under physiological conditions. Because of their large antenna size (largest of all the known PSUs) and the high quantum yield of primary charge separation in the reaction center (RC) (Fetisova and Borisov, 1980) this family of organisms is most promising (Fetisova and Fok, 1984) in the search for the fundamental principles of the PSU structural organization which provide large and highly efficient PSUs in vivo. This work presents the results of the first direct measurements of picosecond excitation energy transfer rates within a light-harvesting antenna in living cells of the green bacterium Chlorobium limicola as well as the results of the investigation of some structural aspects of this energy transfer.

MATERIALS AND METHODS

Cells of the green sulfur bacterium Cb. limicola (Strain C, Moscow University Collection) were grown as described elsewhere (Kondrat'eva and Moshentseva, 1960). Absorption spectra were measured with a Hitachi 557 spectrophotometer. Fluorescence spectra were measured with a Hitachi MPF-4 spectrofluorometer. A phase fluorometer operating at 12.3×10^6 Hz was used for fluorescence lifetime measurements on a nanosecond time scale (Borisov et al., 1977). Picosecond fluorescence kinetics were measured by a picosecond spectrochronograph described in detail by Freiberg (1986). The picosecond pulse source was a Spectra-Physics mode-locked CW oxazine 1 dye laser (pulse duration 3 ps) pumped synchronously at 82 MHz by a krypton-ion laser. The excitation pulses were vertically polarized. The average exciting light intensity was checked by a radiometer NRC Model 880. The emission viewed at an angle of 90° to the exciting beam (in a reflection mode) was filtered by two single-grating LOMO MDR-2 monochromators combined in the subtractive dispersion mount to avoid extra pulse broadening (bandwidth 4 nm) and was detected by a Hamamatsu synchroscan streak camera. The time resolution of the measuring system (full width at half maximum of the response function) was from 16 to 22 ps in different experiments. For polarization measurements two IR polarizers (Carl Zeiss, Jena) were used in the emission beam. The polarizer on the input slit served for correction of polarization of the detection system. Fluorescence polarization values are expressed as $p(t) = I_\parallel - I_\perp)/(I_\parallel + I_\perp)$, where $I_\parallel$ and $I_\perp$ are the signal intensities $I(t)$ associated with the respective fluorescence decay components measured through a polarizer oriented either parallel or perpendicular to the polarization direction of the exciting light. Positioning of the polarizers was optimized using Rayleigh scattering from a glycogen solution. For the ideal case p should be 1.0; typically, p was 0.98. For data recording and processing a Videoton EC-1010 computer-supported B&M Spektronik OSA 500 optical multichannel analyzer with SIT vidicon was used. All fluorescence decay curves were deconvoluted as a sum of exponential components, utilizing the apparatus response function which was stored independently.

RESULTS AND DISCUSSION

All experiments were performed on intact 2-5 day old cells used in their own growth medium under strictly anaerobic conditions. PSUs of these green sulfur bacteria contain about 1000-1500 BChl $\underline{c}$ molecules and about 90 BChl $\underline{a}$ molecules per RC, with their main near-infrared absorption peaks at 730-750 and 810 nm, respectively. The fluorescence spectra show maxima at

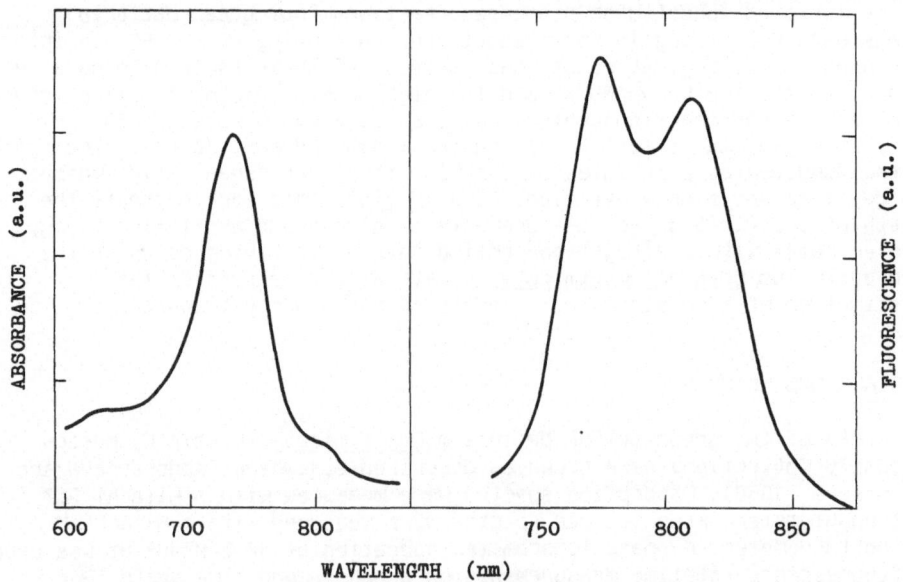

Fig. 1. Room-temperature near-infrared absorption and uncorrected fluorescence spectra of intact cells of <u>Cb. limicola</u>. Excitation at 700 nm was used. A.u., arbitrary units.

770-785 and 820 nm, due to the BChl c and BChl a emissions, respectively. The absorption and fluorescence spectra of the cells used are shown in Fig. 1. The light dependences of BChl c and BChl a fluorescence yields (ϕ_m) and lifetimes (τ_m) measured with the phase fluorometer are shown in Fig. 2. Excitation light intensity was changed from 6×10^{-5} to 6×10^{-3} W/cm².

The character of the light dependences of ϕ_m and τ_m for both BChl c and BChl a fluorescences proved to be analogous, i.e., in the transition from non-saturated to light-saturated photosynthesis the increase in the measured relative fluorescence yields of BChl c and BChl a in intact Cb. limicola cells (under physiological conditions) was accompanied by an increase in their fluorescence lifetimes measured on a nanosecond time scale (from ~1 to ~2 ns). The living cells (under strictly anaerobic conditions) exhibit induction effects in their fluorescence. The absolute values of nanosecond τ_m showed significant sample-to-sample variations (up to 50%). Our data are in good qualitative agreement with earlier data obtained by other investigators (Sybesma and Olson, 1963; Sybesma and Vredenberg, 1963; Clayton, 1965; Govindjee et al., 1972). When cells are aerated, ϕ_m and τ_m decrease steadily to vanishingly small values during the first hour. Aerated cells do not exhibit induction effects in their fluorescence.

In contrast to intact cells photoactive pigment-protein complexes of Cb. limicola (Fig. 3, and Fetisova and Borisov, 1980) showed an antibatic character of the light dependences of ϕ_m and τ_m for both BChl c and BChl a fluorescence. Increasing excitation light intensity caused the photobleaching of the primary electron donor P840 (Fig. 4, and Borisov et

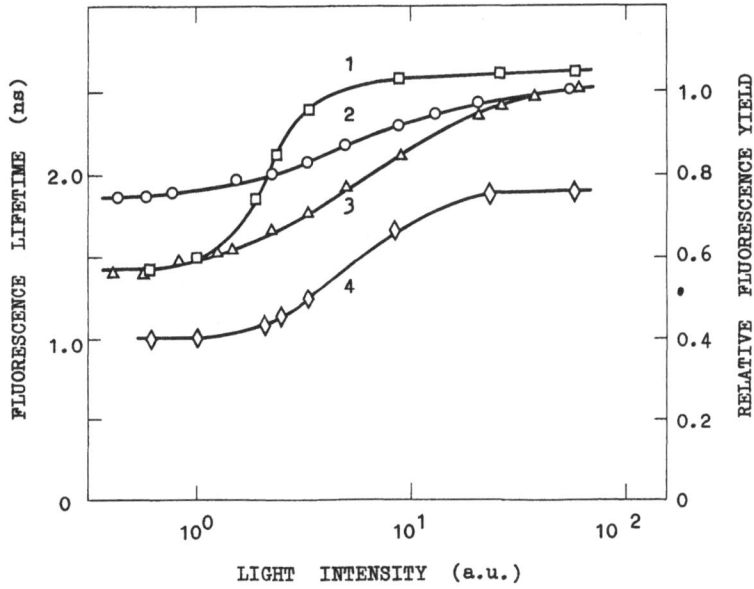

Fig. 2. BChl a fluorescence lifetime (1), relative BChl a fluorescence yield (a.u.) (3), BChl c fluorescence lifetime (4) and relative BChl c fluorescence yield (a.u.) (2) of intact Cb. limicola cells under natural conditions as a function of excitation light intensity. Excitation with mercury lines of 404 and 436 nm was used. Relative BChl c and BChl a fluorescence yields were measured with interference filters at 757 and 819 (or 854) nm, respectively. (The light dependences of ϕ_m measured at 819 and 854 nm were the same.) Maximal excitation intensity was equal to 6×10^{-3} W/cm².

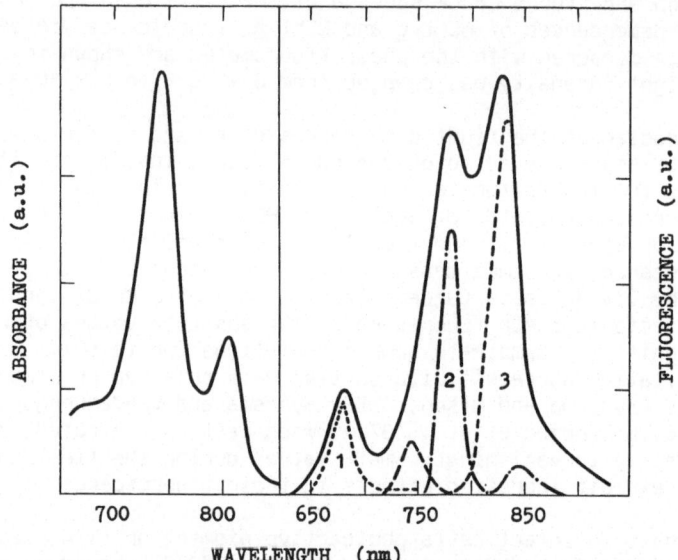

Fig. 3. Absorbance and uncorrected fluorescence spectra of the photochemically active pigment-protein complex obtained from Cb. limicola. Excitation with the 436-nm mercury line was used. Transmittance spectra (1-3), of three filter combinations used for the examination of BChl c and BChl a fluorescence (see Fetisova and Borisov, 1980).

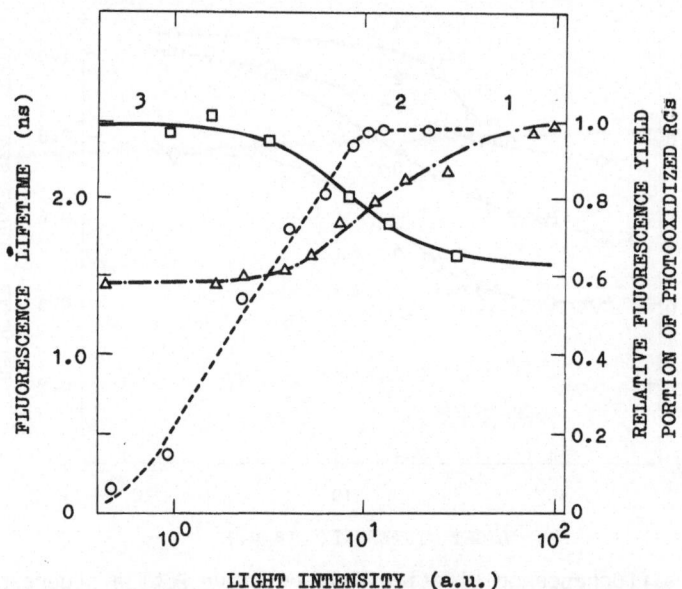

Fig. 4. Relative BChl a fluorescence yield (a.u.) (1), fraction of photo-oxidized P840 (P^+/P_o) (2) and BChl a fluorescence lifetime (3) of the pigment-protein complex obtained from Cb. limicola as a function of light intensity (see Borisov et al., 1977). Maximal excitation intensity was equal to 6×10^{-3} W/cm².

(or equal to) 740 nm, where the contribution from BChl $\underline{a}$ emission is of minor importance. The BChl $\underline{a}$ fluorescence kinetics were detected at 820 nm. The picosecond fluorescence kinetics of BChl $\underline{c}$ and BChl $\underline{a}$ of Cb. limicola cells were measured both without chemical additions and in the presence of 30 μM N-methyl-phenazonium methosulphate (PMS) at the excitation density 0.2 W/cm^2. It was checked with antenna complexes from Rhodospirillum rubrum that PMS did not cause the trivial quenching of fluorescence. The same decay kinetics were observed over an approximately 10^3-fold range in the laser pulse intensity. Thus, under our excitation conditions neither singlet-singlet nor singlet-triplet annihilation took place. Without additions at the excitation power applied (see Figs. 2 and 4) the RCs are in the closed state, probably in the state P$^+$X$_1$X$_2$X$_3^-$, where X$_i$ are the iron-sulfur centers which function serially as secondary acceptors in the electron-transport chain (Swarthoff, 1981a). This is revealed by the steady-state differential absorption changes reflecting the photo-oxidation of P840 and by the appearance of millisecond delayed fluorescence, which is in agreement with earlier observations (Swarthoff, 1981b). In the second case, all RCs are in the open state. This is revealed by the almost complete disappearance of the steady-state differential absorption changes reflecting the photo-oxidation of P840, as well as by the complete disappearance of millisecond delayed fluorescence.

The isotropic fluorescence kinetics for both open and closed RCs are shown in Fig. 5.

When RCs are in the open state the BChl $\underline{a}$ fluorescence decay is well approximated by a single exponential component with time constant $\tau = 66 \pm 5$ ps. This value should be regarded as a lower limit due to the BChl $\underline{c}$ and BChl $\underline{a}$ fluorescence bands overlapping at 820 nm. Probably this overlapping causes also the absence of the expected observable BChl $\underline{a}$ fluorescence rise time. Under the same conditions the BChl $\underline{c}$ fluorescence decay is strongly biphasic and characterized by a fast phase with the lifetime $\tau_1 = 26 \pm 3$ ps and a slow phase having the lifetime $\tau_2 = 65 \pm 10$ ps. The amplitude ratio A_1/A_2 of the fast and slow components is equal to 10 ± 3.

With RCs in the closed state the decay of the excited state is much slower and more complicated. The BChl $\underline{c}$ fluorescence decay is described by a sum of two exponentials with time constants $\tau_1 = 85 \pm 15$ ps and $\tau_2 = 280 \pm 20$ ps ($A_1/A_2 = 5 \pm 1$). The BChl $\underline{a}$ fluorescence decay is but poorly approximated by a sum of two exponentials, a picosecond phase with the time constant $\tau = 290 \pm 30$ ps and a slow phase with the time constant $\tau > 2$ ns. A significant slowdown of the BChl $\underline{a}$ fluorescence rise is also observable.

The ratio of the integrated areas under the BChl $\underline{c}$ and BChl $\underline{a}$ decay curves with closed and open RCs is about 4.

Let us consider the simplest kinetic scheme of the energy transfer within a PSU of Cb. limicola cells applicable at room temperature when RCs are open:

$$\text{BChl } \underline{c} \underset{k_{-1}}{\overset{k_1}{\rightleftharpoons}} \text{BChl } \underline{a} \xrightarrow{k_2} \text{RC}$$
$$\downarrow k_3 \qquad\qquad \downarrow k_3'$$

Under our experimental conditions BChl $\underline{c}$ and BChl $\underline{a}$ excitation decay kinetics $I_c(t)$ and $I_a(t)$, respectively, are described by the following equations:

$$I_c(t) = c_1 \exp(-\lambda_1 t) + c_2 \exp(-\lambda_2 t)$$
$$I_a(t) = a_1 \exp(-\lambda_1 t) + a_2 \exp(-\lambda_2 t)$$

al., 1977), and the increase in ϕ_m was accompanied by a decrease in τ_m (measured on a nanosecond time scale with the same phase fluorometer).

These data revealed that short-lived emissions of both BChl c and BChl a exist on a picosecond time scale. Using the time-lever methods developed for phase fluorometry (Borisov et al., 1977; Fetisova and Borisov, 1980) it was deduced that (i) the main fraction of the BChl c emission decayed with a time constant of 20-50 ps while (ii) the main fraction of the BChl a emission had a decay time of 20-60 ps when all RCs were open. However, the phase fluorometry methods employed are restricted to special cases of two component emissions (see p. 501 in Borisov et al., 1977). In addition, the decay kinetics cannot be observed with these methods, and only an average decay time is obtained for each time scale. Therefore, direct measurements of fluorescence decay kinetics with a picosecond emission spectrochronograph (Freiberg, 1986) are essential.

For the correct quantitative analysis of picosecond experimental data it was first necessary to select BChl c and BChl a absorption and emission wavelengths which are free from overlap. As shown by Fetisova et al. (1986, 1987), in the region of 710-770 nm there is only absorption by BChl c. The region where only BChl c emission occurs is narrower because of significant overlap of the BChl c and BChl a fluorescence bands (see Fig. 1). Therefore, for optimal measurements the BChl c Q_y-transition was excited at 710-725 nm and BChl c fluorescence kinetics were detected at wavelengths shorter than

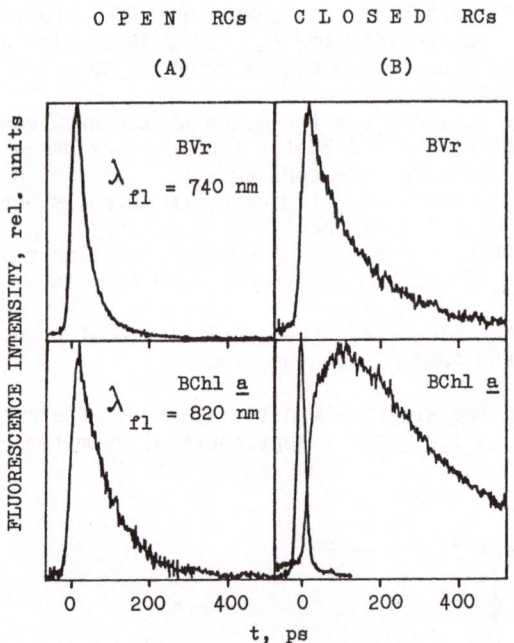

Fig. 5. Room temperature isotropic fluorescence decay kinetics of BChl c at 740 nm and BChl a at 820 nm in living cells of Cb. limicola for open (A) and closed (B) RCs. Excitation at 724 nm was used. Excitation intensity was equal to 0.2 W/cm². The excitation pulse profile is also shown. BVr ≡ BChl c.

8

We assume that the rate constants of trivial losses $k_3, k_3 \ll k_1, k_{-1}, k_2$. Then

$$\lambda_{1,2} = 0.5 \{(k_1 + k_{-1} + k_2) \pm [(k_1 + k_{-1} + k_2)^2 - 4k_1k_2]^{0.5}\}$$

$$c_1/c_2 = (k_1 - \lambda_2)/(\lambda_1 - k_1)$$

Therefore, with the experimental values of λ_1, λ_2 and c_1/c_2 measured under conditions of open RCs, the macroscopic rate constants for the energy transfer from BChl $\underline{c}$ to BChl $\underline{a}$ (k_1) and from BChl $\underline{a}$ to RC (k_2) for active photosynthesis can be determined.

For the sample of Fig. 5 $\tau_1 \equiv 1/\lambda_1 = 26$ ps; $\tau_2 \equiv 1/\lambda_2 = 66$ ps; and $A_1/A_2 \equiv c_1/c_2 = 10$. Then $k_1^{-1} = 28$ ps; $k_2^{-1} = 62$ ps; $k_{-1}^{-1} = 800$ ps and k_1/k_{-1} 29.

Thus, in the framework of the presumably oversimplified model of active photosynthesis at room temperature given here, energy transfer in intact cells of the green bacterium $\underline{Cb. limicola}$ under physiological conditions may be described by the following kinetic scheme

$$
\begin{array}{ccccc}
 & \sim 30 \text{ ps} & & \geqslant 60 \text{ ps} & \\
\text{BChl } \underline{c} & \underset{\sim 800 \text{ ps}}{\overset{\longrightarrow}{\longleftarrow}} & \text{BChl } \underline{a} & \longrightarrow & \text{RC.}
\end{array}
$$

The high rate of the BChl $\underline{c}$ -> BChl $\underline{a}$ energy transfer and of the BChl $\underline{a}$ -> RC energy transfer, in comparison with the respective excitation lifetimes in isolated pigment-protein complexes (for isolated chlorosomes τ ~1 ns, Vos et al., 1987), ensure that (i) the efficiency of the energy transfer from BChl $\underline{c}$ to BChl $\underline{a}$ is ~98% and (ii) the quantum yield of primary charge separation in RCs of living cells of $\underline{Cb. limicola}$ is $\simeq$ 94%.

These data are in good quantitative agreement with previous indirect phase-fluorometry data (Borisov et al., 1977; Fetisova and Borisov, 1980).

Such a high quantum yield of primary charge separation in the RC imposes strong restrictions on the PSU antenna structure (Fetisova and Fok, 1984), which means, in particular, that the latter has to ensure a directed (not random) transfer of excitation energy from an antenna to the RC. It is reasonable to assume that the strong orientational ordering of the transition moments of BChl $\underline{c}$ in the $\underline{Cb. limicola}$ superantenna, discovered recently by Fetisova et al. (1986, 1987) serves the same purpose. It was shown with the linear dichroism method that in each $\underline{Cb. limicola}$ "chromatophore" the Q_y-transition moment vectors of BChl $\underline{c}$ are essentially parallel to each other and practically perfectly oriented along the chlorosome long axis. Comparison with model computer calculations of the optimal antenna structure (Fetisova et al., 1985, 1987) shows that a simultaneous utilization of two optimizing factors in the chlorosome structure may be essential: ordering of mutual orientations of transition moments and specific anisotropy of intermolecular distances.

If the excitation transfer within a chlorosome occurs actually via parallel vectors of BChl $\underline{c}$ transition moments, then the BChl $\underline{c}$ fluorescence anisotropy function may be expected to be approximately constant in time and equal to the limiting value.

In order to investigate this structural aspect of energy transfer, picosecond polarized fluorescence decay kinetics for BChl $\underline{c}$ (730 nm) and BChl $\underline{a}$ (820 nm) emissions in living cells of $\underline{Cb. limicola}$ under physiological conditions were measured upon the selective excitation of the Q_y-transition of BChl $\underline{c}$ at 711 nm. Fig. 6 shows these room-temperature decay kinetics under conditions of closed RCs and the corresponding fluorescence polarization function p(t). According to the data presented in Fig. 6A, the BChl c fluorescence polarization is constant during the BChl c excited state lifetime and is equal to 0.42 ± 0.02 in the time interval 0-120 ps (0.42 ± 0.03 in the time interval 120-270 ps). Preliminary experiments performed with Cb. limicola "chromatophores" showed the same results. The BChl a fluorescence polarization (Fig. 6B) decays from its maximum value (+ 0.42) to zero during the first ~120 ps when the population of BChl c excited states decreases about 3-fold and the population of BChl a excited states achieves its maximum value. It should be noted that the observed population decay is the result of several different processes, in particular, the significant overlap of BChl c and BChl a emission bands.

The value of p obtained for BChl c emission is equal to the limiting value of p achieved in monomeric BChl a (Bolt and Sauer, 1981; Breton and Vermeglio, 1982). This high and constant value of p indicates convincingly that excitation energy transfer within a chlorosome takes place between parallel transition moments. From the structural investigations (Olson, 1980; Zuber, 1985) it is likely that these are transition moments of collective excitations of clusters of strongly coupled BChl c molecules rather than transition moments of the individual chromophores. The excitation energy transfer within the BChl c light-harvesting antenna may then be described as one between these clusters with parallel transition moments, i.e., each cluster may be considered as a single large "molecule" which then may serve as a donor (and acceptor) in the Förster-type excitation transfer, as was recognized for the light-harvesting Chl a/b-protein of higher plants (Knox and Van Metter, 1979). This is the model which is used in computations of the optimal orientation of BChl c dipoles in the PSU of Cb. limicola (Fetisova et al., 1987).

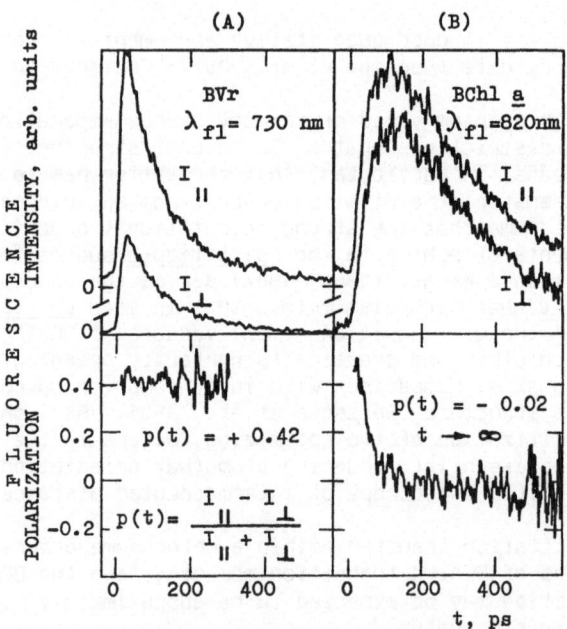

Fig. 6. Room-temperature decay kinetics of polarized fluorescence and the calculated fluorescence polarization function $p/t) = (I_\| - I_\perp)/(I_\| + I_\perp)$ for (A) BChl c at 730 nm and (B) BChl a at 820 nm in living cells of Cb. limicola under physiological conditions. Excitation intensity was 0.2 W/cm² at 711 nm. BVr ≡ BChl c.

In any case the strict orientation of BChl c transition moments is one of the optimizing factors ensuring fast and highly efficient heterogeneous excitation energy transfer from the large BChl c-antenna to the smaller BChl a antenna. Thus the experimental facts, both the optimal orientational ordering of BChl c transition moment vectors (Fetisova et al., 1986) and the high degree of BChl c fluorescence polarization during its excited state lifetime (the present work) confirm the main conclusion of earlier theoretical work (Fetisova and Fok, 1984): the antenna structure should ensure a directed excitation energy transfer from the antenna to the RCs.

ACKNOWLEDGEMENTS

The authors are grateful to V.A. Shuvalov for valuable discussions, to T. Pullerits and S. Savikhin for assistance with some measurements as well as to I. Bukhova for the cultivation of Cb. limicola.

REFERENCES

Amesz, J., 1985, Photosynthesis: structure of the membrane and membrane proteins, Progress in Botany, 47:87.

Amesz, J., and Vasmel, H., 1986, Fluorescence properties of photosynthetic bacteria, in: "Light Emission by Plants and Bacteria," Govindjee, J. Amesz and D. C. Fork, eds., Academic Press, New York.

Betti, J. A., Blankenship, R. E., Natarajan, L. V., Dickinson, L. C., and Fuller, R. C., 1982, Antenna organization and evidence for the function of a new antenna pigment species in the green photosynthetic bacterium Chloroflexus aurantiacus, Biochim. Biophys. Acta, 680:194.

Bolt, J. D., and Sauer, K., 1981, Fluorescence properties of the light-harvesting bacteriochlorophyll protein from Rhodopseudomonas sphaeroides R-26, Biochim. Biophys. Acta 637:342.

Borisov, A. Yu., Fetisova, Z. G., and Godik, V. I., 1977, Energy transfer in photoactive complexes obtained from the green bacterium Chlorobium limicola, Biochim. Biophys. Acta, 461:500.

Breton, J., and Vermeglio, A., 1982, Orientation of photosynthetic pigments in vivo, in: "Photosynthesis: Energy Conversion by Plants and Bacteria," Govindjee, ed., v. 1, Academic Press, New York.

Clayton, R. K., 1965, Characteristics of fluorescence and delayed light emission from green photosynthetic bacteria and algae, J. Gen. Physiol., 48:633.

Fetisova, Z. G., and Borisov, A. Yu., 1980, Picosecond time scale of heterogeneous excitation energy transfer from accessory light-harvesting bacterioviridin antenna to main bacteriochlorophyll a antenna in photoactive pigment-protein complexes obtained from Chlorobium limicola, a green bacterium, FEBS Lett., 114:323.

Fetisova, Z. G., and Fok, M. V., 1984, The ways of optimization of light energy conversion in primary steps of photosynthesis. I. Necessity of photosynthetic unit structure optimization and the method of its efficiency calculation (in Russian), Molek. Biol., 18:1651.

Fetisova, Z. G., Fok, M. V., and Shibaeva, L. V., 1985, The ways of optimization of light energy conversion in primary steps of photosynthesis. VI. The principles of organization of optimal artificial light-harvesting molecular systems (in Russian), Molek. Biol., 19:1489.

Fetisova, Z. G., Kharchenko, S. G., and Abdurakhmanov, I. A., 1986, Strong orientational ordering of the near-infrared transition moment vectors of light-harvesting antenna bacterioviridin in chromatophores of the green photosynthetic bacterium Chlorobium limicola, FEBS Lett., 199:234.

Fetisova, Z. G., Kharchenko, S. G., and Abdurakhmanov, I. A., 1987, in:

"Progress in Photosynthesis Research," J. Biggins, ed., v. 1, Martinus Nijhoff, Dordrecht.

Freiberg, A. M., 1986, Primary processes of photosynthesis studied by fluorescence spectroscopy methods, Laser Chem. 6:233.

Govindjee, Hammond, J. H., and Merkelo, H., 1972, Lifetime of the excited state in vivo. II. Bacteriochlorophyll in photosynthetic bacteria at room temperature, Biophys. J., 12:809.

Knox, R. S., and van Metter, R. L., 1979, Fluorescence of light-harvesting chlorophyll a/b protein complexes: implications for the photosynthetic unit, in: "Chlorophyll Organization and Energy Transfer in Photosynthesis," Ciba Foundation Symposium 61 (new series), Excerpta Medica, Amsterdam.

Kondrat'eva, E. N., 1963, "The Photosynthetic Bacteria" (in Russian), Acad. Sci. USSR Press, Moscow.

Olson, J. M., 1980, Chlorophyll organization in green photosynthetic bacteria, Biochim. Biophys. Acta, 594:33.

Swarthoff, T., Gast, P., Hoff, A. J., and Amesz, J., 1981a, An optical and ESR investigation on the acceptor side of the reaction center of the green photosynthetic bacterium Prosthecochloris aestuarii, FEBS Lett., 130:93.

Swarthoff, T., van der Veek-Horsley, K. M., and Amesz, J., 1981b, The primary charge separation, cytochrome oxidation and triplet formation in preparations from the green photosynthetic bacterium Prosthecochloris aestuarii, Biochim. Biophys. Acta, 635:1.

Sybesma, C., and Olson, J. M., 1963, Transfer of chlorophyll excitation energy in green photosynthetic bacteria, Proc. Natl. Acad. Sci. U.S.A., 49:248.

Sybesma, C., and Vredenberg, W. J., 1963, Evidence for a reaction center P840 in the green photosynthetic bacterium Chloropseudomonas ethylicum, Biochim. Biophys. Acta, 75:439.

van Dorssen, R. J., Gerola, P. D., Olson, J. M., and Amesz, J., 1986a, Optical and structural properties of chlorosomes of the photosynthetic green sulfur bacterium Chlorobium limicola, Biochim. Biophys. Acta, 848:77.

van Dorssen, R. J., Vasmel, H., and Amesz, J., 1986b, Pigment organization and energy transfer in the green photosynthetic bacterium Chloroflexus aurantiacus, Photosynth. Res., 9:33.

Vos, M., Nuijs, A. M., van Grondelle, R., van Dorssen, R. J., Gerola, P. D., and Amesz, J., 1987, Excitation transfer in chlorosomes of green photosytnehtic bacteria, Biochim. Biophys. Acta, 891:275.

Wechsler, T., Suter, F., Fuller, R. C., and Zuber, H., 1984, The complete amino acid sequence of the bacteriochlorophyll c-binding popypeptide from chlorosmes of the green photosynthetic bacterium Chloroflexus aurantiacus, FEBS Lett., 181:173.

Zuber, H., 1985, Structure and function of light-harvesting complexes and their polypeptides, Photochem. Photobiol., 42:821.

PICOSECOND ENERGY TRANSFER KINETICS IN CHLOROSOMES AND BACTERIOCHLOROPHYLL A-PROTEINS OF CHLOROBIUM LIMICOLA

T. Gillbro[a], A. Sandström[a], V. Sundström[a] and J.M. Olson[b]

[a]Department of Physical Chemistry, University of Umeå, Sweden
[b]Institute of Biochemistry, Odense University, Denmark

INTRODUCTION

In this paper we report a picosecond absorption spectroscopy study of the energy transfer and excited state lifetimes of two light-harvesting complexes, namely chlorosomes and the BChl a-protein of the green sulfur bacterium Chlorobium limicola f. thiosulfatophilum (strain Tassajara).

Chlorosomes are supramolecular protein complexes consisting of several rod-like elements each containing about 1000-2000 BChl c chromophores (Olson, 1980). In addition the chlorosomes of Cb. limicola contain a small amount of BChl a so that the molecular ratio BChl c/BChl a is about 90 (Gerola and Olson, 1986). The BChl a-protein is a membrane-bound, water-soluble complex. A related BChl a-protein from Prosthecochloris aestuarii has been crystallized and its structure was determined at high resolution (Matthews et al., 1979). These proteins normally form trimers with 7 BChl a chromophores per monomer unit.

There have been just a few time-resolved investigations on the energy transfer kinetics of chlorosomes and BChl a-proteins of green sulfur bacteria (Fetisova and Borisov, 1980) and to our knowledge no picosecond absorption recovery study has been reported. We believe that this study gives new information on excited state processes not obtained previously by time-resolved fluorescence or by other kinds of optical spectroscopy (van Dorssen et al., 1986; Vos et al., 1987).

MATERIALS AND METHODS

The chlorosomes were prepared according to Gerola and Olson (1986), and were suspended in 2 M NaSCN, 10 mM Ascorbate and 10 mM KP_i buffer (pH 7.4) during the experiments in order to improve the stability (Gerola and Olson, 1986). BChl a-protein was prepared as described by Olson (1978) and dissolved in 10 mM Tris buffer (pH 8) and 0.25 M NaCl. The quality of the preparations were judged by their absorption and fluorescence spectra. Cells with an optical path length of ca. 1 mm were used for the measurements, and the optical density at the excitation wavelength was about 0.5. At 77 K 50% glycerol and a small amount of sucrose was added to the sample in order to avoid cracking.

At room temperature, 23-25°C, we used a rotating sample cell to prevent accumulation of photoproducts in the excitation beam. In some measurements 10 mM dithionite was added to the chlorosomes, since such a treatment has been shown to increase the fluorescence yield (Vos et al., 1987). The picosecond laser system based on a sync-pumped dye laser and the pump-probe technique has been described elsewhere (Sundström and Gillbro, 1983). Here we will just give a brief account of some parameters of importance to this work. The wavelength range 730-825 nm was covered by the two laser dyes DCM and Styryl 8, and a typical pulse length of about 15 ps was obtained. The maximum laser pulse intensity in the excitation beam at the sample was ca. 1 x 10^{14} photons per cm^2. The intensity could be reduced by inserting neutral density filters in the excitation beam.

The angle between the polarization of the excitation light and the probe light was controlled by a Soleil-Babinet compensator in the excitation beam. For isotropic kinetic traces this angle was put at 54.7°. When information about the anisotropy was required, two independent measurements were performed with this angle set at 0° or 90°. From these data the anisotropy, r(t), is calculated as $r(t) = (I_{\parallel} - I_{\perp})/(I_{\parallel} + 2I_{\perp})$.

RESULTS AND DISCUSSION

Chlorosomes

The excited state kinetics of chlorosomes at room temperature were independent of wavelength over the main absorption band of BChl $\underline{c}$, i.e. 730-770 nm. In Fig. 1a we have fitted the experimental curve obtained at 750 nm to the sum of three exponentials. The lifetimes of the best fit were 21 ps, 148 ps and 3.0 ns, respectively, and the ratio of the signal amplitudes at time zero was about 5.5:1:1. Addition of dithionite to the chlorosomes did not change the kinetics to any significant extent. In an attempt to study the effect of excitation annihilation the excitation intensity was reduced by a factor of ten. This did not have any observable effect on the fast kinetics, however. In Fig. 1b the experimental traces for the anisotropy measurements are displayed. A calculation of the anisotropy as a function of time gave r(t) ≈ 0.20 at all times. No initial relaxation of the

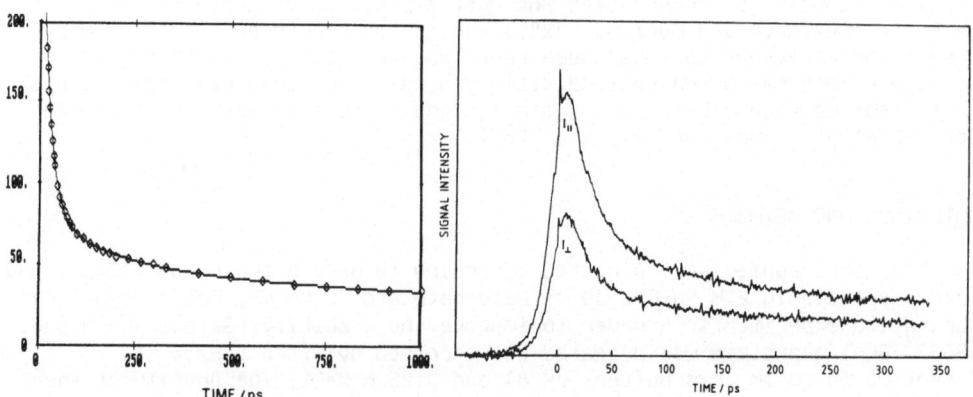

Fig. 1a. Isotropic absorption recovery kinetics of chlorosomes at 750 nm. The solid line shows the best fit to three exponentials. 1b. Decay kinetics of chlorosomes measured at 750 nm with the polarization of the probe light parallel (I) or perpendicular (I) to that of the excitation light.

anisotropy with a lifetime of ca. 20 ps, as found in the isotropic traces, was observed. However, our results show that there is a very fast (<< 10 ps) initial decay of the anisotropy from the theoretical maximum of 0.4 to 0.2. This behaviour is similar to our recent results in some strongly coupled antenna pigment complexes of purple bacteria (Sundström et al., 1986). A possible mechanism for such a fast process might be the loss of coherence in an originally excited dimeric state (or higher aggregate) of BChl c chromophores (Pearlstein et al., 1978; Smith et al., 1983). Such a loss of coherence would lead to a fast depolarization, if the transition dipoles of the monomers are non-parallel to those of the dimeric state. With this model our data predict the angle between the transitions to be ca. 35°. If on the other hand there is a very fast randomization of the excited state between BChl c chromophores in a segment of the rods, the anisotropy at long times should give us the order, S, of the BChl c chromophores in the rods. In this case a simple calculation indicates that the average angle between the long symmetry axis of the rod and the Q_y-transition of the BChl c chromophores is about 26°. This is in fair agreement with steady-state linear dichroism (LD) measurements on oriented chlorosomes that gave an orientation of the transitions parallel to the rod axis (Fetisova et al., 1986; van Dorssen et al., 1986). Since the 20-ps component was not resolved in the anisotropy measurements, we have to conclude that this component is probably not due to energy transfer within the BChl c chromophores of the rods, but rather to a transfer to other chromophores. The best candidate for such a transfer seems to be BChl a. The 3-ns lifetime at 750 nm (mean value 2.5 ± 1.0 ns) is attributed to the excited state lifetime of BChl c in its protein environment. The intermediate lifetime of about 150 ps is more difficult to assign, but it might be due to an imperfect energy transfer in damaged chlorosomes in combination with some energy trapping process.

The absorption recovery kinetics of chlorosomes observed at about 800 nm are shown in Fig. 2a and b. At this wavelength mainly BChl a is excited. In contrast to the experiments at 750 nm dithionite had a dramatic effect on the results as seen in the figure. Without dithionite the excited state lifetime was only about 30 ps (probably due to a scavenger), while in the presence of dithionite at least two components could be resolved. The long ns component is probably due to BChl a fluorescence. The short ca. 100-ps lifetime might come from an energy trapping process within the BChl a chromophores or a transfer of energy to BChl c. The anisotropy (not shown) seems to contain a 40-ps component and relaxes to a very small value (r = 0.04) at longer times.

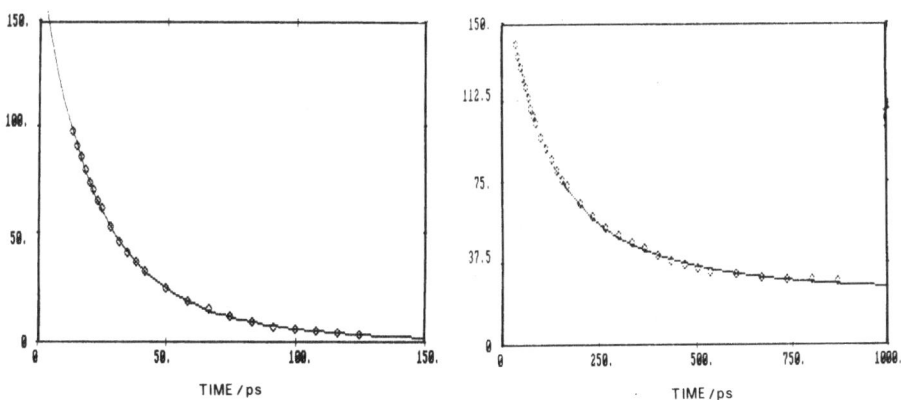

TIME / ps   TIME / ps

Fig. 2a. Absorption recovery of chlorosomes at 796 nm fitted to one exponential. 2b. 10 mM dithionite was added to the chlorosomes. The recording of the kinetics was made at 800 nm.

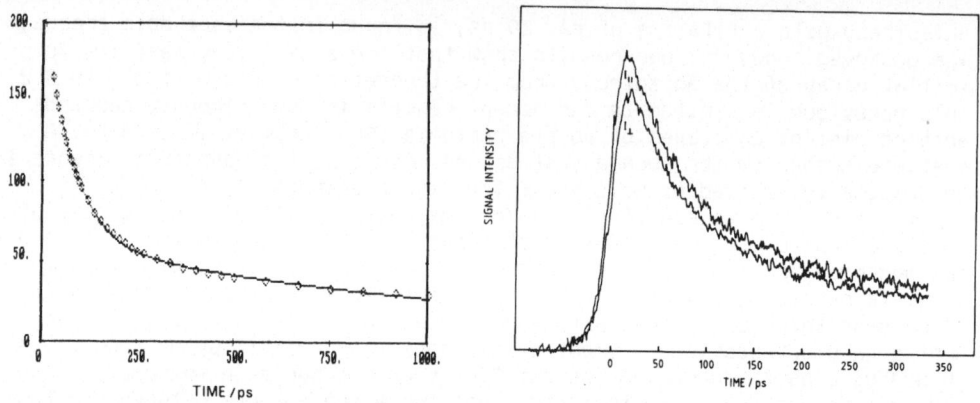

Fig. 3. Absorption recovery kinetics of the BChl a-protein measured at 809
nm. In a) the polarization angle was set at 54.7° and in b) at 0°
and 90°, respectively.

Our attempts to observe excitation annihilation effects at 800 nm were
negative as for 750 nm. Since excitation annihilation was observed recently
by Vos et al. (1987) at similar intensities, we have to conclude that the
annihilation must be strongest in the ns component, which seems to be of
minor importance for the energy transfer.

BChl a-Protein

In Fig. 3a the result of a measurement of the BChl a-protein at 809 nm
is presented. Basically there is a 80-ps decay followed by a 2.5-ns
component, which carries about 25% of the total signal intensity. Within the
experimental error the lifetimes and the signal amplitudes were the same in
the interval 795-825 nm. Also the anisotropy was constant and small (r =

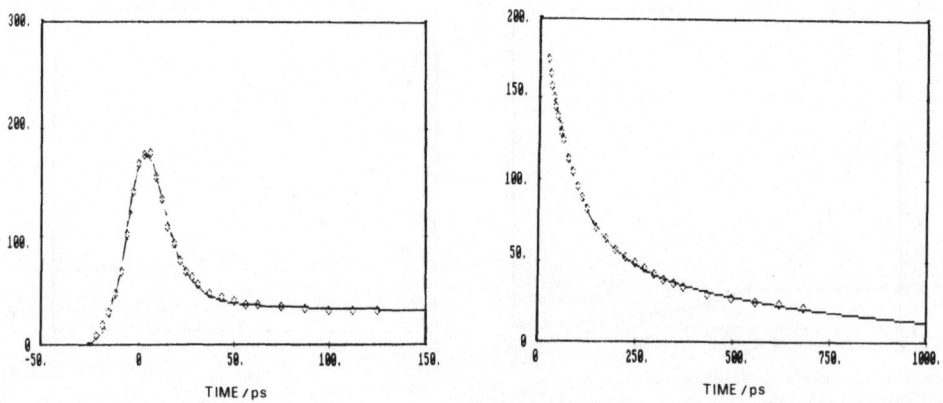

Fig. 4. Kinetics of the ground state recovery of the BChl a-protein at 77 K
recorded at a) 801 nm and b) 823 nm.

0.04) as demonstrated in Fig. 3b. Obviously the initial energy transfer process is very fast (<< 10 ps) just as found for the BChl c chromophores of the rods. The 80-ps component is most likely caused by a scavenging process, since it was not influenced by the excitation light intensity. Attempts to reduce the magnitude of this component through addition of dithionite were unsuccessful, because the BChl a-protein precipitated in the sample. The 2.5-ns lifetime is attributed to BChl a fluorescence.

It is known that the resolution of the absorption spectrum of the BChl a-protein increases significantly upon cooling below 50 K (Olson, 1980), and peaks are resolved at ca. 790, 806, 814.5 and 824 nm. In Fig. 4 results from picosecond measurements at 77 K are shown. The kinetic trace recorded at 801 nm shows a fast 12-ps energy transfer process and a 0.77-ns component. At 823 nm the shortest lifetime that we can resolve is 80 ps, i.e. just as at 23-25°C, and the amplitude of the slow, 0.63-ns, component is much larger than at 801 nm. It thus seems as if there is a distribution of the excited state towards BChl a chromophores absorbing at longer wavelengths. The presence of the 80-ps component at 823 nm shows that the energy trapping occurs in the red-absorbing transition.

We have also studied the anisotropy of BChl a at 77 K and found that it was constant (r = 0.15) and independent of wavelength. Our interpretation of this is that there is a very fast redistribution of excitation energy also at 77 K, but it seems to be temperature-dependent. The redistribution of excitation energy is less complete at 77 K than at 23-25°C. Finally, the 12-ps absorption recovery indicates an over-all transfer of energy from chromophores with blue-shifted spectra to the 824-nm chromophores of the BChl a-protein trimer.

REFERENCES

Gerola, P. D., and Olson, J. M., 1986, A new bacteriochlorophyll a-protein complex associated with chlorosomes of green sulfur bacteria, Biochim. Biophys. Acta, 848:69.
Fetisova, Z. G., and Borisov, A. Yu., 1980, Picosecond time scale of heterogenous excitation energy transfer from accessory light-harvesting bacterioviridin antenna to main bacteriochlorophyll a antenna in photoactive pigment-protein complexes obtained from Chlorobium limicola, a green bacterium, FEBS Lett., 114:323.
Fetisova, Z. G., Kharchenko, S. G., and Abdourakhmanov, I. A., 1986, Strong orientational ordering of the near-infrared transition moment vectors of light-harvesting antenna bacterioviridin in chromatophores of the green photosynthetic bacterium Chlorobium limicola, FEBS Lett., 199:234.
Matthews, B. W., Fenna, R. E., Boglonesi, M. C., Schmid, M. F., and Olson, J. M., 1979, Structure of a bacteriochlorophyll a-protein from the green photosynthetic bacterium Prosthecochloris aestuarii, J. Mol. Biol., 131:259.
Olson, J. M., 1978, Bacteriochlorophyll a-proteins from green bacteria, in: "The photosynthetic bacteria," R. K. Clayton and W. R. Sistrom, eds., Plenum Press, New York.
Olson, J. M., 1980, Chlorophyll organization in green photosynthetic bacteria, Biochim. Biophys. Acta, 594:33.
Pearlstein, R. M., and Hemenger, R. P., 1978, Bacteriochlorophyll electronic transition moment directions in bacteriochlorophyll a-protein, Proc. Natl. Acad. Sci. U.S.A., 75:4920.
Smith, K. M., Kehres, L. A., and Fajer, J., 1983, Aggregation of the bacteriochlorophylls c, d, and e. Models for the antenna chlorophylls of green and brown photosynthetic bacteria, J. Am. Chem. Soc., 105:1387.

Sundström, V., and Gillbro, T., 1983, A discussion of the problem of determining multiple lifetimes from picosecond absorption recovery data as encountered in two carbocyanine dyes, Appl. Phys. B, 31:235.

Sundström, V., van Grondelle, R., Bergström, H., Åkesson, E., and Gillbro, T., 1986, Excitation-energy transport in the bacteriochlorophyll antenna systems of Rhodospirillum rubrum and Rhodobacter sphaeroides, studied by low-intensity picosecond absorption spectroscopy, Biochim. Biophys. Acta, 851:431.

van Dorssen, R. J., Gerola, P. D., Olson, J. M., and Amesz, J., 1986, Optical and structural properties of chlorosomes of the photosynthetic green sulfur bacterium Chlorobium limicola, Biochim. Biophys. Acta, 848:77.

Vos, M., Nuijs, A. M., van Grondelle, R., van Dorssen, R. J., Gerola, P. D., and Amesz, J., 1987, Excitation transfer in chlorosomes of green photosynthetic bacteria, Biochim. Biophys. Acta, 891:275.

ELECTRON TRANSFER IN THE REACTION CENTER OF GREEN SULFUR BACTERIA AND

HELIOBACTERIUM CHLORUM

H.W.J. Smit and J. Amesz

Department of Biophysics
Huygens Laboratory of the State University
P.O. Box 9504, NL-2300 RA Leiden, The Netherlands

INTRODUCTION

During the last years it has become increasingly evident that the photosynthetic mechanisms of the two families that may be called green bacteria: the green sulfur bacteria (Chlorobiaceae) and the green gliding bacteria (Chloroflexaceae) are in many respects different. The first evidence for this was already provided by Pierson and Castenholz (1974) who measured the optical difference spectrum of the photo-oxidation of the primary electron donor P865 of Chloroflexus aurantiacus, which spectrum was found to be much more similar to that of P870 of purple bacteria than to that of P840 in green sulfur bacteria.

Green Sulfur Bacteria

Although relatively little is known about the structure of the membrane and the reaction center of green sulfur bacteria, it is clear that this structure must be basically different from that of purple bacteria and of Cf. aurantiacus. Cytoplasmic membranes of the green sulfur bacterium Prosthecochloris aestuarii, when freed from chlorosomes, contain 80-100 molecules of BChl a per reaction center. Most of this BChl a belongs to the water-soluble BChl a-protein, which is bound to the membrane and can be largely removed by the use of chaotropic agents (Olson et al., 1976). In addition to BChl a the membrane contains BChl c, which is distinguished from the chlorosomal BChl c by its stronger lipophilicity and the nature of the esterifying alcohol, which is phytol instead of farnesol (Braumann et al., 1986). The absorption band near 670 nm in membranes of green sulfur bacteria, first observed by Fowler et al. (1971), may be attributed to the Q_y-transition of this BChl c, which is thus much less red-shifted than BChl c in the chlorosomes. Only traces of bacteriopheophytin a and c were found to be present in the membrane (Braumann et al., 1986).

The presently available evidence indicates that the reaction center of green sulfur bacteria may not exist as a separate entity, as in purple bacteria and Cf. aurantiacus, and that part of the antenna pigments are bound to the same polypeptides which bind the primary electron donor and acceptor. The smallest pigment-protein complex obtained so far contains about 20 BChl a and 15 BChl c per reaction center (Vasmel et al., 1983; Hurt and Hauska, 1984). It contains polypeptide subunits with apparent molecular weight of 68 kDa together with smaller ones (Hurt and Hauska, 1984) which might suggest a structural resemblance to the core of photosystem 1.

Primary electron transport in membranes of P. aestuarii has been studied by Nuijs et al. (1985a) and by Shuvalov et al. (1986). The spectra and kinetics of the absorbance changes brought about by 35-ps flashes could be analyzed in components due to excited antenna pigments, to photo-oxidation of the primary electron donor P840 and to reduction of the primary electron acceptor. The latter was characterized by a bleaching around 670 nm which indicated that one of the BChl c molecules present in the membrane acts as primary electron acceptor. Reduction of BChl c occurred in less than 10 ps (Shuvalov et al., 1986), reoxidation in 600-700 ps (Nuijs et al., 1985a; Shuvalov et al., 1986). Direct spectral evidence on the component which acts as electron acceptor for BChl c is not available; it may be a low-potential iron-sulfur center whose reduction has been observed in photoaccumulation experiments (Swarthoff et al., 1981a).

Heliobacteria

Heliobacterium chlorum does not possess chlorosomes (Gest and Favinger, 1983), nor is there reason to believe that it is phylogenetically related to green sulfur bacteria (Pierson and Olson, 1987). Nevertheless, its primary photochemistry resembles that of P. aestuarii. Flash spectroscopy of membranes of H. chlorum showed that the primary electron acceptor is spectrally and kinetically similar to that of P. aestuarii, indicating that the primary charge separation consists of the transfer of an electron from the primary electron donor P798 (which is presumably a dimer of BChl g; Prince et al., 1985; Brok et al., 1986) to BChl c or a related pigment (e.g., Chl a) with Q_y-absorption near 670 nm (Nuijs et al., 1985b). Little is known about the organization of the antenna in H. chlorum (Van Dorssen et al., 1985), but it is of interest to note that the absorption spectrum shows a conspicuous band near 670 nm, indicating that in addition to the major pigment, BChl g (Brockmann and Lipinski, 1983), significant amounts of the BChl c-like pigment are present in H. chlorum. As in P. aestuarii, the kinetics at 670 nm after a flash showed a rapid reversal, with a time constant of about 500 ps (Nuijs et al., 1985b), presumably due to a reoxidation by a secondary electron acceptor. The identity of this electron acceptor is not known, but ESR measurements (Brok et al., 1986) indicated the photoaccumulation of a reduced iron-sulfur center upon cooling of illuminated membranes, suggesting that the secondary electron transport chain too may be similar in H. chlorum and green sulfur bacteria.

The present communication reports a spectroscopic study of the kinetics of secondary electron transport, back reactions and triplet formation as a function of the redox potential in membranes of H. chlorum. The results are compared with similar measurements on membranes of P. aestuarii.

MATERIALS AND METHODS

Membranes of H. chlorum were obtained by sonication, followed by repeated centrifugation; membranes of P. aestuarii as described by Swarthoff and Amesz (1979). All experiments were done in the presence of ascorbate and Tris or phosphate buffer and, in the case of H. chlorum, in the presence of glucose, glucose oxidase and catalase to remove traces of oxygen.

Flash-induced absorbance kinetics were measured with a single beam spectrophotometer, provided with a Q-switched frequency-doubled Nd-YAG laser (532 nm , 15 ns), a rhodamine B dye laser pumped by the YAG laser (590 nm). Redox titrations were carried out in the presence of sodium dithionite and redox mediators. The magnetic field dependence of the absorbance changes was measured as described by Kingma (1983). The absorbance of the samples was approximately 1.1 at 788 or 1.5 at 810 nm, unless otherwise indicated.

RESULTS

Difference Spectra and Kinetics of P798 Oxidation and Triplet Formation

Fig. 1 shows the absorbance difference spectrum of membranes of H. chlorum obtained with non-saturating 15-ns flashes in the presence of ascorbate. This spectrum is largely due to photo-oxidation of P798; in the red and near-infrared region it is similar to that obtained with continuous light (Fuller et al., 1985; Nuijs et al., 1985b) and in the region 440-620 nm to that obtained by flash illumination by Prince et al. (1985). The absorbance changes at 800 nm reversed with components of 6 and 30 ms which together comprised about 90% of the total decay. Most of this decay was due to a back reaction with one or more acceptors; analysis of the spectrum of these decay components indicated that only in 15% of the reaction centers a photo-oxidation of Cyt c_{553} (Fuller et al., 1985; Prince et al., 1985) occurred, with a time constant of 6 ms.

The difference spectrum of Fig. 1 in principle should contain a contribution due to the photoreduction of one or more electron acceptors. Comparison with the spectrum obtained in continuous light (Nuijs et al., 1985b) shows that above about 600 nm such a contribution must be very small, but at shorter wavelength a comparison is difficult, because of interfering cytochrome absorbance changes. Therefore we also measured the difference spectrum in the presence of ascorbate and N-methylphenazonium methosulfate (PMS). With 20 µM PMS present a monoexponential decay of 6 ± 1 ms was obtained. The spectrum of this 6 ms component was very similar to that of Fig. 1. Assuming that PMS directly reduces $P798^+$, we conclude that also below 600 nm the spectrum of Fig. 1 is almost entirely due to photo-oxidation of P798.

The 6-ms decay component observed in the presence of PMS disappeared almost completely at lower redox potentials and was replaced by faster components. Fig. 2 shows the kinetics obtained in the presence of dithionite

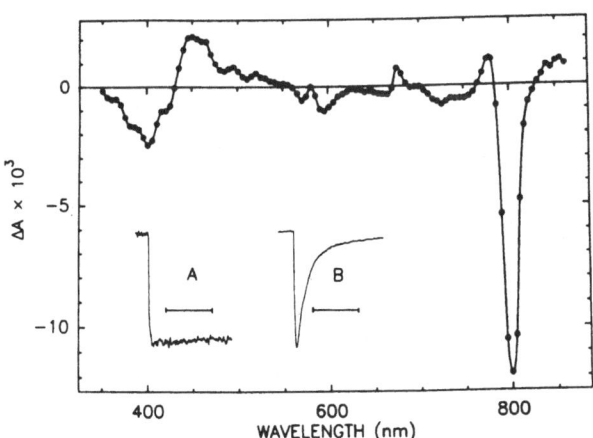

Fig. 1. Difference spectrum induced in membranes of H. chlorum by 15-ns laser flashes. Excitation at 532 nm or 590 nm (for the region 500-550 nm). The membranes were suspended in 10 mM Tris, 10 mM ascorbate, 10% (w/v) sucrose and 2 mM dithiothreitol, pH = 8.0. Inset A: kinetics at 400 nm; B at 800 nm. The bars indicate 10 µs and 50 ms, respectively.

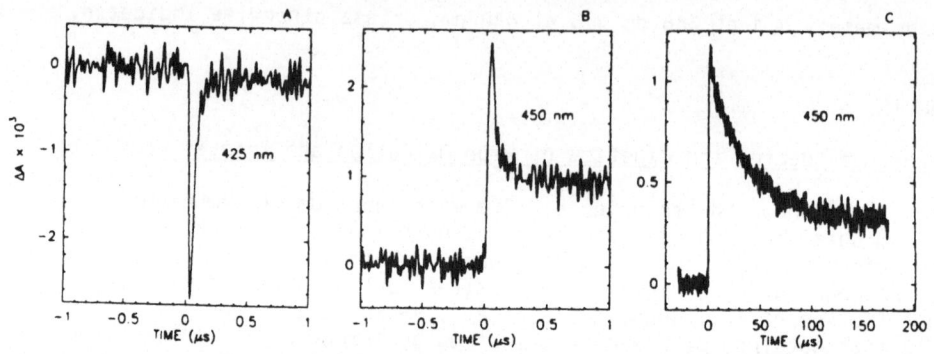

Fig. 2. Flash-induced kinetics at pH = 9.5, with 10 mM dithionite and 20 µM
PMS. For tracing c the rapid initial rise and decay were removed by
a high-frequency cut-off filter.

(pH = 9.5). Two decay components were now observed of about 30 ns and 35 µs.
The amplitude of the 35-µs component decreased in the presence of a magnetic
field. The maximum value of the depression was 50%; half the effect occurred
at about 30 mT. Similar effects have been seen in reaction centers of purple
bacteria, and have been discussed in terms of the radical pair mechanism of
triplet formation (Hoff, 1981).

The spectrum of the 35-µs component is given in Fig. 3. It shows a
bleaching centered at 793 nm and is presumably due to the triplet of P798
formed in a back reaction of $P798^+$ with a reduced electron acceptor, which
we call X_1, with an overall time constant of 30 ns. The spectrum is clearly
different from that of $P798^+$. The bleaching band in the near-infrared is
blue-shifted and broader, presumably due to the absence of the band shift
that causes the maximum at 776 nm in the $P798^+$ spectrum, and there is only a
single bleaching band at 575 nm in the Q_y-region. The yield of triplet

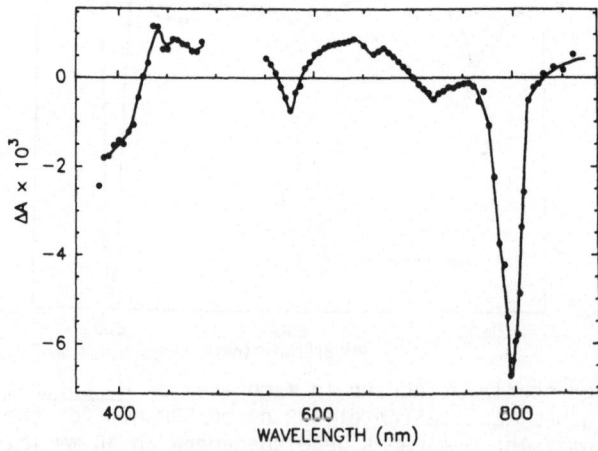

Fig. 3. Spectrum of the formation of the triplet of P798. Conditions as in
Fig. 2.

formation in the presence of dithionite was about 30% as judged from the amplitude of the 35 μs-component.

Electron Transport and Triplet Formation in Membranes of P. aestuarii

Formation of the triplet of P840 in green sulfur bacteria has only been studied in solubilized pigment-protein complexes obtained from the cytoplasmic membrane (Swarthoff et al., 1981b,c; van Bochove et al., 1984; Vasmel et al., 1984). These preparations, however, are all to some extent deficient in the electron transport chain (Swarthoff et al., 1981a,b; van Bochove et al., 1984), and therefore we decided to extend these studies to isolated membranes in order to compare the results with those obtained with membranes of H. chlorum. Moreover, in view of the seemingly conflicting results reported by Swarthoff et al. (1981b) and Prince and Olson (1976) it was of interest to obtain information on the rate of flash-induced Cyt c_{553} oxidation in these preparations. Fig. 4 shows the flash-induced difference spectrum of such a preparation, measured at 10 μs after a flash in the presence of ascorbate. It can be seen that the spectrum is quite complicated, and kinetic and spectral analysis showed that it contained at least four components: the oxidation of P840 and Cyt c_{553} and formation of the triplet states of P840 and carotenoid. At 840 nm two decay components were observed, of 70 ± 5 μs and 70 ms. At the time scale of the kinetics of Fig. 5 the latter is seen as a constant component. In the visible region an 8-10 μs component was also present (Fig. 5).

The difference spectrum of the "constant" component is shown in Fig. 6, and can be attributed to photo-oxidized P840 and Cyt c_{553}. In the Q_y-region the difference spectrum of P840 oxidation shows significant differences with the spectrum obtained in continuous light (Swarthoff and Amesz, 1979). Like the spectrum obtained by van Bochove et al. (1984), it shows a single band at 840 nm, and the band shift near 800 nm is smaller. The same is true for the band shift of BChl c near 670 nm. These differences may be caused by the presence of a different reduced electron acceptor and by the presence of oxidized cytochrome c in reaction centers subjected to continuous illumination, causing more extensive electrochromic band shifts of neighboring pigments.

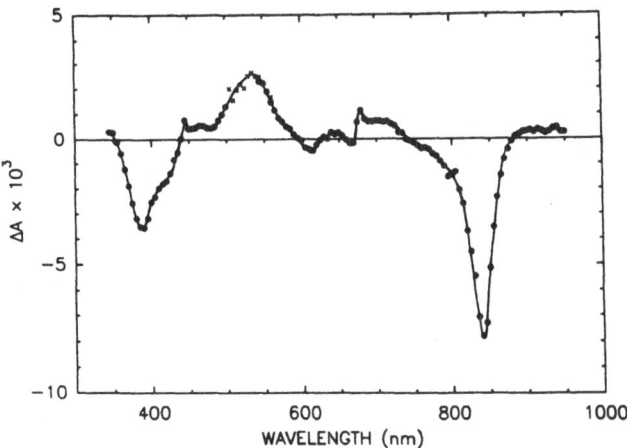

Fig. 4. Difference spectrum of membranes of P. aestuarii, measured at 10 μs after a flash. The membranes were suspended in 10 mM phosphate, 10 mM ascorbate, pH 7.4, and 40% (w/v) sucrose. Excitation at 532 nm or 590 nm (crosses). The amplitudes at 550 nm were normalized for the two types of flashes used.

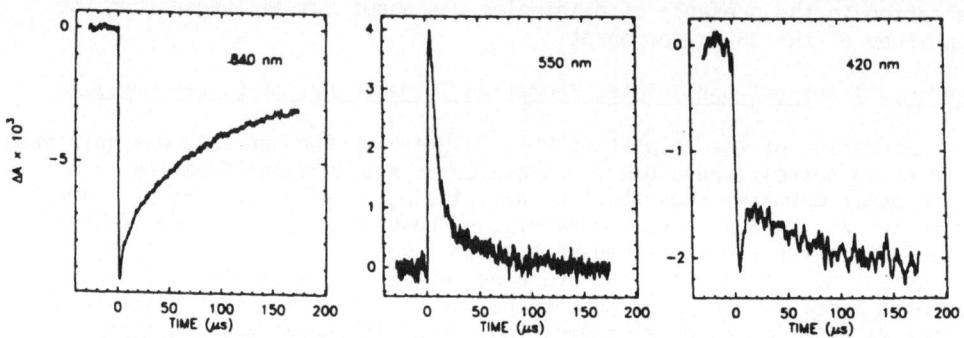

Fig. 5. Kinetics at various wavelengths for membranes of P. aestuarii.
Conditions as for Fig. 4.

Comparison of the difference spectrum of Fig. 6 with those obtained at
10 μs (Fig. 4) and at 5 μs after the flash showed that cytochrome c
oxidation was completed within 5 μs. This indicates that the time constant
for cytochrome oxidation must be less than 3 μs. A time constant of less
than 5 μs was reported for membranes of Clorobium limicola by Prince and
Olson (1976). In the presence of 20 μM PMS a 30-μs component in cytochrome c
oxidation became visible, which may correspond to the 50-μs component
observed in membranes of Cb. limicola (Prince and Olson, 1976), and the
total amplitude at 420 nm increased by about two-fold. From the amplitude of
the absorbance changes of cytochrome c and P840 it followed that only in
about 25-30% of the reaction centers did a stable charge separation
occur. This is less than earlier observed in a similar preparation
(Swarthoff et al., 1981b). Cytochrome c oxidation occurred in 1/2 to 2/3 of
these reaction centers. With continuous illumination P840+ was formed in all
reaction centers.

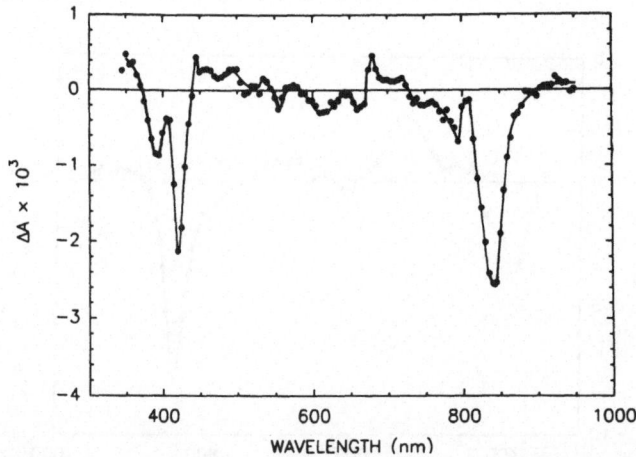

Fig. 6. Spectrum of P840 and Cyt c₅₅₃ oxidation, obtained by plotting the
signal at 170-μs after a flash and correcting for the contribution
by P840 triplet formation (see Fig. 7).

The spectrum of the component which decayed in 70 μs (Fig. 7) may be ascribed to the formation of the triplet of P840. The difference spectrum shows a minimum at 840 nm, in agreement with earlier measurements (van Bochove et al., 1984). Additional minima are seen at 605 nm (Q_x-region) and at 380 nm. The feature around 670 nm probably reflects a change in interaction with a neighboring BChl c molecule, possibly the primary acceptor. A similar red shift is seen in the triplet spectrum of P798 (Fig. 3). The amount of triplet formed was about 0.3 per reaction center.

The spectrum of the 8-10 μs (Fig. 8) component shows the characteristics of a carotenoid triplet (Kingma et al., 1985) as indicated by the absorbance decrease in the region below 500 nm and the positive band at 545-550 nm. In contrast to those at 840 nm, the absorbance changes at 550 nm were not saturated at the highest flash intensities used. Formation of the carotenoid triplet occurred upon excitation at 590 as well as at 532 nm. These observations indicate that the triplet is generated in the antenna, rather than in the reaction center, and that at least part of it is produced by a process which involves excited BChl a, presumably intersystem crossing, followed by triplet energy transfer (Kingma, 1983; Kingma et al., 1985).

Electron Transport at Low Redox Potentials

Illumination of H. chlorum membranes with continuous light at an apparent redox potential of -480 mV produced absorbance changes which can be ascribed to the photoaccumulation of a reduced iron-sulfur center (Fig. 9). A similar difference spectrum has been obtained with P. aestuarii (Swarthoff et al., 1981a). As with P. aestuarii, the spectrum shows negative bands in the blue region, together with negative and positive absorbance changes near 670 nm and in the near-infrared region, which may be explained by electrochromic effects on the absorption bands of neighboring pigment molecules. The iron-sulfur center may be identical to that found by Brok et al. (1986) in the ESR spectrum of illuminated membranes at low temperature.

Fig. 10 shows the redox titration curve of the fraction of P798⁺ which decayed on a ms time scale. The data can be fitted by a one-electron Nernst curve centered at -440 mV, significantly higher than the value of -510 mV obtained by Prince et al. (1985). At redox potentials above -430 mV the

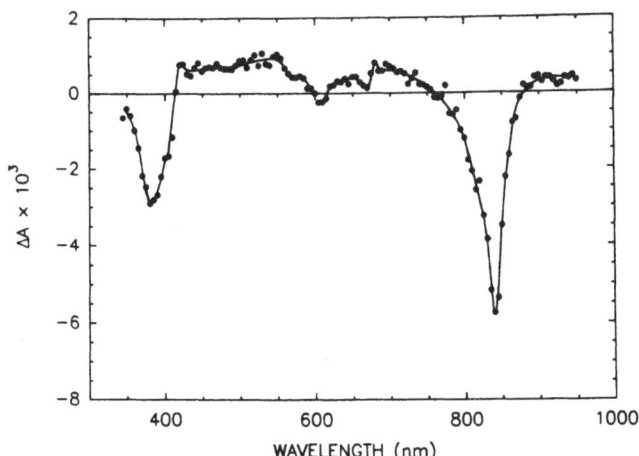

Fig. 7. Difference spectrum of the formation of the triplet of P840, obtained by plotting the 70-μs decay component (see Fig. 5) as a function of wavelength.

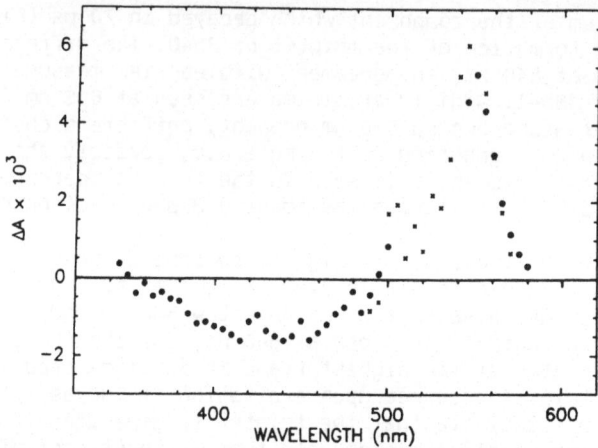

Fig. 8. Difference spectrum of carotenoid triplet formation, obtained by
subtracting the spectra at 5 and 30 μs after a flash and correcting
for the contribution by the triplet of P840.

curve for the formation of the triplet of P798 was complementary to that of
P798$^+$, but below that value the yield did not increase any further in the
presence of redox mediators. The reason for this effect is not clear. A
"normal" yield of 30% was only obtained in the absence of redox mediators
(solid square), but under these conditions a proper titration was not
possible. Nevertheless, our results indicate that the disappearance of the
6-ms component of P798$^+$ is correlated with triplet formation, indicating
that the electron acceptor for X_1 has a midpoint potential of -440 mV,
considerably higher than that of the iron-sulfur center. The yield of
triplet formation in a flash did not increase any further when the

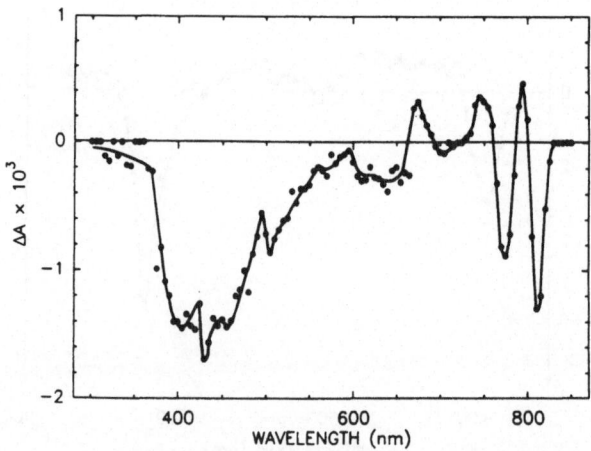

Fig. 9. Difference spectrum obtained with membranes of H. chlorum upon
illumination with strong continuous light in the presence of 15 mM
dithionite and 20 μM PMS at pH = 8.0.

iron-sulfur center was brought in the reduced state by continuous background light. These results make it difficult to assign a role for this center in the electron transport chain.

Results obtained with P. aestuarii were different. Although our measurements indicate that in our preparation the electron acceptor chain was deficient or partially uncoupled in most reaction centers, the "active" reaction centers retained their activity down to much lower redox potentials than those of H. chlorum. Fig. 11 shows experiments in the presence of 10 mM dithionite at an apparent redox potential of -440 mV. The kinetics of cytochrome c oxidation (420 and 555 nm) again showed the 30-μs component. The "constant" component of P840 and the amount of P840 triplet produced, measured at 380 nm, where contributions by P840+ and cytochrome c are small, remained essentially the same. This indicates that no back reaction was induced in the "active" reaction centers by lowering the redox potential. Similar results were obtained at pH = 10.0. Addition of dithionite at high pH gave a decrease of only about 10% in the fraction of reaction centers in which a "stable" charge separation occurred, and the kinetics at 390 nm indicated only a small increase in the amount of triplet produced in a flash.

DISCUSSION

At least two different electron acceptors appear to be involved in the kinetics of the flash-induced absorbance changes of membranes of H. chlorum, which we call X_1 and X_2. X_2 is responsible for the decay of P798+ in the ms region observed in the presence of ascorbate, whereas X_1 causes the approximately 30-ns decay at low redox potentials, which is associated with the formation of the triplet of P798. Neither X_1 nor X_2 can be equated with the spectroscopically identified electron acceptors BChl c or the iron-sulfur center. BChl c reoxidation in the presence of dithionite occurs in about 2 ns and is not coupled to the rereduction of P798+ (Nuijs et al., 1985b). Moreover, the difference spectrum of the 30-ns decay component (not shown) gave no evidence for the involvement of BChl c. Our titration

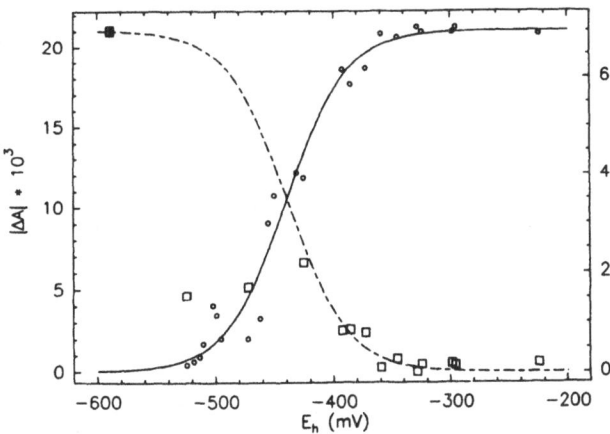

Fig. 10. Titration curves of P798+ reduction and of triplet formation in H. chlorum membranes. Circles, left hand scale: P798+ reduction on a ms time scale. Squares, right hand scale: P798 triplet formation. pH = 9.5. Mediators used: methyl and benzyl viologen, 2-hydroxy-1,4-naphthoquinone, PMS and neutral red, except for the solid square.

experiments indicate that X_2 (which we assume to be a single substance giving rise to somewhat different rates of back reaction in different reaction centers) has a midpoint potential of -440 mV. This indicates that it is not identical to the iron-sulfur center. Our observation that the yield of triplet formation was not dependent on the redox state of the iron-sulfur center indicates that this center is not identical to either X_1 or its electron acceptor and makes it difficult to assign a role for the center in the main electron acceptor chain. X_1 may be identical to the substance that produces the quinone-like ESR signal observed by Brok et al. (1986).

The results obtained with P. aestuarii can be accomodated in the presently accepted scheme for electron transport (Pierson and Olson, 1987). In agreement with results obtained with Cb. limicola the reaction centers were found to retain their normal activity down to much lower redox potentials than in H. chlorum and the yield of formation of the triplet of P840 was not significantly enhanced.

It is of interest to note that, in contrast to what is observed in purple bacteria, in H. chlorum and P. aestuarii the triplet state of the primary electron donor is not transferred to a carotenoid molecule. In H. chlorum no carotenoid triplet was observed at all, whereas in P. aestuarii the triplet state of the carotenoid is apparently generated in the antenna.

Our results indicate that, in the presence of ascorbate and PMS, Cyt c_{553} in P. aestuarii is oxidized by a flash in about 15-20% of the reaction centers. In another 10% of the reaction centers P840+ is eventually rereduced in 70 ms, perhaps by a back reaction with a reduced iron-sulfur center. Most of the remaining reaction centers appear to be deficient in the electron transport chain and produce the triplet of P840 with a yield of roughly 35%. The rate of cytochrome c oxidation is comparable to that observed in many species of purple bacteria. A puzzling phenomenon is the low rate (6 ms) of cytochrome oxidation (Prince et al., 1985) observed in membranes of H. chlorum. We obtained essentially the same results with intact cells (not shown) indicating that the phenomenon is not due to an artifact of preparations.

Taken together our results indicate that, although there are obvious similarities, electron transport in the reaction center of heliobacteria may in some significant respects differ from that in green sulfur bacteria. Further experiments will be needed to explore this question.

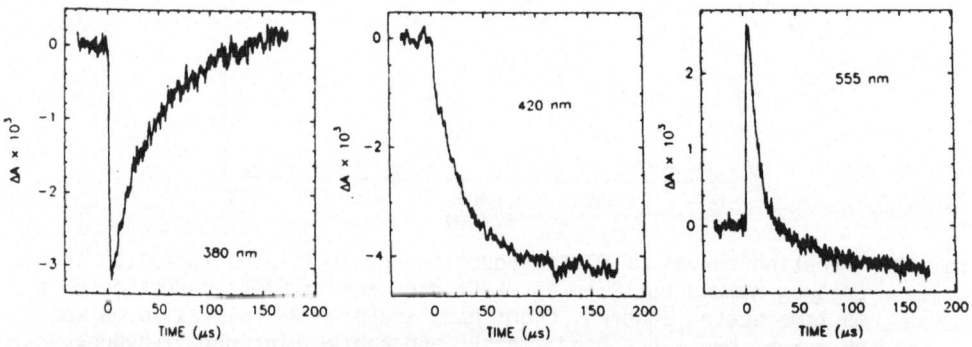

Fig. 11. Kinetics of flash-induced absorbance changes in membranes of P. aestuarii in the presence of 10 mM dithionite, pH = 7.4.

REFERENCES

Braumann, T., Vasmel, H., Grimme, L. H., and Amesz, J., 1986, Pigment composition of the photosynthetic membrane and reaction center of the green bacterium Prostecochloris aestuarii, Biochim. Biophys. Acta, 848:83.

Brockmann, H., and Lipinski, A., 1983, Bacteriochlorophyll g. A new bacteriochlorophyll from Heliobacterium chlorum, Arch. Microbiol., 136:17.

Brok, M., Vasmel, H., Horikx, J. T. G., and Hoff, A. J., 1986, Electron transport components of Heliobacterium chlorum investigated by EPR spectroscopy at 9 and 35 GHz, FEBS Lett., 194:322.

Fowler, C. F., Nugent, N. A., and Fuller, R. C., 1971, The isolation and characterization of a photochemically active complex from Chloropseudomonas ethylica, Proc. Natl. Acad. Sci. U.S.A., 68:2278.

Fuller, R. C., Sprague, S. G., Gest, H., and Blankenship, R. E., 1985, Unique photoreaction center from Heliobacterium chlorum, FEBS Lett., 182:345.

Gest, H., and Favinger, J. L., 1983, Heliobacterium chlorum, an anoxygenic brownish-green photosynthetic bacterium containing a "new" form of bacteriochlorophyll, Arch. Microbiol., 136:11.

Hoff, A. J., 1981, Magnetic field effects on photosynthetic reactions, Quart. Rev. Biophys., 14:599.

Hurt, E. C., and Hauska, G., 1984, Purification of membrane-bound cytochromes and a photoactive P840 protein complex of the green sulfur bacterium Chlorobium limicola f. thiosulfatophilum, FEBS Lett., 168:149.

Kingma, H., 1983, Redox states of the reaction center in relation to energy transfer and mechanism of carotenoid triplet formation in photosynthetic bacteria. Effects of magnetic field, Thesis, University of Leiden, The Netherlands.

Kingma, H., van Grondelle, R., and Duysens, L. N. M., 1985, Magnetic-field effects in photosynthetic bacteria. II. Formation of triplet states in the reaction center and the antenna of Rhodospirillum rubrum and Rhodopseudomonas sphaeroides. Magnetic-field effects, Biochim. Biophys. Acta, 808:383.

Nuijs, A. M., Vasmel, H., Joppe, H. L. P., Duysens, L. N. M., and Amesz, J., 1985a, Excited states and primary charge separation in the pigment system of the green photosynthetic bacterium Prostecochloris aestuarii as studied by picosecond absorbance difference spectroscopy, Biochim. Biophys. Acta, 807:24.

Nuijs, A. M., van Dorssen, R. J., Duysens, L. N. M., and Amesz, J., 1985b, Excited states and primary photochemical reactions in the photosynthetic bacterium Heliobacterium chlorum. Proc. Natl. Acad. Sci. U.S.A., 82:6865.

Olson, J. M., Giddings, T. H., and Shaw, E. K., 1976, An enriched reaction center preparation from green photosynthetic bacteria, Biochim. Biophys. Acta, 449:197.

Pierson, B. K., and Castenholz, R. W., 1974, Studies of pigments and growth in Chloroflexus aurantiacus, a phototrophic filamentous bacterium, Arch. Mikrobiol., 100:283.

Pierson, B. K., and Olson, J. M., 1987, Photosynthetic bacteria, in "Photosynthesis," J. Amesz, ed., Elsevier, Amsterdam.

Prince, R. C., and Olson, J. M., 1976, Some thermodynamic and kinetic properties of the primary photochemical reactants in a complex from a green photosynthetic bacterium, Biochim. Biophys. Acta, 423:357.

Prince, R. C., Gest, H., and Blankenship, R.E., 1985, Thermodynamic properties of the photochemical reaction center of Heliobacterium chlorum, Biochim. Biophys. Acta, 810:377.

Shuvalov, V. A., Amesz, J., and Duysens, L. N. M., 1986, Picosecond spectroscopy of isolated membranes of the photosynthetic green sulfur bacterium <u>Prosthecochloris aestuarii</u> upon selective excitation of the primary electron donor, <u>Biochim. Biophys. Acta</u>, 851:1.

Swarthoff, T., and Amesz, J., 1979, Photochemically active pigment-protein complexes from the green photosynthetic bacterium <u>Prosthecochloris aestuarii</u>, <u>Biochim. Biophys. Acta</u>, 548:427.

Swarthoff, T., Gast, P., Hoff, A. J., and Amesz, J., 1981a, An optical and ESR investigation on the acceptor side of the reaction center of the green photosynthetic bacterium <u>Prosthecochloris aestuarii</u>, <u>FEBS Lett.</u>, 130:93.

Swarthoff, T., van der Veek-Horsley, K. M., and Amesz, J., 1981b, The primary charge separation, cytochrome oxidation and triplet formation in preparations from the green photosynthetic bacterium <u>Prosthecochloris aestuarii</u>, <u>Biochim. Biophys. Acta</u>, 635:1.

Swarthoff, T., Gast, P., and Hoff, A. J., 1981c, Photooxidation and triplet formation of the primary electron donor of the green photosynthetic bacterium <u>Prosthecochloris aestuarii</u>, observed with EPR spectroscopy, <u>FEBS Lett.</u>, 127:83.

van Bochove, A. C., Swarthoff, T., Kingman, H., Hof, R. M., van Grondelle, R., Duysens, L. N. M., and Amesz, J., 1984, A study of the primary charge separation in green bacteria by means of flash spectroscopy, <u>Biochim. Biophys. Acta</u>, 764:343.

van Dorssen, R. J., Vasmel, H., and Amesz, J., 1985, Antenna organization and energy transfer in membranes of <u>Heliobacterium chlorum</u>, <u>Biochim. Biophys. Acta</u>, 809:199.

Vasmel, H., Swarthoff, T., Kramer, H. J. M., and Amesz, J., 1983, Isolation and properties of a pigment-protein complex associated with the reaction center of the green photosynthetic sulfur bacterium <u>Prosthecochloris aestuarii</u>, <u>Biochim. Biophys. Acta</u>, 723:361.

Vasmel, H., den Blanken, H. J., Dijkman, J. T., Hoff, A. J., and Amesz, J., 1984, Triplet-minus-singlet absorbance difference spectra of reaction centers and antenna pigments of the green photosynthetic bacterium <u>Prosthecochloris aestuarii</u>, <u>Biochim. Biophys. Acta</u>, 767:200.

COMPARATIVE STUDY OF SPECTRAL AND KINETIC PROPERTIES OF ELECTRON TRANSFER IN

PURPLE AND GREEN PHOTOSYNTHETIC BACTERIA

A.O. Ganago, V.S. Gubanov, A.V. Klevanik, A.N. Melkozernov,
A.Ya. Shkuropatov and V.A. Shuvalov

Institute of Soil Science and Photosynthesis
USSR Academy of Sciences
Pushchino, Moscow region, U.S.S.R.

INTRODUCTION

The spectral properties in the visible and near infrared regions of reaction centers of the purple bacterium Rhodobacter sphaeroides and the green nonsulfur bacterium Chloroflexus aurantiacus are determined by special interaction between 4 molecules of BChl a and 2 molecules of bacteriopheophytin (BPh) a in Rb. sphaeroides (Straley et al., 1973) and 3 molecules of BChl a and 3 molecules of BPh a in Cf. aurantiacus (Vasmel et al., 1986). These molecules are in L- and M-protein subunits of Rb. sphaeroides (Allen et al., 1986) and bound to the two protein subunits with molecular mass of 26 kDa in Cf. aurantiacus RCs (Feick and Fuller, 1984). Two BChl a molecules form the special pair, the primary electron donor P, absorbing at 870 nm (900 nm at 77 K) in both species. In Rb. sphaeroides RCs the L- as well as M-protein subunits contain one BChl a and one BPh a , BL and HL, and BM and HM, respectively (Allen et al., 1986). The pigments form two prosthetic chains: $P-BL-HL-Q_a$ and $P-BM-HM-Q_b$ where Q_a and Q_b are the primary and secondary quinones. Only one chain $P-BL-HL-Q_a$ is photochemically active (Shuvalov et al., 1986a).

In Cf. aurantiacus RCs the photoactive chain is probably similar to that of Rb. sphaeroides RCs (Shuvalov et al., 1986a). The non-active chain in Cf. aurantiacus contains two H molecules. A similar structure of the M-chain is observed in Rb. sphaeroides RCs after treatment with $NaBH_4$ (Shuvalov et al., 1986a). In this case the BM molecule was modified into H or removed while the photoactive L-chain was preserved. In Cf. aurantiacus RCs at room and low temperature a small band around 790 nm is observed. This band was assigned to the second excitonic band of P (Vasmel et al., 1986). A similar band is not observed in intact Rb. sphaeroides RCs in agreement with its calculated negligible dipole strength.

Picosecond and femtosecond measurements of absorbance changes (A) in RCs have shown that the excited state of P has a lifetime of 3-4 ps and 14 ps in Rb. sphaeroides and Cf. aurantiacus RCs, respectively (Breton et al., 1986; Shuvalov and Duysens, 1986; Shuvalov et al., 1986b). The deactivation of P* is accompanied by the formation of P+ and HL-. The participation of BL in the electron transfer is a subject of debate. The formation of some amount (8-35%) of the state P+BL- prior to P+HL- was observed in picosecond measurements using selective excitation of P for RCs isolated from both species (Shuvalov and Duysens, 1986; Shuvalov et al., 1986b). Femtosecond kinetic measurements were not able to reveal a difference between the kinetics of the bleachings of the B- and H-bands in Rb. sphaeroides (Breton et al., 1986). This was interpreted as evidence for the absence of the formation of BL^-. In intact RCs the blue shift of the BM-band is observed in the femtosecond and picosecond time domains. This shift can be intense enough to obscure the absorbance changes related to the BL-band. The recent femtosecond measurements of kinetics and spectra of ΔA in modified Rb. sphaeroides RCs have shown that at 6 ps after excitation the spectrum of A shows the formation of a mixed state of P^+BL^- and P^+HL^- (Chekalin et al., 1987; see below).

As mentioned above the excited state P^* is deactivated with kinetics similar to those of the formation of HL^-. This was registered by measurements of stimulated emission from P^* at 920 nm in Rb. sphaeroides and Cf. aurantiacus RCs (Breton et al., 1986; Chekalin et al., 1987; Shuvalov et al., 1986). However, hole-burning experiments have shown that in Rb. sphaeroides and Rhodopseudomonas viridis RCs there are no narrow bleachings of the P-band at 900 nm at 2-4 K corresponding to the picosecond lifetime of P^* (Meech et al., 1985; 1986; Boxer et al., 1986a,b). This was interpreted as evidence for the formation of an earlier state (probably PL^-PM^+) within 20 fs.

In this paper we compare the absorption spectra, femtosecond and picosecond measurements and hole-burning experiments for Cf. aurantiacus RCs and modified Rb. sphaeroides (R-26) RCs with similar pigment composition and arrangement.

MATERIALS AND METHODS

Reaction centers from Rb. sphaeroides (R-26) and Cf. aurantiacus were isolated as described earlier (Shuvalov et al., 1986a). For the modification the R-26 RCs suspended in 100 mM Tris-HCl buffer (pH 8.0) containing 0.1% lauryldimethylamine N-oxide (LDAO), were treated with a relatively large amount of NaBH₄ (the pH shifted from 8.0 to 10.6) for 8 hours. Then the RCs were twice dialyzed, against 100 mM and then 10 mM Tris-HCl (pH 8.0) buffer containing 0.1% LDAO and purified by DEAE-cellulose chromatography. For low temperature measurements the samples were diluted with 50% glycerol.

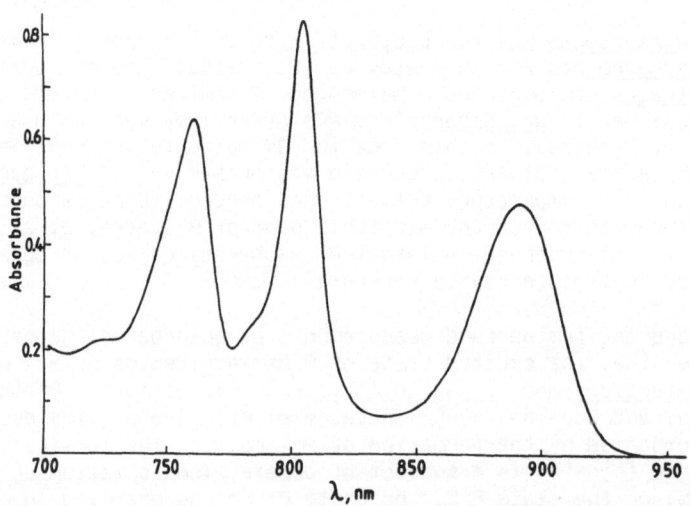

Fig. 1. Absorption spectrum of modified Rb. sphaeroides (R-26) RCs at 77 K.

Light-induced absorbance changes in the fs and ps time domains were measured as described earlier (Chekalin et al., 1987; Shuvalov and Duysens, 1986). For the μs region a ruby laser (duration 20 ns, wavelength 694 nm) was used for excitation.

Hole-burning experiments were done using a parametric generator pumped by 30-ps pulses at 532 nm from a Nd-YAG laser (Shuvalov and Klevanik, 1983). The generator bandwidth was 2-3 nm in the region of 820-910 nm. The spectrum of ΔA was measured at 6 ms after excitation for 1 ms in two regions: 790-840 nm and 840-930 nm using a fast scanning computer-controlled spectrophotometer made in the laboratory. Each spectrum represents an average of ~200 sweeps.

RESULTS

Fig. 1 shows the spectrum of R-26 RCs at 77 K after modification with NaBH₄. The decrease of the ratio of A_{805}/A_{900} as well as the increase of A_{760}/A_{900} with respect to those for intact RCs are observed. An absorption band at 790 nm appears which is similar to that observed in Cf. aurantiacus RCs and is assigned to the second excitonic band of P (Vasmel et al., 1986). The absorption spectrum of modified R-26 RCs in the near infrared is very similar to that of Cf. aurantiacus RCs. These RCs have 3 molecules of BChl a and 3 molecules of BPh a (Vasmel et al., 1986).

Fig. 2 shows kinetic measurements of ΔA induced by 20-ns flashes in Cf. aurantiacus RCs at 870 nm (P-band) and 550 nm (cytochrome band). In the presence of ascorbate (1 mM) and Vitamin K₃ (20 μM) the slow relaxation (~150 ms) of the bleaching at 870 nm is observed due to the electron transfer to Vitamin K₃ from Q_b^- and stabilization of P⁺. The addition of Cyt c (20 μM) to RCs induces a relatively fast rereduction of P⁺ due to the electron donation by cytochrome (time constant of ~15 ms). The amount of cytochrome oxidation in one flash is shown by ΔA at 550 nm. The value of the

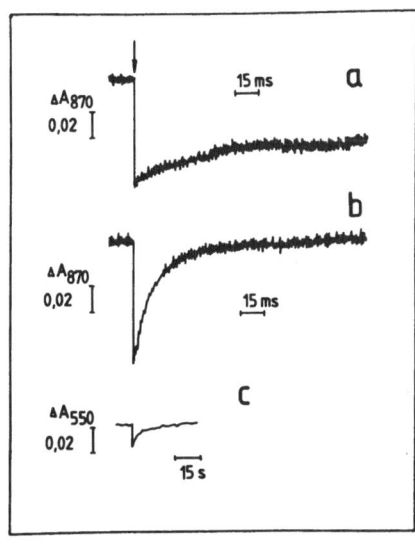

Fig. 2. The kinetics of ΔA at 870 nm (P-band) and 550 nm (cytochrome band) in Cf. aurantiacus RCs (2 μM) at 293 K in the presence of Vitamin K₃ (20 μM) and ascorbate (1 mM) (a), the same plus cytochrome c (20 μM) (b and c).

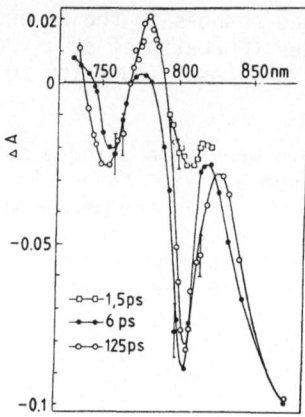

Fig. 3. Difference absorption spectra of modified R-26 RCs at 293 K excited
by non-saturating 300-fs pulses at 620 nm, measured at various
delays. Absorbance at 870 nm was 1.2.

difference extinction coefficient ($\Delta\epsilon$) of Cyt $\underline{c}$ at 550 nm is 19.7 nM^{-1} cm^{-1}
in agreement with earlier work. Comparison of Fig. 2b and c shows that $\Delta\epsilon$
for P at 870 nm is 115 ± 15 mM^{-1} cm^{-1}, close to that for Rb. sphaeroides RC
(Straley et al., 1973).

Fig. 3 shows the spectrum of ΔA in modified R-26 RCs measured at 6 ps
delay after the center of 300-fs pulses at 625 nm (Chekalin et al., 1987).
In the absence of the band shift of the BM molecule in modified RCs, a clear
bleaching of the BL-band at 805 nm without an absorbance increase at 780 nm
as well as the bleaching of the HL-band at 755 nm and of the P-band at 870
nm are observed. The kinetics of the bleaching at 870 nm correspond to the
autocorrelation function of the excitation pulse, but those at 750 nm and
805 nm are delayed with time constant of ~3 ps (Fig. 4). Stimulated emission
of P* at 930 nm has a lifetime of ~3 ps in agreement with measurements on
intact RCs (Breton et al., 1986).

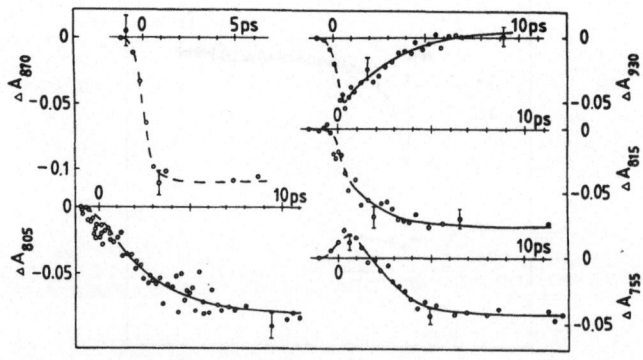

Fig. 4. Kinetics of absorbance changes at 870 nm (P-band), 805 nm and 815 nm
(BL-bands), 755 nm (HL-band and 930 nm stimulated emission from P*).
Other conditions as for Fig. 3.

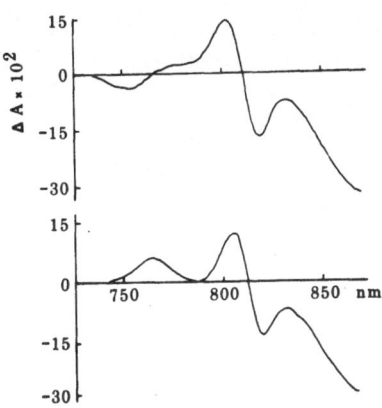

Fig. 5. Absorption difference spectra in Cf. aurantiacus induced by 33-ps
pulses at 870 nm, measured at 71 ps (upper, the state P+H−) and 2.9
ns (lower, the state P+Q$_a$−).

In contrast to R-26 RCs picosecond spectroscopy of Cf. aurantiacus RCs
has shown that the lifetime of P* and the time constant of the formation of
P+H− are close to ~14 ps (Shuvalov et al., 1986b). The spectrum of P+H−
includes the bleachings of the H-band at 760 nm and the P-band at 870 nm. In
the region of the B-band a blue shift of this band at 815 nm is observed
(Fig. 5).

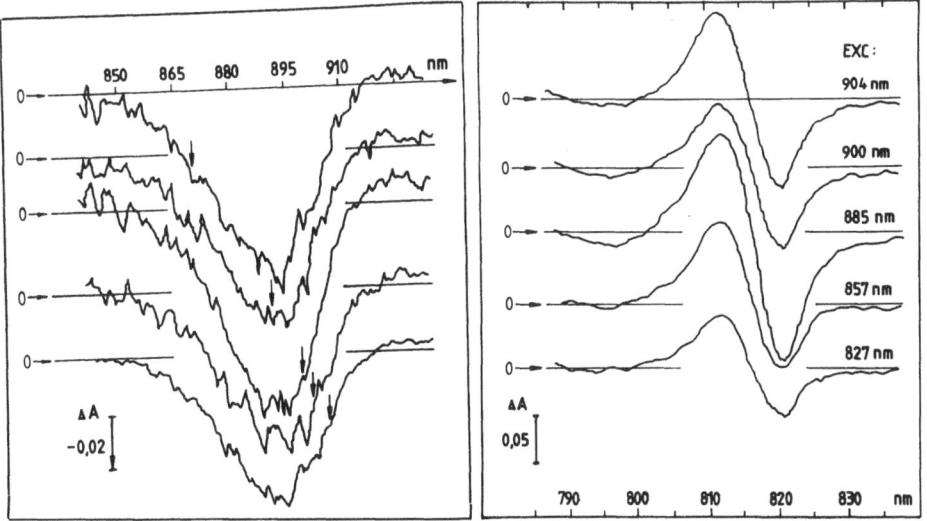

Fig. 6. Spectra of ΔA of P+ formation in Cf. aurantiacus RCs at 1.9 K
induced by the 30-ps pulses at different wavelengths (shown by
arrows in Fig. 6A and by numbers in Fig. 6B), measured at 6 ms.

Fig. 6 shows the spectra of ΔA of P⁺ formation in the region of the P-band at 900 nm at 1.9 K in <u>Cf. aurantiacus</u> RCs induced by the 30-ps pulses at different wavelengths: 870 nm, 891 nm, 899 nm, 901.5 nm and 906 nm. In all cases the bleaching of the broad band (linewidth 350 cm⁻¹) is observed with a shape close to Gaussian. In the region of the blue shift of the B band at 815 nm the shape of spectra is independent of the wavelength of excitation.

DISCUSSION

The spectrum of modified R-26 RCs (Fig. 1) was interpreted in terms of the modification (pheophytinization) of a BM molecule which resulted in the formation of an M-chain in which two BPh <u>a</u> molecules are present (Shuvalov et al., 1986). The absorption and circular dichroism spectra of modified R-26 RCs are similar to those of <u>Cf. aurantiacus</u> RCs suggesting that the M-chain of modified R-26 RCs is similar to the non-active chain of <u>Cf. aurantiacus</u> RCs (Shuvalov et al., 1986). Furthermore Fig. 1 shows the presence of a weak band at 790 nm in modified R-26 RCs which is also clearly seen in <u>Cf. aurantiacus</u> RCs. Excitonic calculations for <u>Cf. aurantiacus</u> RCs reveal a second exciton band of P near 790 nm (Vasmel et al., 1986). The presence of the band at 790 nm in modified R-26 RCs supports the idea that the pigment content and arrangement are similar to those in <u>Cf. aurantiacus</u> RCs. The difference extinction coefficient for P at 870 nm for <u>Cf. aurantiacus</u> RCs (115 mM⁻¹ cm⁻¹) is close to that for R-26 RCs (108 mM⁻¹ cm⁻¹) (Straley et al., 1973).

Picosecond spectroscopy of the electron transfer reveals some difference between modified R-26 and <u>Cf. aurantiacus</u> RCs. The time constants for the formation of P⁺H⁻ are different for modified R-26 RCs (~3 ps) and <u>Cf. aurantiacus</u> RCs (~14ps). The simultaneous bleachings of the BL- and HL-bands in modified R-26 RCs at 6 ps delay suggest that the electron density is delocalized between BL and HL, and equal populations of P⁺BL⁻ and P⁺HL⁻ are observed (Chekhalin et al., 1987). It was shown that the absorption of a second photon at 870 nm in this state leads to transient kinetics at 800 nm and redistribution of the bleachings between the 760 and 800-nm bands (Akhmanov et al., 1980). This was interpreted as evidence for an electron density shift from B to H induced by second photon absorption. No transient kinetics at 800 nm are observed when RCs are excited by non-saturating pulses (Shuvalov and Duysens, 1986).

In contrast to R-26, in <u>Cf. aurantiacus</u> RCs only the bleachings of the H- and P-bands related to the formation of P⁺H⁻ are observed. The difference between two types of RCs can be due to the difference in energy (ΔE) of P⁺H⁻ and P⁺B⁻. For R-26 RCs ΔE is probably close to zero while the energy of P⁺H⁻ is lower than that of P⁺B⁻ for <u>Cf. aurantiacus</u> RCs. In both cases the state P⁺B⁻ appears to be formed earlier than P⁺H⁻ (Shuvalov and Duysens, 1986; Shuvalov et al., 1986b) in agreement with the X-ray structure of the RCs which shows that BL is located between P and HL in <u>Rb. sphaeroides</u> as well as in <u>Rbs. viridis</u> RCs (Allen et al., 1986; Deisenhofer et al., 1985).

Hole-burning studies of <u>Cf. aurantiacus</u> RCs give results similar to those obtained for <u>Rb. sphaeroides</u> (Boxer et al., 1986; Meech et al., 1985). In both cases the bleaching of the largely homogeneous broad band (linewidth of 350 cm⁻¹) for <u>Cf. aurantiacus</u> is observed. Such results were earlier interpreted as evidence for an ultra-fast decay of P*, probably to a charge transfer state such as PM⁺PL⁻ or PM⁺PLBL(Boxer et al., 1986a,b; Meech et al., 1985,1986).

Another possible explanation of these results is that the zero-phonon line is suppressed in spectra of ΔA. The Debye-Waller factor (α) gives the

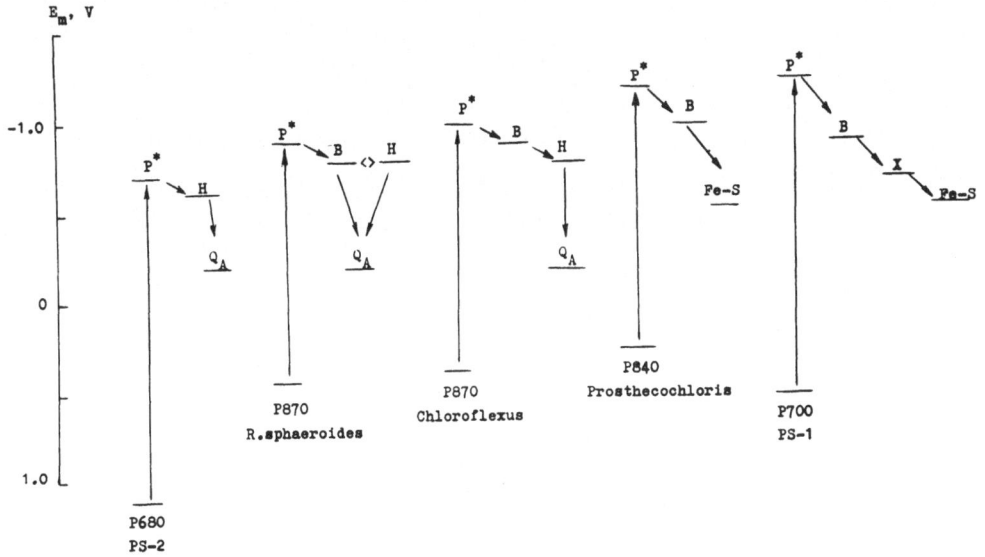

Fig. 7. Energy scheme of the electron carriers in RCs of photosystems 1
(PS-1) and 2 (PS-2) of green plants, of the purple bacterium Rb.
sphaeroides, green nonsulfur bacterium Cf. aurantiacus and green
sulfur bacterium Prosthecochloris aestuarii. P is the primary donor;
B is BChl a, BChl c, or Chl a; H is BPh a or pheophytin a; X is an
Fe-S center.

relative intensity of the zero-phonon line normalized by comparison with the
total intensity of absorption. This factor at T = 0 is equal to (Friederich
and Haarer, 1984):

$$\alpha(T = 0) = \exp(-S)$$

where S corresponds to the number of vibrational quanta which are excited in
the maximum of the phonon sideband. For S ⩾ 5 there is a strong
electron-phonon coupling. The zero-phonon line is entirely suppressed and
only the phonon sideband is observed.

This explanation is supported by the considerable red shift ($\sim$300 cm^{-1})
of the fluorescence maximum with respect to the absorption maximum and by
the mirror-symmetry of the two spectra in the crossing area for Rps. viridis
RCs at 1.9 K (Maslov et al., 1983). For these RCs a similar bleaching of the
broad band is observed in hole-burning experiments (Boxer et al., 1986b;
Meech et al., 1986). In the case of the fast formation of a new state from
P* (e.g., charge transfer state) the mirror-symmetry of the fluorescence and
absorption spectra should be absent in contrast to the experimental data.

Fig. 7 shows the energy scheme of the electron transfer in different
RCs, including photosystems 1 (PS-1) and 2 (PS-2) of green plants and the RC
of green sulfur bacteria. The increase of the energy of P* leads to the
involvement of BChl a, BChl c, or Chl a (for PS-1) between P and H or
instead of H as an intermediate electron carrier and to the increase of the
energy of vacant orbitals of terminal acceptors.

REFERENCES

Akhmanov, S. A., Borisov, A. Yu., Danielius, R. V., Gadonas, R. A.,
    Kozlovski, V. S., Piskarskas, A. S., Razjivin, A. P., and Shuvalov,
    V. A., 1980, One and two-photon picosecond processes of electron
    transfer among the porphyrin molecules in bacterial reaction centers,
    FEBS Lett., 114:149.
Allen, J. P., Feher, G., Yeats, T. O., Rees, D. C., Deisenhofer, J., Michel,
    H., and Huber, R., 1986, Structural homology of reaction centers from
    Rhodopseudomonas sphaeroides and Rhodopseudomonas viridis as
    determined by x-ray diffraction, Proc. Natl. Acad. Sci. USA,
    83:8589.
Boxer, S. G., Lockhart, D. J., Middendorf, T. R., 1986a, Photochemical
    hole-burning in photosynthetic reaction centers, Chem. Phys. Lett.,
    123:476. Boxer, S. G., Middendorf, T. R., and Lockhart, D. J., 1986b,
Reversible photochemical hole-burning in Rhodopseudomonas viridis reaction
    centers, FEBS Lett., 200:237.
Breton, J., Martin, J.-L., Petrich, J., and Antonetti, A., 1986, The absence
    of spectroscopically resolved intermediate state P$^+$B$^-$ in bacterial
    photosynthesis, FEBS Lett., 209:37.
Chekalin, S. V., Matveetz, Yu. A., Shkuropatov, A. Ya., Shuvalov, V. A., and
    Yartzev, A. P., 1987, Femtosecond spectroscopy of primary charge
    separation in modified reaction centers of Rhodobacter sphaeroides
    (R-26), FEBS Lett., 216:245.
Deisenhofer, J., Epp, O., Miki, K., Huber, R., and Michel, H., 1985,
    Structure of the protein subunits in the photosynthetic reaction
    centre of Rhodopseudomonas viridis at 3 A resolution, Nature,
    318:618.
Feick, R. G., and Fuller, R. C., 1984, Topography of the photosynthetic
    apparatus of Chloroflexus aurantiacus, Biochemistry, 23:3693.
Friederich, J., and Haarer, D., 1984, Photochemical hole burning: a
    spectroscopic study of relaxation processes in polymers and glasses,
    Angew. Chem. Int. Ed. Engl., 23:113.
Kirmaier, C., Holten, D., Feick, R., and Blankenship, R. E., 1983,
    Picosecond measurements of the primary photochemical events in
    reaction centers isolated from the facultative green photosynthetic
    bacterium Chloroflexus aurantiacus, FEBS Lett., 158:73.
Maslov, V. G., Klevanik, A. V., Ismailov, M. A., and Shuvalov, V. A., 1983,
    About the nature of long wavelength absorption band of reaction
    centers of Rhodopseudomonas viridis in connection with the primary
    charge separation (in Russian), Dokl. Akad. Nauk SSSR, 269:1217.
Meech, S. R., Hoff, A. J., and Wiersma, D. A., 1985, Evidence for a very
    early intermediate in bacterial photosynthesis. A photon-echo and
    hole-burning study of the primary donor band in Rhodopseudomonas
    sphaeroides, Chem. Phys. Lett., 121:287.
Meech, S. R., Hoff, A. J., and Wiersma, D. A., 1986, Role of charge-transfer
    states in bacterial photosynthesis, Proc. Natl. Acad. Sci. USA,
    83:9464.
Shuvalov, V. A., and Duysens, L. N. M., 1986, Primary electron transfer
    reactions in modified reaction centers from Rhodopseudomonas
    sphaeroides, Proc. Natl. Acad. Sci. USA, 83:1690.
Shuvalov, V. A., and Klevanik, A. V., 1983, The study of the state P$^+$B$^-$ in
    bacterial reaction centers by selective picosecond and low
    temperature spectroscopies, FEBS Lett., 160:51.
Shuvalov, V. A., Shkuropatov, A. Ya., Kulakova, S. M., Ismailov, M. A., and
    Shkuropatova, V. A., 1986a, Photoreactions of bacteriopheophytins and
    bacteriochlorophylls in reaction centers of Rhodopseudomonas
    sphaeroides and Chloroflexus aurantiacus, Biochim. Biophys. Acta,
    849:337.
Shuvalov, V. A., Vasmel, H., Amesz, J., and Duysens, L. N. M., 1986b,
    Picosecond spectroscopy of the charge separation in reaction centers

of _Chloroflexus aurantiacus_ with selective excitation of the primary electron donor, _Biochim. Biophys. Acta_, 851:350.

Straley, S. C., Parson, W. W., Mauzerall, D., and Clayton, R. K., 1973, Pigment content and molar extinction coefficient of photochemical reaction centers from _Rhodopseudomonas sphaeroides_, _Biochim. Biophys. Acta_, 305:597.

Vasmel, H., Amesz, J., and Hoff, A., 1986, Analysis by exciton theory of the optical properties of the _Chloroflexus aurantiacus_ reaction center, _Biochim. Biophys. Acta_, 852:159.

TRIPLET-MINUS-SINGLET OPTICAL DIFFERENCE SPECTROSCOPY OF SOME GREEN

PHOTOSYNTHETIC BACTERIA

A.J. Hoff, H. Vasmel, E.J. Lous and J. Amesz

Department of Biophysics
Huygens Laboratory of the State University
P.O. Box 9504, 2300 RA Leiden, The Netherlands

SUMMARY

Prosthecochloris aestuarii

Several pigment-protein complexes have been isolated (H. Vasmel et al., 1983, 1984) to wit: 1) the core complex (CC), 2) the photosystem-pigment complex (PP), 3) the reaction-center-pigment protein complex (RCPP), 4) the light-harvesting water-soluble antenna complex (LHP). The RCPP complex contains the reaction center (RC), the core complex and subunits of LHP, all membrane-bound; the PP complex contains in addition two loosely bound LHP molecules. The CC is not photoactive. We summarize triplet data obtained by absorbance-detected magnetic resonance (ADMR) at 1.2 K, and present ADMR-monitored triplet-minus-singlet spectra of these four complexes.

The RCPP complex contains three triplet states, one characteristic of $P840^T$, one identical to that of the core complex and one similar to that of isolated LHP. The triplet-minus-singlet (T - S) spectra of the complexes were measured using ADMR. This allows one to discriminate between T - S spectra of the various triplet states when present in one complex by selecting the appropriate microwave transition frequency. The T - S spectra of PP, RCPP and CC complexes are shown in Figs. 1 and 2. The corresponding linear dichroic (LD-(T - S)) spectra of $P840^T$ are shown in Fig. 3.

The bleaching at 837 nm of the T - S spectra of $P840^T$ is due to bleaching of the primary donor absorption band. The bleachings at 827 and 834 nm correspond to bands in the singlet absorption spectrum. The bleachings have the same sign and similar intensity in the two LD-(T-S) spectra, indicating that their transition moments are approximately parallel. This means that they cannot be the exciton components of a strongly coupled dimer. They are either both due to a weakly coupled dimer with its triplet state localized on the lowest energy component absorbing at 837 nm, or only one of the two (presumably the 837-nm one) is due to a strongly coupled dimer, the other band arising from interactions with a neighboring pigment absorbing around 826 nm and a pigment absorbing at 814 nm. Features in the 660-nm region of the T - S spectrum (not shown) indicate coupling to a BChl c molecule that presumably functions as an early acceptor (van Bochove et al., 1984).

Table 1. Microwave Transitions of the Triplet States of P. aestuarii

	ν_1	ν_2 (MHz)	Remarks
PP	464*	not resolved	LHP triplet
RCPP	515	735	reaction-center triplet
CC	450	785	isolated CC
LHP	465	795	isolated LHP
BChl a in MTHF	516	864	

*PP shows in addition the resonances due to the RCPP and CC triplet.

The T - S spectrum of the core complex is typical for a dimeric antenna pigment that is strongly coupled in the singlet state. The 837-nm bleaching is due to a localized triplet state; the 808-nm band, to an appearing monomer band.

The third T - S spectrum is observed in the PP, RCPP and LHP complexes and is due to a triplet state of the water-soluble LHP complex. Its interpretation is difficult as the LHP molecule contains 7 electrostatically coupled BChl a molecules. Spectral simulations with exciton theory suggest that the triplet state is predominantly localized on the BChl a that is mainly responsible for the longest wavelength absorption (827 nm), with perhaps some delocalization on the 806-nm BChl a. This gives rise to a bleaching of the 829- and 806-nm bands, and to a 3-nm red shift of the strong 815-nm band of the singlet absorption spectrum.

Chloroflexus aurantiacus

The triplet state of Cf. aurantiacus is typically due to a reaction center triplet. Its EPR characteristics are shown in Table 3.

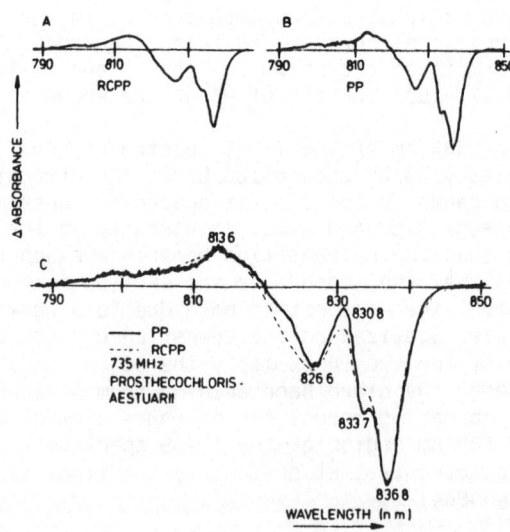

Fig. 1. T - S spectra of PP and RCPP complexes from P. aestuarii.

Table 2. Triplet Data of <u>P. aestuarii</u> Pigment-Protein Complexes

	$|D|$	$|E|$	k_x	k_y	k_z
	(x 10^{-4} cm^{-1})			(s^{-1})	
RCPP	208.3 ± 0.7	36.7 ± 0.7	6790 ± 500	3920 ± 300	1275 ± 100
CC	205.8 ± 0.7	55.8 ± 0.7			
LHP	209.8 ± 0.7	55.2 ± 0.7			
BChl <u>a</u> in MTHF	230.0 ± 2.0	58.0 ± 2.0	11950 ± 750	15900 ± 1300	1635 ± 50

The T - S spectrum is shown in Fig. 4, the LD-(T - S) spectra in Fig.
5. From the extrapolation to zero microwave power of the intensity ratio
LD-(T - S)/T - S at the isolated 887-nm band it follows that $\alpha_x = 19 \pm 3°$,
$\alpha_y = 65 \pm 2°$, α_i being the angle of the optical transition moment with the
x, y triplet spin axes. It follows that the 887-nm transition moment lies
approximately in the plane spanned by the x and y spin axes ($\alpha_x + v_y \approx 90°$).
Similarly, from the 790-nm band $\alpha_x = 36 \pm 2°$, $\alpha_y = 41 \pm 2°$. The angle
between the 887- and 790-nm transition moments is then 65° or 19°, which may
be compared with the value obtained from fluorescence polarization (37°) and
from exciton calculations (49°) (Vasmel et al., 1986). The calculations
indicate that the 790-nm band is predominantly the high energy exciton band
of P865. The 805-nm band, with $\alpha_x = 0 \pm 10°$ and $\alpha_y = 65 \pm 2°$, is ascribed to
the appearing monomer band of P865^T. The (T - S) spectra can be well
understood assuming that the crystal structure of the RC of <u>Cf. aurantiacus</u>
is the same as that of <u>Rhodopseudomonas viridis</u>, provided one of the
accessory BChls is replaced by a bacteriopheophytin (Scherer and Fischer,
1987).

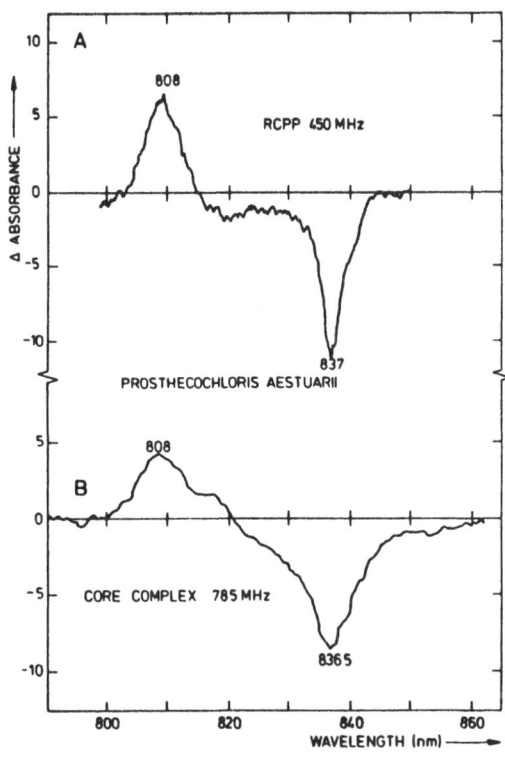

Fig. 2. T - S spectra of RCPP and CC complexes from <u>P. aestuarii</u>.

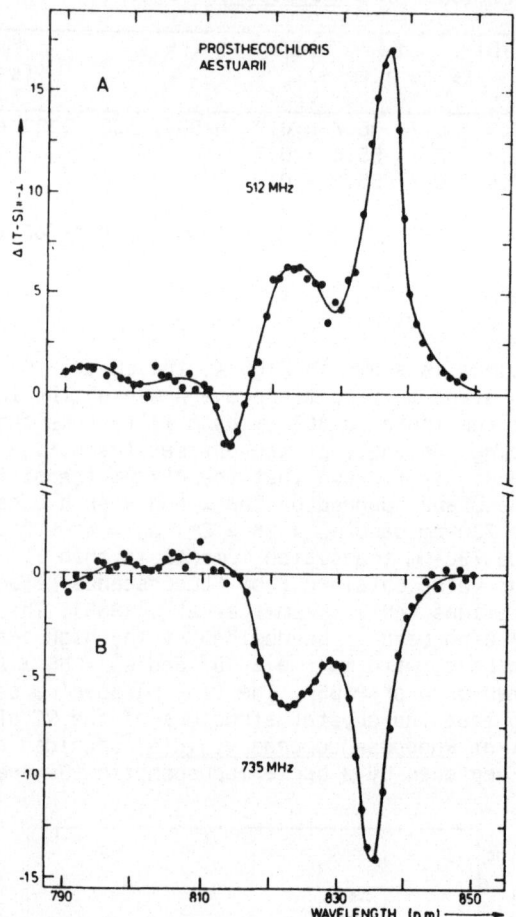

Fig. 3. LD-(T - S) spectra of P840^T from _P. aestuarii_.

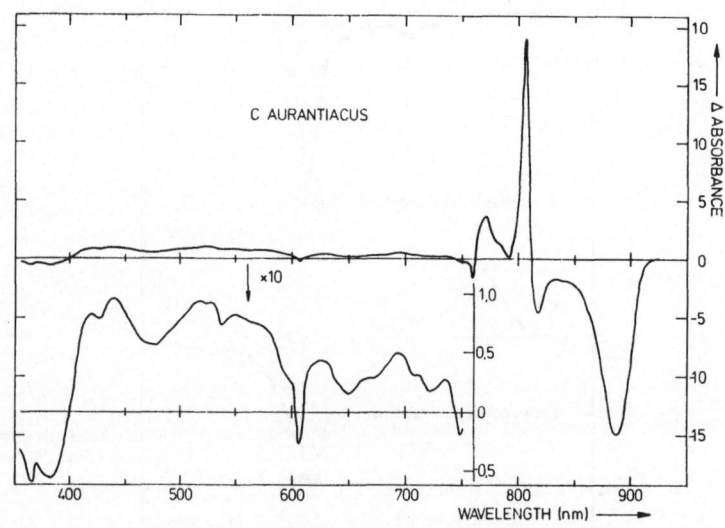

Fig. 4. T - S spectrum for _Cf. aurantiacus_

Table 3. Triplet Data of <u>Cf. aurantiacus</u> Reaction Centers

ν_1 ν_2 (MHz)	$\|D\|$	$\|E\|$	k_x	k_y	k_z
	($\times 10^{-4}$ cm^{-1})			(s^{-1})	
451 735	197.7 ± 0.7	47.3 ± 0.7	12660 ± 750	14290 ± 800	1690 ± 50

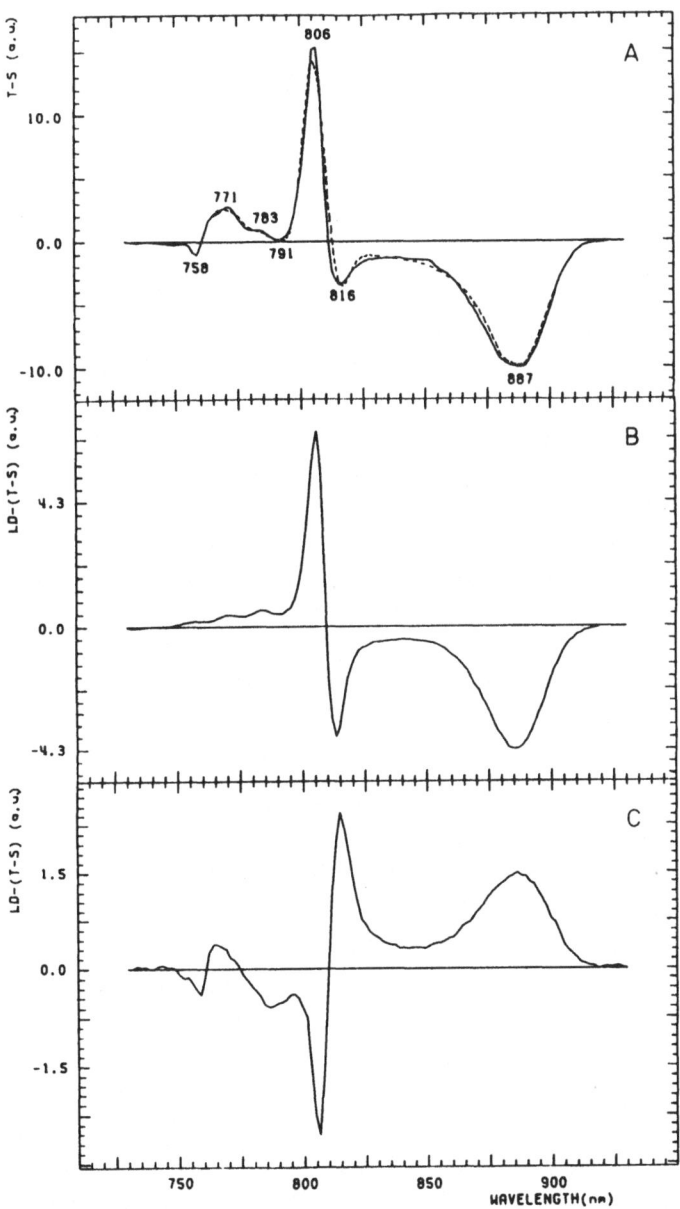

Fig. 5. LD-(T - S) spectra for <u>Cf. aurantiacus</u>.

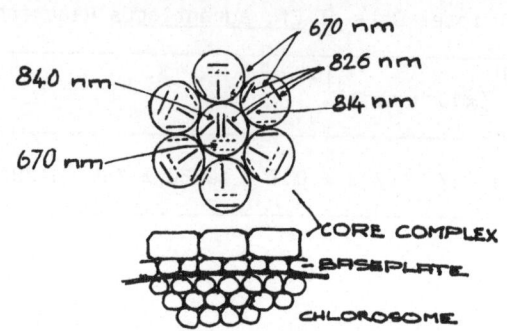

Fig. 6. Geometry of the various components of the RCPP. See text for details.

Reaction-Center Pigment Protein

The RCPP contains about 35 BChl a molecules. From these, about 14 are lost upon making the CC. These probably constitute two subunits (2 x 7 BChl a) of the LHP belonging to the base-plate proteins. The RCPP and the CC both contain about 15 BChl c molecules. The CC is not photoactive because the BChl c primary acceptor is inactivated. RCPP shows a CD signal at 670 nm characteristic of a dimeric BChl c molecule. This signal is lost upon CC formation and may be attributed to the BChl c acceptor. The T - S spectra show that the primary donor P840 has an interaction with at least one, and probably two, BChl a molecules absorbing at 826 and 833 nm, and with the BChl c dimer at 670 nm. In addition there is an interaction with a BChl a absorbing at 814 nm. Thus, from about 15 BChl c one dimer shows interaction with P840; the others do not interact appreciably with each other or with BChl a. Fig. 6 shows a geometry that is compatible with the BChl a and BChl c stoichiometries and with geometrical constraints, and takes into account all the spectroscopic data discussed above. Note that a heterodimer of BChl a and BChl c does not show much interaction because of the different energies involved.

REFERENCES

Scherer, P. O. J., and Fischer, S. F., 1987, Model studies of low-temperature optical transitions of photosynthetic reaction centers. II. Rb. sphaeroides and Cf. aurantiacus, Biochim. Biophys. Acta, 891:157.

van Bochove, A. C., Swarthoff, T., Kingma, H., Hof, R. M., van Grondelle, R., Duysens, L. N. M. and Amesz, J., 1984, A study of the primary charge separation in green bacteria by means of flash spectroscopy, Biochim. Biophys. Acta, 764:343.

Vasmel, H., Swarthoff, T., Kramer, H. J. M., and Amesz, J., 1983, Isolation and properties of a pigment-protein complex associated with the reaction center of the green photosynthetic bacterium Prosthecochloris aestuarii, Biochim. Biophys. Acta, 725:361.

Vasmel, H., den Blanken, H. J., Dijkman, J. T., Hoff, A. J., and Amesz, J., 1984, Triplet-minus-singlet absorbance difference spectra of reaction centers and antenna pigments of the green photosynthetic bacterium Prosthecochloris aestuarii, Biochim. Biophys. Acta, 767:200.

Vasmel, H., Amesz, J., and Hoff, A. J., 1986, Analysis by exciton theory of the optical properties of the *Chloroflexus aurantiacus* reaction center, *Biochim. Biophys. Acta*, 852:159.

SOLUBLE ELECTRON-TRANSFER PROTEINS OF CHLOROBIACEAE

U. Fischer

Fachbereich Biologie, AG Geomikrobiologie
Universität Oldenburg
Postfach 2503, D-2900 Oldenburg, F.R.G.

SUMMARY

Green sulfur bacteria (Chlorobiaceae) utilize mainly sulfide or
thiosulfate as electron donors for their anoxygenic photosynthesis and
carbon dioxide reduction (Trüper and Fischer, 1982). Depending on how sulfur
compounds are metabolized by anoxyphotobacteria, they can either be oxidized
or reduced. Electron-transfer proteins are thus necessary for these redox
reactions. Green sulfur bacteria contain various soluble cytochromes and
iron-sulfur proteins like ferredoxins and rubredoxins as well as
membrane-bound cytochromes. Occurrence and distributions of soluble
electron-transfer proteins of Chlorobiaceae so far examined are summarized
in Table 1 (Steinmetz, 1984; Fischer, 1986).

As one can see from Table 1, ferredoxin, rubredoxin and the small Cyt
c-555 occur in nearly all species of green sulfur bacteria examined. Only
the thiosulfate-utilizing species of Chlorobium limicola and Cb. vibrioforme
contain the monomeric, acidic hemoprotein Cyt c-551 (Meyer et al., 1968;
Steinmetz and Fischer, 1982a). Cyt c-551 serves as the first endogenous
electron acceptor in thiosulfate oxidation catalyzed by a Cyt c-551
oxidoreductase, before the electrons flow to oxidized BChl a via the small
Cyt c-555. The reduction rate of Cyt c-551 is even stimulated when catalytic
amounts of Cyt c-555 are present. The other cytochromes (c-555 or
flavocytochrome c-553) cannot replace Cyt c-551 in this function (Steinmetz
and Fischer, 1981; 1982a). Since Cyt c-551 is absent in the
non-thiosulfate-utilizing species of green sulfur bacteria (Table 3), this
might be one explanation of why these organisms cannot use thiosulfate as an
additional electron source for their photosynthesis. While Cyt c-551
participates in thiosulfate oxidation, the flavocytochromes of Chlorobiaceae
are involved in anaerobic sulfide oxidation and act as sulfide:cytochrome c
reductases. The enzymatic activity of these flavocytochromes is strongly
inhibited by heat treatment or by the presence of cyanide (Steinmetz and
Fischer, 1981; 1982a). Flavocytochrome c-552 of Chromatium vinosum not only
has sulfide:cytochrome c reductase activity but also catalyzes the reduction
of elemental sulfur to sulfide with reduced benzyl-viologen as electron
donor (Steinmetz and Fischer, 1981). This could not be confirmed for the
flavocytochromes of Chlorobiaceae (Steinmetz and Fischer, 1981; 1982a).
Concerning the redox potentials of -240 mV for the reaction of elemental
sulfur to sulfide and of +90 mV for the flavocytochromes one cannot expect
this reaction to be realized in vivo by these hemoproteins without being

Table 1. Distribution of C-type Cytochromes and Fe-S Proteins in Chlorobiaceae

Organisms	Cytochromes			Flavo-cyt.	Ferre-doxin	Rubre-doxin
	c-551	c-553	c-555			
Cb. limicola f. thiosulfatophilum	+		+	+	+	+
Cb. vibrioforme f. thiosulfatophilum	+		+	+	+	+
Cb. vibrioforme			+		+	+(2)
Cb. limicola			+	+	+	+
Cb. phaeobacteroides			+	+	ND	ND
P. luteolum		+	+		+	+
Pr. aestuarii			+		+	+

ND = not determined

coupled to an energy producing process (Steinmetz and Fischer, 1982a). This assumption is supported by the fact that sulfide formation from elemental sulfur in Cb. limicola f. thiosulfatophilum is a light-dependent process (Paschinger et al., 1974).

Cytochrome c-555 not only serves as positive effector for the thiosulfate-Cyt c-551-oxidoreductase but also acts as electron acceptor for the sulfide:cytochrome c reductase, a flavocytochrome (Fischer, 1986; Steinmetz et al., 1983). A direct photo-oxidation in Chlorobiaceae was only demonstrated for a membrane-bound Cyt c-555 but not for the soluble Cyt c-555 (Knaff et al., 1973). One might therefore assume that soluble Cyt c-555 leads the electrons obtained during anaerobic sulfide (or thiosulfate) oxidation via the membrane-bound cytochrome to the reaction centre (Knaff et al., 1973; Fischer, 1986).

It has been reported that cells of Cb. limicola f. thiosulfatophilum grown with sulfide as the main electron donor contained more flavocytochrome c-553 than those which were cultivated with thiosulfate (Steinmetz and Fischer, 1982). We could not confirm this finding when we made similar experiments with Cb. vibrioforme f. thiosulfatophilum. The cytochromes (c-551, c-553 and c-555) in the latter organism remained nearly in a constant ratio to each other (3.5:1:2.7 with thiosulfate and 4:1:1.7 with sulfide). Cytochrome formation in this organism is independent of the sulfur source offered (Steinmetz and Fischer, 1982a).

A comparison of molecular properties of soluble cytochromes of thiosulfate-utilizing and non-thiosulfate-utilizing species of Chlorobiaceae is summarized in Table 2 and 3.

Among phototrophic bacteria the iron-containing protein rubredoxin occurs only in the green sulfur bacteria (Steinmetz and Fischer, 1982b; Steinmetz et al., 1983). All Chlorobiaceae rubredoxins are acidic proteins of low molecular weight, with negative redox potentials, and they possess per protein molecule only one iron atom, which is tetrahedrally chelated to the sulfur atoms of the four cysteine residues in the amino acid chain (Woolley and Meyer, 1987). The oxidized form exhibits maxima at 570, 490 and 370 nm and the visible part of the oxidized spectrum is completely bleached after the addition of dithionite. The molecular properties of rubredoxins from green sulfur bacteria are listed in Table 4.

Table 2. Comparison of Molecular Properties of Soluble Cytochromes of the Thiosulfate-Utilizing Strains of Cb. limicola and Cb. vibrioforme

Molecular Properties	Organism and Cytochromes					
	Cb. limicola f. thiosulfatophilum[1]			Cb. vibrioforme f. thiosulfatophilum[2]		
	c-551	c-553	c-555	c-551	c-553	c-555
Molecular weight	45,000	50,000	10,000	32,000	63,000	12,500
Redox potential (mV)	+ 135	+ 98	+ 145	+ 150	+ 90	+ 155
Flavin group	-	FAD	-	-	FAD	-
Isoelectric point	6.0	6.7	10.5	6.0	6.3	10-10.5
Purity index ($A_{280}/A_{\gamma-band}$)	0.39	0.94*	0.16	0.39	0.8	0.16
Maxima (nm) oxidized (γ-band)	410.5	410	412.5	410	410	412
Reduced (α- and γ-band)	551	553.5	555	551	553.5	555
			(551)			(551)
	416	416.7	418.5	416	417	418

[1] Data from Meyer et al. (1968) and Bartsch (1978)
[2] Data from Steinmetz and Fischer (1982a)
*= oxidized

Table 3. Comparison of Molecular Properties of Soluble Cytochromes of Non-Thiosulfate-Utilizing Species of Chlorobiaceae: Molecular weight (kDa) (1), Redox potential (mV) (2), Flavin groups (3), Isoelectric point (4), Purity index (red) ($A_{280}/A_{\gamma-band}$) (5), Maxima (nm) oxidized ($\gamma-band$) (6), and Reduced (α-band and γ-band) (7).

Molecular Properties	Organism and Cytochromes c							
	Cb. limicola[1]		Cb. phaeobacteroides[2]		P. luteolum[3]		Cb. vibrioforme[4]	Pr. aestuarii[5]
	553	555	553	555	553	555	555	555
1.	56	10	ND	9.5	10.7 (13)	9.5	11.5 11	12 11.1
2.	+ 65	+ 140	+ 63	+ 105	+ 220	+ 160	+ 80	+ 103
3.	FAD	-	FAD	-	-	-	-	-
4.	9.0	9.5-10.0	9.5-10.0	9.5-10.0	10.2	10.5	7.3	4.65
5.	0.96	0.13	0.65	0.176	0.18	0.17	0.15	0.18
6.	410	412	410	411	413	413	412.5	412
7.	553.5	555	552	554.5	553	555	555	555
		(551)		(551)		(551)	(551)	(550)
	417	417.5	417	418	417	418	418	417.5

Data taken from [1] Steinmetz and Fischer (1981), [2] Fischer (unpublished), [3] Steinmetz and Fischer (1982b), [4] Steinmetz et al. (1983), [5] Shioi et al. (1972).
ND = not determined
Cb. = Chlorobium, P. = Pelodictyon, Pr. = Prosthecochloris

Table 4. Comparison of Molecular Properties of Rubredoxins of Chlorobiaceae

| Molecular Properties | Organism and Rubredoxin | | | |
| | Cb. limicola f. thiosulf.[1] Rub | Cb. vibrioforme[2] | | P. luteolum[3] |
		Rub I	Rub II	Rub
Molecular weight	7,400	8,100	8,100	6,300
				27,000
Nonheme iron (nmol/nmol of protein)	ND	1	1	1
Isoelectric point	ND	2.9	2.7	2.8
Redox potential (mV)	-61	ND	ND	-100
Purity index (A_{280}/A_{370})* or (A_{280}/A_{490})	2.65*	3.27	3.57	3.5
Maxima (nm) (oxidized form)	372	370	370	370
	492	492	492	490
	570	575	557	570

Data taken from: [1] Meyer et al. (1971); [2] Steinmetz et al. (1983);
[3] Steinmetz and Fischer (1982b)
ND = not determined

Rubredoxins are mainly found in anaerobic bacteria (sulfate-reducing bacteria, Clostridia or Chlorobiaceae). Only two rubredoxins have been isolated from the obligate aerobes Pseudomonas oleovorans and Acinetobacter calcoaceticus (Woolley and Meyer, 1987). To my knowledge, only Clostridium thermoaceticum and Cb. vibrioforme contain two distinct rubredoxins (Steinmetz et al., 1983; Woolley and Meyer, 1987). Nothing is known about the function of this protein in green sulfur bacteria. In Desulfovibrio gigas rubredoxin is necessary for the hydrogenase reaction and in Clostridium thermoaceticum this protein is the preferred electron donor for carbon monoxide dehydrogenase. In Pseudomonas oleovorans it participates in hydroxylation reactions of alkanes and fatty acids (Fischer, 1986; Woolley and Meyer, 1987).

REFERENCES

Bartsch, R. G., 1978, Cytochromes, in: "The Photosynthetic Bacteria," R. K. Clayton and W. R. Sistrom, eds., Plenum Press, New York.

Fischer, U., 1986, Schwefelstoffwechsel und Elektronentransportproteine in Anoxyphutobakterien, Habilitationsschrift, Univ. Oldenburg, West-Germany.

Knaff, D. B., Buchanan, B. B., and Malkin, R. C., 1973, Effect of oxidation-reduction potential on light-induced cytochrome and bacteriochlorophyll reactions in chromatophores from the photosynthetic green bacterium Chlorobium, Biochim. Biophys. Acta, 325:94.

Meyer, T. E., Bartsch, R. G., Cusanovich, M. A., and Mathewson, J. H., 1968, The cytochromes of Chlorobium thiosulfatophilum, Biochim. Biophys. Acta, 153:854.

Meyer, T. E., Sharp, J. J., and Bartsch, R. G., 1971, Isolation and properties of rubredoxin from the photosynthetic green sulfur bacteria, Biochim. Biophys. Acta, 234:266.

Paschinger, H., Paschinger, J., and Gaffron, H., 1974, Photochemical disproportionation of sulfur into sulfide and sulfate by Chlorobium limicola f. thiosulfatophilum, Arch. Microbiol., 96:341.

Shioi, Y., Takamiya, K., and Nishimura, M., 1972, Studies on energy and

electron transfer systems in the green photosynthetic bacterium
Chloropseudomonas ethylica strain 2-K. I. Isolation and
characterization of cytochromes from _Chloropseudomonas ethylica_
strain 2-K, _J. Biochem. (Tokyo)_, 71:285.

Steinmetz, M. A., 1984, Cytochrome und Eisenschwefelproteine in
Chlorobiaceae, Doctoral thesis, Univ. Bonn, F.R.G.

Steinmetz, M. A., and Fischer, U., 1981, Cytochromes of the
non-thiosulfate-utilizing green sulfur bacterium _Chlorobium limicola_,
Arch. Microbiol., 130:31.

Steinmetz, M. A., and Fischer, U., 1982a, Cytochromes of the green sulfur
bacterium _Chlorobium vibrioforme_ f. _thiosulfatophilum_. Purification,
characterization and sulfur metabolism, _Arch. Microbiol._, 131:19.

Steinmetz, M. A., and Fischer, U., 1982b, Cytochromes, rubredoxin, and
sulfur metabolism of the non-thiosulfate-utilizing green sulfur
bacterium _Pelodictyon luteolum_, _Arch. Microbiol._, 132:204.

Steinmetz, M. A., Trüper, H. G., and Fischer, U., 1983, Cytochrome c-555 and
iron-sulfur proteins of the non-thiosulfate-utilizing green sulfur
bacterium _Chlorobium vibrioforme_, _Arch. Microbiol._, 135:186.

Trüper, H. G., and Fischer, U., 1982, Anaerobic oxidation of sulphur
compounds as electron donors for bacterial photosynthesis, _Phil.
Trans. Roy. Soc. Lond. B_, 298:529.

Wooley, K. J., and Meyer, T. E., 1987, The complete amino acid sequence of
rubredoxin from the green phototrophic bacterium _Chlorobium
thiosulfatophilum_ strain PM, _Eur. J. Biochem._, 163:161.

COMPLEX FORMATION BETWEEN CHLOROBIUM LIMICOLA F. THIOSULFATOPHILUM C-TYPE CYTOCHROMES

M.W. Davidson[a], T.E. Meyer[b], M.A. Cusanovich[b] and D.B. Knaff[a]

[a]Department of Chemistry and Biochemistry
 Texas Tech University, Lubbock, TX 79409, U.S.A.
[b]Department of Biochemistry, University of Arizona
 Tucson, AZ 85721, U.S.A.

SUMMARY

Chlorobium contains three soluble c-type cytochromes that are involved in the oxidation of sulfur-containing electron donors by this bacterium. Flavocytochrome "c553", which consists of a 47-kDa subunit containing a single covalently bound FAD and a 11-kDa subunit containing a single heme c (Meyer et al., 1986; Yamanaka, 1976; Yamanaka et al., 1979), has been demonstrated to catalyze electron flow from sulfide to Cyt "c555" (Kusai and Yamanaka, 1973, 1978; Kusai and Fakumori, 1980). It has been proposed that elemental sulfur is the product resulting from sulfide oxidation (Kusai and Fukumori, 1980).

Cytochrome "c555", which has been shown by X-ray crystallography to be structurally similar to Rhodospirillum rubrum Cyt "c2" and mitochondrial Cyt c (Salemme, 1977), has been found in all photosynthetic green sulfur bacteria studied to date (Trüper and Fischer, 1982). In addition to its likely role as an electron acceptor in the flavocytochrome "c555"-catalyzed oxidation of sulfide, Cyt "c555" can also serve as an acceptor of electrons from thiosulfate. The oxidation of thiosulfate in Chlorobium and in other thiosulfate-oxidizing green sulfur bacteria involves a third soluble c-type cytochrome, Cyt "c551" (Kusai and Yamanaka, 1973, 1978; Trüper and Fischer, 1982).

It has been possible to demonstrate, using affinity chromatography, that Cyt "c555" forms electrostatically stabilized complexes with both Cyt "c551" and flavocytochrome "c553" (Davidson et al., 1986). The ionic strength dependence of the flavocytochrome "c553" catalyzed electron transfer reaction from sulfide to Cyt "c555" suggests that the complex between these two proteins is catalytically significant. The binding site for Cyt "c555" on flavocytochrome "c553" appears to be on its heme-containing rather than on the FAD-containing subunit.

The photosynthetic purple sulfur bacterium Chromatium vinosum contains a soluble flavocytochrome "c552" that is similar to the Chlorobium flavocytochrome "c553" in many ways, including the ability to catalyze electron flow from sulfide to c-type cytochromes (Davidson et al., 1986). The Chromatium flavocytochrome "c552" is able to form an electrostatically stabilized complex with the Chlorobium cytochrome "c555" (Davidson et al., 1986), suggesting that the binding sites for Cyt c-electron acceptors on the two flavocytochromes may be similar. Equine Cyt $\underline{c}$, which can function as an electron accepting analog for Cyt "$\underline{c}555$" in the flavocytochrome "$\underline{c}553$"-catalyzed oxidation of sulfide, loses its ability to act as an electron acceptor and to form a complex with the flavocytochrome when its lysine residues were trifluoroacetylated (Davidson et al., 1986; Gray and Knaff, 1982). These results are consistent with a role for positive charges on lysine residues that surround the exposed heme edges of these cytochromes (Salemme, 1977) in complex formation with the flavocytochromes.

ACKNOWLEDGEMENTS

This work was supported by grants from the U.S. National Science Foundation (PCM 84-16649 to M.A.C.) and the Robert A. Welch Foundation (D-710 to D.B.K.).

REFERENCES

Bosshard, H. R., Davidson, M. W., Knaff, D. B., and Millett, F., 1986, Complex formation and electron transfer between mitochondrial cytochrome $\underline{c}$ and flavocytochrome $\underline{c}552$ from Chromatium vinosum, J. Biol. Chem., 261:190.

Davidson, M. W., Meyer, T. E., Cusanovich, M. A., and Knaff, D. B., 1986, Complex formation between Chlorobium limicola f. thiosulfatophilum $\underline{c}$-type cytochromes, Biochim. Biophys. Acta, 850:396.

Gray, G. O., and Knaff, D. B., 1982, The role of a cytochrome $\underline{c}552$: cytochrome $\underline{c}$ complex in the oxidation of sulfide in Chromatium vinosum, Biochim. Biophys. Acta, 680:290.

Kusai, A., and Yamanaka, T., 1973, Cytochrome $\underline{c}$ (553 Chlorobium thiosulfatophilum) is a sulphide-cytochrome $\underline{c}$ reductase, FEBS Lett., 34:235.

Kusai, A., and Yamanaka, T., 1973, A novel function of cytochrome $\underline{c}$ (555, Chlorobium thiosulfatophilum) in oxidation of thiosulfate, Biochem. Biophys. Res. Comm., 51:107.

Kusai, A., and Yamanaka, T., 1978, The oxidation mechanisms of thiosulphate and sulphide in Chlorobium thiosulfatophilum: roles of cytochrome $\underline{c}$-551 and cytochrome $\underline{c}$-553, Biochim. Biophys. Acta, 325:304.

Meyer, T. E., Bartsch, R. G., Cusanich, M. A., and Mathewson, J. S., 1968, The cytochromes of Chlorobium thiosulfatophilum, Biochim. Biophys. Acta, 153:854.

Salemme, R. F., 1977, Structure and function of cytochromes $\underline{c}$, Ann. Rev. Biochem., 46:299.

Trüper, H. G., and Fischer, U., 1982, Anaerobic oxidation of sulphur compounds as electron donors for bacterial photosynthesis, Phil. Trans. Roy. Soc. London B, 298:529.

Yamanaka, T., 1976, The subunits of Chlorobium flavocytochrome $\underline{c}$, J. Biochem., 79:655.

Yamanaka, T., Fukumori, Y., and Okunuki, K., 1979, Preparation of subunits of flavocytochromes $\underline{c}$ derived from Chlorobium limicola f. thiosulfatophilum and Chromatium vinosum, Anal. Biochem., 95:209.

Yamanaka, T., and Fukumori, Y., 1980, A biochemical comparison between Chlorobium and Chromatium flavocytochromes $\underline{c}$, in: "Flavins and Flavoproteins," K. Yaqi, and T. Yamano, eds., Japan Scientific Societies Press, Tokyo.

THE MECHANISM OF PHOTOSYNTHETIC ELECTRON TRANSPORT AND ENERGY TRANSDUCTION

BY MEMBRANE FRAGMENTS FROM <u>CHLOROFLEXUS AURANTIACUS</u>

D. Zannoni and G. Venturoli

Department of Biology, Institute of Botany
University of Bologna
I-40126 Bologna, Italy

INTRODUCTION

The photosynthetic apparatus of the thermophilic, facultatively aerobic green bacterium <u>Chloroflexus aurantiacus</u> is located in two cytologically distinct structures. This arrangement and the presence of two bacteriochlorophyll types, BChl <u>a</u> and species-dependent BChl <u>c</u>, <u>d</u> or <u>e</u> are characteristic for all green bacteria (Chlorobiaceae and Chloroflexiaceae). In <u>Cf. aurantiacus</u>, the photochemical reaction centers and a BChl <u>a</u>-containing light-harvesting complex with absorption maxima at 866 nm and 808 nm are located in the cytoplasmic membrane closely intermingled with the secondary electron transport system (Zannoni, 1986a). Recently, a method has been described for the isolation of the purified cytoplasmic membrane (CM), completely devoid of BChl <u>c</u> and chlorosomes (Feick et al., 1982). Spectral and thermodynamic characterization of the cytoplasmic membrane isolated by this method indicated that the reaction center P870 of <u>Cf. aurantiacus</u> has a midpoint potential at pH 8.1 of + 360 mV whereas a membrane-bound photooxidizable <u>c</u>-type cytochrome (<u>c</u>554) that serves as the electron donor to P870 has its E_m 8.1 around + 260 mV. This cytochrome has subsequently been isolated and found to contain two spectrally identical but thermodynamically distinct hemes with E_m of + 265 mV and + 140 mV (Blankenship et al., 1985). An electron acceptor with E_m = 50 mV (pH 8.1) was detected by the failure to observe Cyt <u>c</u>554 photo-oxidation at low potentials (Bruce et al., 1982). Subsequent studies with isolated reaction centers have also established that this acceptor is menaquinone (Vasmel and Amesz, 1983; Blankenship et al., 1984). The presence of a two-quinone acceptor system is also suggested by the fact that o-phenantroline, an inhibitor of electron transfer between the primary (Q_a) and the secondary (Q_b) quinone acceptor, greatly increases the rate constant for the recovery of P870 photobleaching in whole membranes (Bruce et al., 1982; Zannoni and Venturoli, in preparation). In this connection it is important to note that in isolated CM, on the contrary, the rate constant for the recovery of P870 photobleaching was found to be largely independent of o-phenantroline and equal to that measured in whole membranes in the presence of saturating o-phenantroline concentrations (Bruce et al., 1982). In purple bacteria, o-phenantroline is known to cause a positive shift in the E_m of Q_a (Prince et al., 1978). Since reversible analogous shifts have been observed in isolated reaction centers (Wraight, 1982) and in chromatophores (Venturoli, unpublished results) upon extraction of Q_b, it can be concluded that o-phenantroline reverses the effect that Q_b has on the midpoint potential of

Q_a. These considerations suggest that isolated CMs are devoid of Q_b and therefore the actual E_m of Q_a might be considerably lower than the previously reported value of - 50 mV at pH 8.0 (Bruce et al., 1982). In this work we show that this latter suggestion is indeed correct, because in whole membranes isolated from photosynthetically grown Cf. aurantiacus, the primary acceptor titrates at approximately - 100 mV (pH 7.0) with a pH-dependency of - 60 mV/pH unit. Furthermore, we have been able to demonstrate that the two hemes of Cyt c554 undergo photo-oxidation and can be thermodynamically resolved following a single saturating flash of actinic light. Two different cytochromes c with apparent E_m 7.0 of approx. + 220 mV and + 20 mV seem to be rapidly photo-oxidized in the subsequent flashes in a train. The former of these hypothetical c-type components is re-reduced in the dark in a reaction which is sensitive to HQNO (heptylhydroxy-quinoline-N-oxide) and largely insensitive to antimycin A, mucidin, myxothiazol and UHDBT (undecylhydroxy-dioxobenzothiazole). Interestingly, cytochromes c with similar redox midpoint potentials have previously been detected by dark equilibrium redox titrations in both photosynthetic and aerobic membranes (Zannoni and Ingledew, 1985; Zannoni, 1986b; Wynn et al., 1987).

In this report preliminary data on the energy transduction by membrane fragments isolated from photosynthetically grown cells of Chloroflexus are also presented. In contrast to the general belief that phosphorylation is difficult to demonstrate in membranes from green bacteria due to the fact that chlorosomes tend to prevent the formation of sealed plasma membrane vesicles, we demonstrate that under appropriate conditions (E_h = 0 mV, pH 8.2 and temperature 42°C), photophosphorylation activity can readily be observed in membranes isolated from the green bacterium Cf. aurantiacus.

MATERIALS AND METHODS

Organism Cultivation and Membrane Isolation

The medium used for phototrophic growth of Cf. aurantiacus strain J-10-fl was that described by Pierson and Castenholz (1974). Cells were cultivated in a 14l fermentor (Microferm; New Brunswick Scientific, NJ) at 55°C with an incident light intensity of 2000 W m^{-2} (high-light cells) or 200 W m^{-2} (low-light cells) for 20 h with a stirring at 400 rpm. Membranes were prepared as previously described (Zannoni and Ingledew, 1985).

Kinetic Spectrophotometry

The kinetics of flash-induced redox changes of cytochromes were measured using a single-beam spectrophotometer with a time resolution of 0.5 μs and a bandwidth of 1.5 nm. Flash excitation was provided by a xenon lamp filtered by a Wratten 88A glass filter, giving a saturating (> 90%) flash of 15 μs duration at half maximal intensity. The photomultiplier was protected by a Corning glass filter No. 9782. Rapid digitisation of the photo-multiplier linear amplifier output was done by a Datalab DL905 Transient Recorder, interfaced to an Olivetti M24 computer. Redox poise was done as in Zannoni and Ingledew (1985).

ATPase and Phosphorylation

Standard photophosphorylation and ATPase activities were assayed according to Baccarini-Melandri et al. (1970). Photophosphorylation measurements at a controlled ambient potential were carried out under strict anaerobic conditions as previously described (Baccarini-Melandri et al., 1979).

Protein and Pigment Determination

Proteins were assayed by using the method of Lowry et al. (1951). BChl _a_ and _c_ concentrations were measured in acetone/methanol extracts (7:2, v/v) at 769 and 666 nm using extinction coefficients of 68.6 and 74 mM^{-1} cm^{-1}, respectively.

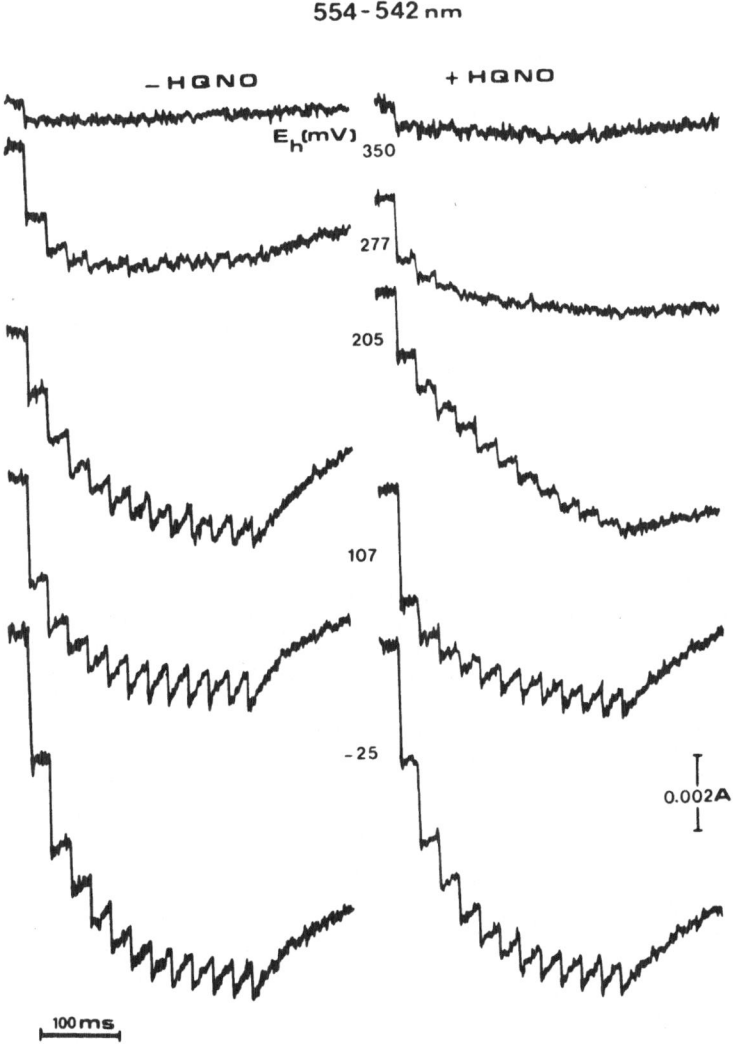

554-542 nm

Fig. 1. Flash-induced cytochromes _c_ oxidation measured at 554-542 nm in the presence of in the absence of HQNO in membranes from low-light cells of _Cf. aurantiacus_ at different E$_h$. Membranes were suspended in 50 mM MOPS/100 mM KCl (pH 7.0) containing 2 μM each of PMS, PES, DAD and pyocianine; 10 μM each of 1,4-naphthoquinone; 1,2-naphthoquinone, 2,3,5,6-tetramethyl-1,4-benzoquinone and p-benzoquinone; 20 μM valinomycin and 2 myM nigericin. HQNO was present at a concentration of 6 μM. Traces were average of 4; sweep, 500 ms; filter RC, 500 μs; time between each measurement, 60 s.

Light-Induced Cytochrome(s) c Oxidation

It has previously been shown that photoheterotrophically grown cells of
Cf. aurantiacus contain both the b and c type elements which are diagnostic,
together with the Rieske iron-sulphur center, of a putative b/c₁ complex
(Zannoni and Ingledew, 1985). It has also been demonstrated that the g =
1.90 centre (Rieske iron-sulphur protein) can be photo-oxidized under
steady-state illumination at controlled ambient redox potential so as to
provide a strong indication of its involvement in photosynthetic electron
transport (Zannoni and Ingledew, 1985). To characterize the photoinduced
electron flow further, we have therefore studied the oxidation of
cytochromes c at 554-542 nm following a train of 12 closely spaced flashes
of light in the absence or in the presence of HQNO at a controlled ambient
redox potential. Traces in Fig. 1 indicate that around + 200 mV, HQNO
stimulates by approx. 15% the total extent of Cyt c oxidation induced by a
train of 12 flashes fired 30 ms apart. This stimulation, however, is not
seen at all in the extent of Cyt c oxidation induced by the first flash, but
becomes more and more evident following the subsequent flashes in the train.
The marked inhibitory effect of HQNO on the re-reduction rate of
flash-oxidized Cyt c is particularly evident in the recovery kinetics of the
traces. It appears that following a single flash of light the observed Cyt c
photo-oxidation is not determined at any potential by the re-reduction rate
in the dark, whereas the extent of Cyt c seen between the 2nd and the 12th
flash in a train can be varied by an electron transport inhibitor such as
HQNO. This finding suggests that different c type species are involved in
cyclic electron flow by membrane fragments from Cf. aurantiacus. This type
of reasoning has been applied to the analysis of the redox titration of the
extent of flash-induced Cyt c oxidation. Experimental points were obtained
from the extent of the absorbance changes induced by a series of 12 actinic
flashes at a series of E_h values at different pHs. The results of these
titrations are summarized in Fig. 2, in which the closed circles indicate
the extent of the signal at 554-542 nm before the flash (0 time) and 30 ms
after the flash while the closed squares and empty triangles indicate the
extent of the signal between the 2nd and the 12th flash in the presence and
in the absence of HQNO, respectively. At the first flash, the data nicely
fit with a two-component n = 1 Nernst curve with E_m 7.0 of approx. + 295 mV
and + 140 mV. It is also evident that the extent of the signal at 554-542 nm
is progressively reduced below - 50 mV, when the primary quinone acceptor
(Q_a) which titrates with an E_m 7.0 of - 100 mV, starts to be reduced before
the flash. The pH dependency of the midpoint potential of Q_a between pH 7.0
and pH 9.0 is detailed in panel B of Fig. 2. The analysis of the titration
of the extent of Cyt c photo-oxidized between the 2nd and the 12th flash is
complex. As expected from the traces of Fig. 1 the extent of the signal is
increased by HQNO between + 250 mV and 100 mV while it is unaffected by this
inhibitor below 100 mV. Both in the presence and in the absence of HQNO,
when the E_h is lowered from + 250 mV to + 150 mV, most of the signal which
becomes observable between the 2nd and the 12th flash, can be accounted for
by the presence of a redox component titrating with an apparent E_m of
approx. + 220 mV. It is therefore tempting to attribute this signal to a
c-type component, already identified in dark equilibrium titration by
Zannoni and Ingledew (1985), distinct from the diheme Cyt c554 which goes
oxidized following the first flash. A striking feature of titration curves 2
and 3 is the reduction of the extent of the signal between + 150 mV and + 70
mV. This phenomenon might be due to the removal of a redox rate-limiting
step affecting the re-reduction level of cytochromes c photo-oxidized
between the 2nd and the 12th flash. Since the decrease in the observed
extent of Cyt c oxidation titrates with an apparent E_m around + 120 mV it
might be speculated that this effect is related in some way to the initial
redox state of the Rieske iron-sulphur centre, previously titrated around +

100 mV (Zannoni and Ingledew, 1985), or unidentified electron transport carrier. Both hypotheses, although suggestive, require however a very high electron-carrier reaction-center ratio and this particular point has to be clearly established. From the data of Fig. 2 it is also evident that below + 70 mV the extent of cytochromes c photo-oxidized by a train of 12 flashes increases further. This increment tends to suggest that a low potential c-type species (E_m 7.0 = + 20 mV), in addition to Cyt $c554$ and Cyt c_1, can readily be photo-oxidized. At present the role of this cytochrome in photoactivated electron transport is rather obscure.

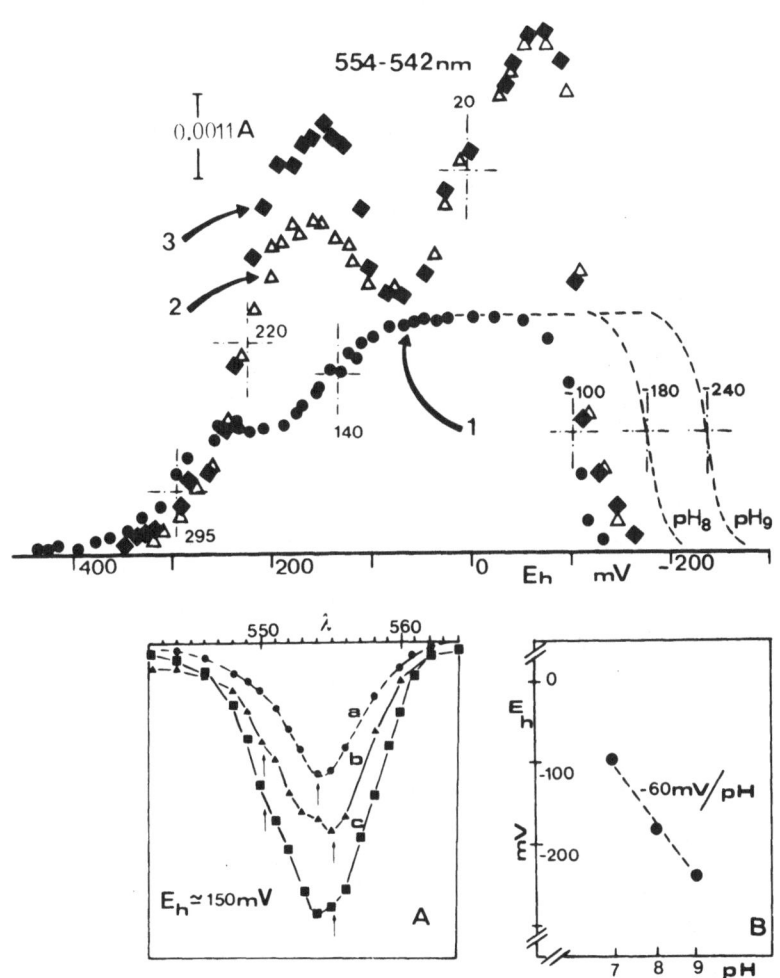

Fig. 2. Redox titrations of the flash-induced cytochromes c oxidation measured at 554-542 nm in membranes from low-light cells of Cf. aurantiacus. Experimental conditions as in Fig. 1. Symbols: (●) changes occurring between 0-30 ms; (Δ) changes occurring between 30-330 ms in the absence of HQNO; (◆) changes occurring between 30-330 ms in the presence of HQNO; (---) titration curves for the changes occurring between 0-30 ms at pH 8.0 and pH 9.0. The midpoint potentials corresponding to n = 1 Nernst components are indicated. Panels A and B show the time-resolved spectra of the signals seen at + 150 mV and the pH dependency of the redox midpoint potential of the primary acceptor, respectively. See text for further details.

In Fig. 2, panel A, the time-resolved spectra are shown of the components photo-oxidized at the first flash (trace a) and between the 2nd and 12th flash in the absence (trace b) and in the presence (trace c) of HQNO (ambient redox potential of + 150 mV). It is apparent that the spectrum of the change occurring from just before the flash (0 time) to approx. 30 ms after the flash has a peak at 554 nm whereas the absorption from 30 ms and 0.33 s gives broader spectra, possibly including minor contributions from bands peaking at different wavelength (550, 553 and 555 nm). Comparison of these spectra with those obtained during dark-equilibrium titrations suggests that the peak at 554 nm corresponds to the α-band of Cyt c295 while the shoulders at 550 nm and 555 nm correspond to the split α-band of Cyt c220 (Cyt c₁) (Zannoni et al., in preparation).

ATPase and Photophosphorylation Activities

Under experimental conditions in which the Mg^{++}/ATP ratio is close to unity at 60°C and pH 8.2 the K_m of the Mg^{++}dependent ATPase in membranes from Cf. aurantiacus is 0.18 mM ATP (not shown). This latter value is very similar to previous determinations in facultative photosynthetic bacteria (Baccarini-Melandri et al., 1975). No activation of the ATPase by light could be observed. In Fig. 3 the temperature dependence of the ATPase activity (Arrhenius-plot) is shown. Quite interestingly, the Arrhenius plot shows a dramatic break at 50°C. At this temperature the activation energy shifts from approx. 0.7 kcal/mol to 17 kcal/mol. The significance of this definite change is now under investigation.

In Figs. 4 and 5 the photophosphorylation activity as a function of temperature, pH and ambient redox potential is shown. Under aerobic conditions in the presence of Na succinate (Fig. 4) the photophosphorylation, although present, is very low. This activity is slightly affected by the pH and very sensitive to the temperature. The data indicate that at pH 8.0 a temperature around 42°C maximizes the rate of phosphorylation in vitro. Under these experimental conditions, however, the

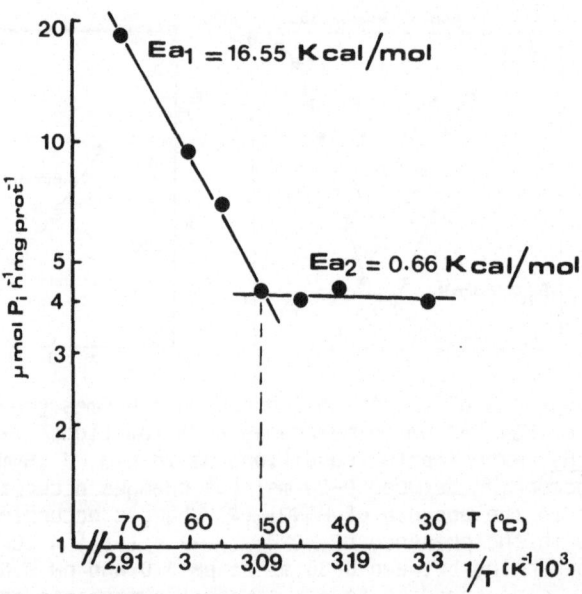

Fig. 3. Temperature dependence of the ATPase activity in membranes from high-light cells from Cf. aurantiacus. Reaction time, 1 min; pH 7.8; ATP, 3 mM; Mg^{++}, 3 mM.

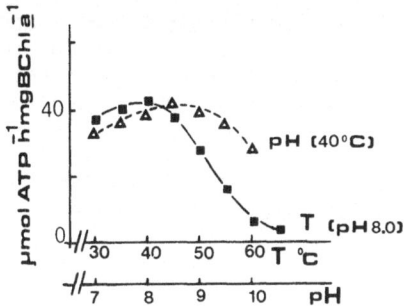

Fig. 4. Temperature and pH dependence of the light-induced ATP synthesis by membranes from high-light cells of Cf. aurantiacus. The redox poise was obtained by adding Na succinate under aerobic conditions. Reaction time, 1 min.

ATP synthesis can be drastically activated by changing the ambient midpoint potential (see Fig. 5).

As shown in Fig. 5 photophosphorylation is tightly controlled by the (initial) redox conditions of the system as previously observed in other photosynthetic organisms (Baccarini-Melandri et al., 1979; Van der Berg et al., 1983). However, while in ubiquinone-containing species the phosphorylation is fastest around 100-130 mV, in Cf. aurantiacus (which contains menaquinone as sole quinone species) the photophosphorylation activity requires a lower ambient redox potential (around 0 mV). In this respect it is interesting to note that in facultative photosynthetic bacteria the optimum redox poise for photophosphorylation coincides with the operating redox midpoint potential of the ubiquinol/ubiquinone couple at the myxothiazol-sensitive site (Q_z) of the bc$_1$ complex (Van der Berg et al., 1983). Assuming that this coincidence reflects a general property and that

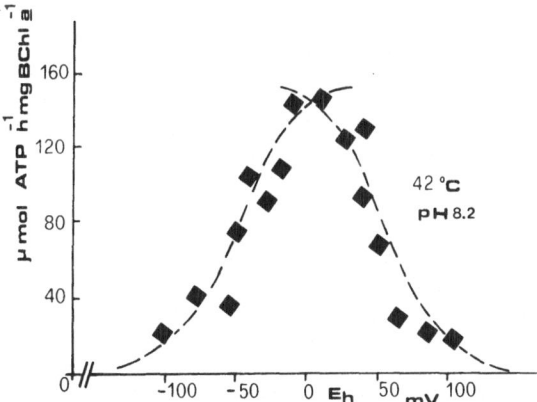

Fig. 5. E_h dependence of photophosphorylation in membranes from high-light cells of Cf. aurantiacus. In order to avoid artificial electron flow, all the redox mediators (see Fig. 1) were used at very low concentrations (2 μM each). Temperature, 42°C; pH 8.2.

the putative bc_1 complex of Chloroflexus operates according to a Q-cycle mechanism (Meinhardt and Crofts, 1983), thermodynamic considerations based on the E_m of the Rieske-centre and of the low-potential cytochrome b (b-70) suggest that the Q_z counterpart in Cf. aurantiacus should have an E_m 7.0 around 0 mV.

Experiments aiming to define the redox behaviour of the menaquinone pool in membranes from Chloroflexus are in progress.

ACKNOWLEDGEMENTS

We wish to thank Dr. P. Minardi for her advice in performing some of the ATPase activity experiments. This work was supported by the National Research Council of Italy (CNR), grant No. 850563.

REFERENCES

Baccarini-Melandri, A., Fabbri, E., and Melandri, B. A., 1976, Energy transduction in photosynthetic bacteria. VIII. Action of the energy-transducing ATPase by inorganic phosphate, Biochim. Biophys. Acta, 376:82.

Baccarini-Melandri, A., Gest, H., and San Pietro, A., 1970, A coupling factor in bacterial photophosphorylation, J. Biol. Chem., 245:1224.

Baccarini-Melandri, A., Melandri, A., and Hauska, G., 1979, The stimulation of photophosphorylation and ATPase by artificial redox mediators in chromatophores of Rhodopseudomonas capsulata at different redox potentials, J. Bioenerg. Biomemb., 11:1.

Blankenship, R. E., Huynch, P., Gabrielson, H., and Mancino, L. J., 1985, Purification physical properties and kinetic behaviour of cytochrome c554 from Chloroflexus aurantiacus, Abstracts 29th Ann. Meet. Biophys. Soc., Biophys. J., 47:2a.

Blankenship, R. E., Mancino, L. J., Feick, R., Fuller, R. C., Machnicki, J., Frank, H. A., Kirmaier, C., and Holten, D., 1984, Primary photochemistry and pigment composition of reaction centers isolated from the green photosynthetic bacterium Chloroflexus aurantiacus, in: "Advances in Photosynthesis Research," C. Sybesma, ed., vol. I, Martinus Nijhoff/Dr. W. Junk Publishers, The Hague.

Bruce, B. D., Fuller, R. C., and Blankenship, R. E., 1982, Primary photochemistry in the facultatively aerobic green photosynthetic bacterium Chloroflexus aurantiacus, Proc. Natl. Acad. Sci. U.S.A., 79:6532.

Feick, R. G., Fitzpatrick, M., and Fuller, R. C., 1982, Isolation and characterization of cytoplasmic membranes and chlorosomes from the green bacterium Chloroflexus aurantiacus, J. Bacteriol., 150:905.

Lowry, O. H., Rosebrough, N. J., Farr, A. L., and Randall, R. J., 1951, Protein measurements with the Folin phenol reagent, J. Biol. Chem., 193:265.

Meinhardt, S. W., and Crofts, A. R., 1983, The role of cytochrome b-556 in the electron transfer chain of Rhodopseudomonas sphaeroides, Biochim. Biophys. Acta, 723:219.

Pierson, B. K., and Castenholz, R. W., 1974, A phototrophic gliding filamentous bacterium of hot springs, Chloroflexus aurantiacus, Arch. Microbiol., 100.5.

Prince, R. C., and Dutton, P. L., 1978, Protonation and the reducing potential of the primary electron acceptor, in: "The Photosynthetic Bacteria," R. K. Clayton and W. R. Sistrom, eds., Plenum Press.

Van den Berg, W. H., Bonner, W. D., and Dutton, P. L., 1983, Redox potential dependence of photophosphorylation and electron transfer in continuous illumination of Rhodopseudomonas sphaeroides chromatophores, <u>Arch. Biochem. Biophys</u>., 222:299.

Vasmel, H., and Amesz, J., 1983, Photoreduction of menaquinone in the reaction center of the green photosynthetic bacterium <u>Chloroflexus aurantiacus</u>, <u>Biochim. Biophys. Acta</u>, 724:118.

Wraight, C. A., 1982, The involvement of stable semiquinone in the two-electron gates of plant and bacterial photosystems, <u>in</u>: "Function of Quinones in Energy Conserving Systems," B. F. Trumpower, ed., Academic Press, New York.

Wynn, R. M., Redlinger, T. E., Foster, J. M., Blankenship, R. E., Fuller, R. C., Shaw, R. W., and Knaff, D., 1987, Electron-transport of phototrophically and chemotrophically grown <u>Chloroflexus aurantiacus</u>, <u>Biochim. Biophys. Acta</u>, 891:216.

Zannoni, D., and Ingledew, W. J., 1985, A thermodynamic analysis of the plasma membrane electron transport components in photoheterotrophically grown cells of <u>Chloroflexus aurantiacus</u>. An optical and electron paramagnetic resonance study, <u>FEBS Lett</u>., 193:93.

Zannoni, D., 1986a, respiration vs. photosynthesis in membranes from the thermophilic green photosynthetic bacterium <u>Chloroflexus aurantiacus</u>, <u>EBEC Reports</u>, 4:139.

Zannoni, D., 1986b, The branched respiratory chain of heterotrophically dark-grown <u>Chloroflexus aurantiacus</u>, <u>FEBS Lett</u>., 198:119.

ELECTRON TRANSPORT CHAINS OF PHOTOTROPHICALLY AND CHEMOTROPHICALLY GROWN

CHLOROFLEXUS AURANTIACUS

D.B. Knaff[a], R.M. Wynn[a], T.E. Redlinger[b], R.E. Blankenship[c],
J.M. Foster[d], R.W. Shaw[a] and R.C. Fuller[b]

[a]Department of Chemistry and Biochemistry
 Texas Tech University, Lubbock, TX 79409, U.S.A.
[b]Department of Biochemistry
 University of Massachusetts, Amherst, MA 01003, U.S.A.
[c]Department of Chemistry, Arizona State University
 Tempe, AZ 85287, U.S.A.
[d]School of Natural Science, Hampshire College
 Amherst, MA 01002, U.S.A.

SUMMARY

    The photosynthetic gliding green bacterium Chloroflexus aurantiacus is
capable of both phototrophic and chemotrophic growth (Sprague et al., 1981).
Cells grown under phototrophic conditions contain BChl $\underline{a}$ associated with the
cytoplasmic membrane plus antenna BChl $\underline{c}$ located in chlorosomes attached to
the cytoplasmic side of the cell membrane (Pierson and Castenholz, 1978).
During chemotrophic growth chlorophyll synthesis is repressed and the number
of chlorosomes decline (Sprague et al., 1981). Synthesis of pigment-binding
polypeptides and of electron carrier proteins such as Cyt "$\underline{c}$554" the
immediate electron donor to the reaction center P870$^+$, has been shown to be
controlled by O_2 tension in Chloroflexus (Foster et al., 1986; Bruce et al.,
1982; Wynn et al., 1987). Other aspects of the cytochrome content of
Chloroflexus, particularly the Cyt $\underline{b}$:Cyt $\underline{c}$ ratio and the presence of heme $\underline{a}$,
have been shown to vary with growth conditions (Bruce et al., 1982; Wynn et
al., 1987; Pierson, 1985; Zannoni and Ingledew, 1986; Zannoni, 1986). We
have characterized the membrane-bound components of the electron transfer
chains of both phototrophically and chemotrophically grown Chloroflexus.
Particular attention has been placed on the cytochrome and iron-sulfur
center content of chemotrophically grown cells, cultured under conditions of
vigorous aeration in the dark (Foster et al., 1986; Wynn et al., 1987).
Cells grown under these conditions are completely free of BChl $\underline{c}$ and contain
< 0.02% of the BChl $\underline{a}$ found in cells grown anaerobically in the light.

    Oxidation-reduction titrations, optical difference spectra,
immunoblotting and SDS-PAGE indicate that chemotrophically grown cells
contain neither the reaction center (Redlinger and Fuller, 1985) nor the E_m
= +260 mV Cyt "$\underline{c}$554" (Foster et al., 1986; Wynn et al., 1987). However,
membranes isolated from these cells appear to contain at least three $\underline{c}$-type
cytochromes, all with α-band maxima at 551 nm, and with E_m values of +310
mV, +120 mV and 0 V, respectively (Wynn et al., 1987). The E_m = +310 mV
component was not found in phototrophically grown cells and may thus play a
role unique to aerobic electron flow (Wynn et al., 1987). The other two

cytochromes have also been observed in redox titrations by Zannoni (1986). Two major heme c-containing peptides with M_r = 36 and 24 kDa were detected in these membranes. Three b-cytochromes (E_m = +260 mV, +60 mV and -60 mV) were detected (Wynn et al., 1987) in redox titrations similar to those reported by Zannoni (1986). The overall protoheme (Cyt b):heme c (Cyt c) ratio of 1:2.8 in membranes isolated from chemotrophically grown cells was found to be 10-fold greater than that of membranes isolated from phototrophically grown cells, in agreement with previous reports (Bruce et al., 1982; Pierson, 1985). Neither heme a nor a Cu^{+2} signal of the type found in copper-containing Cyt "$aa3$" oxidases was detected. The absence of heme a, which is in contrast to reports from other laboratories of Cyt "$aa3$" oxidases in aerobically-grown Chloroflexus (Pierson, 1985; Zannoni, 1986), may result from the more vigorous aeration used to grow the cells used in these studies.

CO-difference spectra of membranes isolated from chemotrophically grown cells indicate the presence of a CO-binding component that resembles the proto-heme containing "Cyt $a1$". The E_m = +260 mV Cyt b-like component may be a cytochrome o-type oxidase, while it is likely (Wynn et al., 1987) that the E_m = +60 and -60 mV Cyt bs represent components of a Cyt "$bc1$" complex (Hauska et al., 1983). While we were unable to obtain definitive evidence for the presence of the other components of a Cyt "$bc1$" complex in these membranes (neitehr a g = 1.90 EPR signal characteristic of a reduced Rieske iron-sulfur protein (Hauska et al., 1983) nor any cross-reaction with an antibody raised against Rhodobacter sphaeroides Cyt "$c1$" was observed), Zannoni (1986) has reported that electron flow catalyzed by membranes isolated from chemotrophically grown Chloroflexus is inhibited by two specific inhibitors of Cyt "$bc1$" complexes (Hauska et al., 1983), antimycin A and myxothiazol.

A g = 1.89 EPR signal, probably arising from a reduced Rieske iron-sulfur protein, was observed in membranes isolated from phototrophically grown Chloroflexus (Wynn et al., 1986, 1987). A similar observation was made by Zannoni and Ingledew (1986), who also determined a E_m value (at pH 7.0) of +100 mV for this Rieske protein and showed that it could be photo-oxidized. Cells grown under both conditions exhibited EPR signals consistent with the presence of iron-sulfur centers associated with succinate and NADH dehydrogenases (Wynn et al., 1987; Zannoni and Ingledew, 1986). Membranes isolated from phototrophically grown cells were shown (Wynn et al., 1987) to contain small amounts of protoheme, in agreement with previous reports (Pierson, 1985; Wynn et al., 1986, but see Zannoni and Ingledew, 1986 for reports of larger amounts) and four different heme c-containing peptides, with M_r values of 43-45, 36, 24 and 19 kDa, respectively. Immunoblotting identified the 43-45 kDa component as Cyt "$c554$". Although the other heme c-containing peptides have not yet been identified, one corresponds to a low potential (Wynn et al., 1987) CO-binding Cyt c (Pierson, 1985, Zannoni and Ingledew, 1986). In contrast to the low potential, membrane-bound cytochromes c of bacteria such as Chromatium vinosum and Rhodopseudomonas viridis, and Chloroflexus cytochrome was not photo-oxidized at 77 K.

ACKNOWLEDGEMENTS

This work was supported by grants from the U.S. National Science Foundation (PCM-840854) to D.B.K. and DMB-05077 to R.C.F.) and Department of Agriculture (84-CRCR-1-1523 to R.E.B.).

REFERENCES

Bruce, B. D., Fuller, R. C., and Blankenship, R. E., 1982, Primary
   photochemistry in the facultatively aerobic green photosynthetic
   bacterium Chloroflexus aurantiacus, Proc. Natl. Acad. Sci. U.S.A.,
   79:6532.
Foster, J. M., Redlinger, T. E., Blankenship, R. E., and Fuller, R. C.,
   1986, Oxygen regulation of development of the photosynthetic membrane
   system in Chloroflexus aurantiacus, J. Bact., 167:631.
Hauska, G., Hurt, E., Gabellini, N., and Lockau, W., 1983, Comparative
   aspects of quinol-cytochrome c/plastocyanin oxidoreductases.,
   Biochim. Biophys. Acta, 726:97.
Pierson, B. K., 1985, Cytochromes in Chloroflexus aurantiacus grown with and
   without oxygen, Arch. Microbiol., 143:260.
Pierson, B. K., and Castenholz, R. W., 1978, Photosynthetic apparatus and
   cell membranes of the green bacteria, in: "The Photosynthetic
   Bacteria," R. K. Clayton, and W. R. Sistrom, eds., Plenum Press, New
   York.
Redlinger, T. E., and Fuller, R. C., 1985, Protein processing as a
   regulatory mechanism in the synthesis of the photosynthetic antenna
   in Chloroflexus, Arch. Microbiol., 141:344.
Sprague, S. G., Staehelin, A., and Fuller, R. C., 1981, Semiaerobic
   induction of bacteriochlorophyll synthesis in the green bacterium
   Chloroflexus aurantiacus, J. Bact., 147:1021.
Wynn, R. M., Gaul, D. F., Choi, W.-K., Shaw, R. W., and Knaff, D. B., 1986,
   Isolation of cytochrome "bc1" complexes from the photosynthetic
   bacteria Rhodopseudomonas viridis and Rhodospirillum rubrum,
   Photosyn. Res., 9:181.
Wynn, R. M., Redlinger, T. E., Foster, J. M., Blankenship, R. E., Shaw, R.
   W., and Knaff, D. B., 1987, Electron transport chains of
   phototrophically and chemotrophically grown Chloroflexus aurantiacus,
   Biochim. Biophys. Acta 891:216.
Zannoni, D., 1986, The branched respiratory chain of heterotrophically
   dark-grown Chloroflexus aurantiacus, FEBS Lett., 198:119.
Zannoni, D., and Ingledew, W. J., 1986, A thermodynamic analysis of the
   plasma membrane electron transfer components in
   photoheterotrophically grown cells of Chloroflexus aurantiacus, FEBS
   Lett., 193:93.

A NEW CO₂ FIXATION MECHANISM IN <u>CHLOROFLEXUS AURANTIACUS</u> STUDIED BY ¹³C-NMR

H. Holo[a] and D. Grace[b]

[a]Dept. of Biology, University of Oslo
 P.O. Box 1050, N-0316 Oslo 3, Norway
[b]Dept. of Chemistry, University of Oslo
 P.O. Box 1033, N-0315 Oslo 3, Norway

INTRODUCTION

Among autotrophs three different mechanisms for autotrophic CO_2 fixation are known. Of these the reductive pentose phosphate cycle (Calvin cycle) is the best studied. It is found in all phototrophic eukaryotes and in a number of autotrophic prokaryotes. This cycle has been considered universal among autotrophs (Quayle and Ferenci, 1978), but during recent years several autotrophs have been shown to lack the key enzymes of the reductive pentose phosphate cycle and thus use other mechanisms for CO_2 fixation.

The alternative mechanisms known are the reductive tricarboxylic acid cycle which is found in the phototrophic green sulfur bacteria and the acetyl-CoA pathway found in several anaerobic chemotrophs (for a review see Fuchs and Stupperich, 1985). The thermophilic phototrophic green bacterium <u>Chloroflexus aurantiacus</u> does not seem to use any of the known pathways when grown autotrophically (Holo and Sirevåg, 1986). Its mechanism for CO_2 fixation is especially interesting because of its thermophilic properties and its distant phylogenetic relationship to other phototrophs (Woese et al., 1985).

In a recent work we concluded that acetyl-CoA is a central intermediate in the CO_2 fixation pathway in <u>Cf. aurantiacus</u>, and this compound is reductively carboxylated to form pyruvate by the enzyme pyruvate synthase (Holo and Sirevåg, 1986). The pyruvate synthase step appears to be universal to autotrophs that do not use the reductive pentose phosphate cycle, but the pathways differ with respect to the way acetyl-CoA is formed from CO_2. Whereas this compound is formed by condensation of two C_1 compounds in the acetyl-CoA pathway, in the reductive tricarboxylic acid cycle acetyl-CoA synthesis is less direct and the product itself may also act as a substrate for further production. In this study we have undertaken a long term labelling study to investigate the metabolism of acetyl-CoA in autotrophically grown cells. The results indicate that the cells are able to change the labelling pattern of acetyl-CoA, suggesting that acetyl-CoA is involved as an intermediate in its own synthesis from CO_2.

In this work we have taken advantage of the fact that metabolism in <u>Cf. aurantiacus</u> can be restricted and directed towards polyglucose (PG) synthesis by adding fluoroacetate (Holo and Sirevåg, 1986). Because of the large amounts of PG formed, metabolism can be studied by an insensitive method like ^{13}C-NMR. ^{13}C-NMR has the advantage that the extent of labelling of all the individual carbon atoms in a molecule can be determined.

MATERIALS AND METHODS

Organism and Culture Conditions

Cf. aurantiacus strain OK-70 fl. was grown photoautotrophically with hydrogen as the electron donor as described by Holo and Sirevåg (1986). Labelling experiments were conducted in the presence of 0.2 mM fluoroacetate with cultures containing 75-150 µg protein/ml as described by Holo and Grace (1987). Normally, the 200 ml cultures were incubated in 1-liter rubber-stoppered infusion bottles with H_2/CO_2 (95:5) in the gas phase. Where indicated, the cultures were stirred by a magnetic stirring bar at 300 r.p.m. and continuously gassed with H_2/CO_2 (95:5) at a rate of 100 ml/min. All incubations were carried out at 55°C with a 25 or 40W light bulb kept at a distance of about 50 cm as the light source.

At the end of the labelling experiments the cells were harvested by centrifugation and washed twice with 0.5 M H_2SO_4.

Analytical Methods

The cell pellets were hydrolysed in 0.5 M H_2SO_4 and glucose was determined according to Sirevåg (1975). Glucose was analysed by ^{13}C-NMR as described by Holo and Grace (1987).

Formation of $^{14}CO_2$ from ^{14}C-acetate by cultures treated with fluoroacetate was measured as described by Holo and Grace (1987).

Total cell protein was determined on methanol extracted samples with the method of Lowry et al. (1951) using bovine serum albumin as standard.

Chemicals

$^{13}C1$-Na-acetate (90% ^{13}C) was from Stohler Isotope Chemicals (Rutherford, N.J., U.S.A.). $^{13}C3$-alanine (98% ^{13}C) and $^{13}C2$-Na-acetate (98% ^{13}C) were from Cambridge Isotope Laboratories (Woburn, Mass., U.S.A.).

RESULTS AND DISCUSSION

Fig. 1 shows ^{13}C-NMR spectra of glucose obtained from the cultures incubated in the presence of fluoroacetate and ^{13}C-labelled compounds. The labelling of the individual carbon atoms are presented in Table 1. In all cases the labelling of C3 exceeds that of C4. This may be due to impurity peaks in the 73 ppm area since a lower enrichment was always found if the overlapping C3β-C5β peaks were used to calculate the enrichment of C3 instead of using the C3 α peak. Taking this into account, the labelling patterns in Table 1 can be considered symmetric. This shows that Cf. aurantiacus forms hexose by condensation of two C_3 units, as in a reversed Embden-Meyerhof pathway. This is supported by Kondratieva et al. (1985) who demonstrated the enzymes of the Embden-Meyerhof pathway in other Chloroflexus strains.

The amount of carbon derived from a tracer molecule can be estimated from the sum of labelling of all the six carbon atoms of the glucose formed. From Table 1 it can thus be calculated that about one third of the carbon of

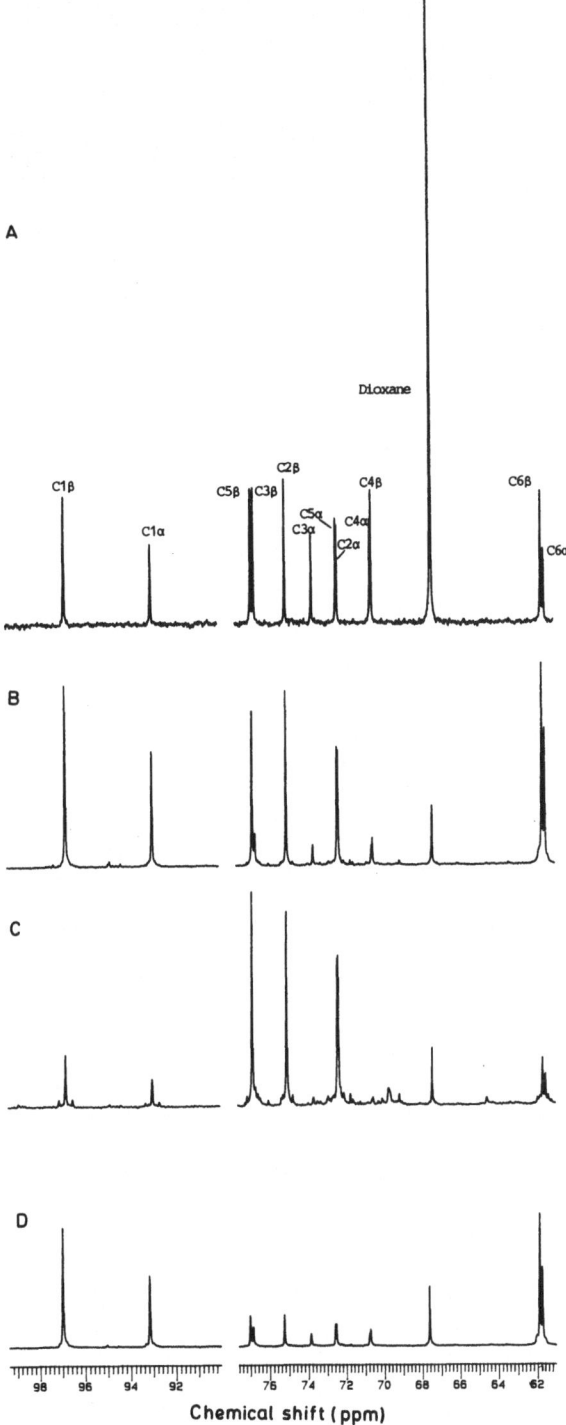

Fig. 1. ¹³C-NMR spectra of glucose from <u>Chloroflexus aurantiacus</u>:
A, unenriched glucose; B, formed in the presence of ¹³C1-acetate;
C, formed in the presence of ¹³C2-acetate; and D, formed in the
presence of ¹³C3-alanine. (From Holo and Grace (1987) with
permission.)

151

Table 1. Labelling of Carbon Atoms in Glucose from <u>Chloroflexus aurantiacus</u>
Incubated in the Presence of H_2, CO_2, Fluoroacetate and
[13]C-Labelled Compounds (Data from Holo and Grace (1987)).

Tracer	Culture	Incubation time (h)	Polyglucose (PG) content (μmol glucose/ml culture) initial	final	Labelling of glucose formed (% of tracer)[1] C1	C2	C3	C4	C5	C6
1-[13]C Acetate, 3mM	1	40	0.22	0.75	30	21	7	4	24	34
	2[2]	24	0.09	0.68	25	20	3	2	22	29
	3[3]	50	0.08	0.79	33	19	8	6	19	34
2-13C Acetate, 3mM	4	43	0.08	0.56	13	32	7	2	37	14
	5[2]	24	0.10	0.80	9	32	2	0	31	9
	6[3]	51	0.07	1.08	6	42	4	1	47	6
3-[13]C Alanine, 1mM	7	51	0.07	0.90	16	3	2	0	3	17

[1] $\dfrac{\text{PG (final)} \cdot (\%^{13}C \text{ (sample)} - 1.1\%) \cdot 100}{(\text{PG (final)} - \text{PG (initial)}) \cdot (\%^{13}C \text{ (tracer)} - 1.1\%)}$

[2] Continuously stirred and gassed with $H_2:CO_2$ (95:5)
[3] 0.2 mM KCN present

glucose formed in the presence of labelled acetate has been derived from the
acetate. The rest of the carbon is from CO_2 since PG synthesis from
fluoroacetate is negligible (Holo and Sirevåg, 1986). The low degree of
labelling of C3 and C4 in glucose formed in the presence of ^{13}C-acetate
shows that these carbon atoms are derived almost entirely from CO_2 which
supports the claim of pyruvate synthase operation in CO_2 fixation and
acetate metabolism in <u>Cf. aurantiacus</u>. The observed labelling of C3 and C4
can probably be ascribed to fixation of [13]CO_2 formed during the experiments
(Table 2) because this labelling was reduced in cultures continuously gassed
with H_2/CO_2 to remove the [13]CO_2 formed.

Autotrophically grown <u>Cf. aurantiacus</u> can also metabolize acetate by
the glyoxylate cycle (Holo and Sirevåg, 1986). With C1-labelled acetate as
tracer, the glyoxylate cycle would cause labelling of C3 and C4 in glucose
exclusively. The data clearly excludes this cycle as a major route in our
experiments. The glyoxylate cycle involves aconitase, and the results
therefore support the idea that the inhibitory effect of fluoroacetate on
growth of <u>Cf. aurantiacus</u> is due to aconitase inhibition as in many other

Table 2. Polyglucose and [14]CO_2 Formation by <u>Chloroflexus aurantiacus</u>
Incubated in the Presence of H_2, CO_2, Fluoroacetate and
[14]C-Labelled Acetate (Data from Holo and Grace (1987)).

Tracer	Incubation time (h)	Polyglucose formed (μmol glucose/ml culture)	[14]CO_2 formed (μmol/ml culture)
1 [14]C Acetate (2 mM)	18	0.22	0.18
2-[14]C Acetate (2 mM)	16	0.33	0.13

organisms (Peters et al., 1953). This makes it difficult to explain how CO_2 is formed from acetate in these experiments (Table 2). Acetate cannot be oxidized via the tricarboxylic acid cycle when aconitase is blocked. Neither can all the CO_2 formation be ascribed to the oxidative pentose phosphate cycle; this cycle would cause extensive labelling of C3 in glucose and the labelling patterns would be highly asymmetric (Cohen et al., 1979a). The CO_2 formation is probably not due to acetate oxidation, because this would produce reducing equivalents that could replace hydrogen in PG production. However, no carbohydrate synthesis was observed in fluoroacetate-treated cells in the presence of acetate when hydrogen was replaced by argon. Thus at present we cannot explain how CO_2 can be formed from acetate in these experiments.

Glucose formed from C1-labelled acetate via pyruvate synthase and the reversed Embden-Meyerhof pathway would be labelled in carbon atoms 2 and 5, whereas C2-labelled acetate would give rise to C1 and C6 labelled glucose. However, when ^{13}C1-acetate was used as the tracer, glucose became more labelled in C1 and C6 than in C2 and C5. Similarly, in the presence of ^{13}C2-acetate, the glucose became more labelled in C2 and C5 than in C1 and C6. This cannot be explained by a futile loop causing label scrambling like the shuttle between phosphoenolpyruvate and fumarate (Cohen et al., 1979b), since the "inverse" labelling is dominating. Apparently the cells form acetyl-CoA with a labelling pattern opposite to that of the acetate tracer used. In addition, the apparent formation of C1-labelled acetyl-CoA from C2-labelled is faster than the conversion of C1-labelled acetate to C2-labelled acetyl-CoA. Furthermore, the overall rate of both processes appears to be higher than that of PG synthesis from added acetate.

Thus by adding together enrichment levels of individual carbon atoms of glucose formed in the presence of the two acetate tracers, it can be seen that more carbon derived from the added acetate is found in C2 and C5 than in C1 and C6. The data from comparable experiments in Table 1 indicate that the content of carbon derived from added acetate is at least 20% higher in C2 of the glucose formed than in C1.

Pyruvate is the first intermediate in carbohydrate syntehsis from acetyl-CoA via pyruvate synthase. The metabolism of pyruvate was studied by using ^{13}C-alanine as tracer in the labelling experiments. This amino acid is converted directly to pyruvate by the cells. The results of the alanine labelling experiment indicate pyruvate metabolism by established mechanisms; C3-labelled alanine labels glucose at C1 and C6. The low degree of labelling seen in C2 and C5 of glucose may be caused by the phosphoenolpyruvate-fumarate shuttle discussed above or pyruvate oxidation to acetyl-CoA by pyruvate synthase. Thus the exceptional labelling patterns of glucose formed in the presence of acetate are caused by reactions prior to pyruvate on the route leading from acetate to PG.

Table 3. Effect of KCN on Polyglucose Formation by
Autotrophically Grown Chloroflexus aurantiacus
in the Presence of Fluoroacetate.

Additions	Incubation time (h)	Polyglucose formed (μmol glucose/ml culture)
none	18	0.22
0.2 mM KCN	18	0.11
0.2 mM KN + 4 mM acetate	18	0.23

Acetyl-CoA is formed directly from acetate by Cf. aurantiacus (Holo and Sirevåg, 1986) and it is likely to be the first intermediate in acetate assimilation by the cells. Thus the results suggest that most of the acetyl-CoA formed is metabolized by another enzyme before it is converted to pyruvate by pyruvate synthase. This leads to the suggestion that acetyl-CoA is metabolized via a metabolic cycle operating at a higher rate than pyruvate synthase.

The results of this and previous work (Holo and Sirevåg, 1986) indicate that all the C3-units used for carbohydrate synthesis in the presence of acetate have been formed via pyruvate synthase. The finding that carbon derived from added acetate is found with some preference in C2 and C5 of glucose, indicates that acetyl-CoA has been formed in which the carboxyl group is derived from added acetate and the methyl group is derived from CO_2. The data suggest that at least 10% (at least 26% in the presence of KCN) of all the acetyl-CoA used for carbohydrate synthesis had such a composition. In fact the numbers might have been much higher if the cells also had made acetyl-CoA in which only the methyl group was derived from acetate, but this is impossible to tell from these experiments. Nevertheless, our data show that the C-C bond of acetate is broken by the cells and that the product ultimately acts as a receptor molecule in the CO_2 fixation process. This indicates that CO_2 fixation is an integral part of the new metabolic pathway causing apparent label rearrangement reactions of acetyl-CoA.

As is shown in Table 3, PG synthesis from CO_2 is inhibited by 0.2 mM KCN. The inhibition, however, is overcome by adding acetate. This suggests that acetyl-CoA synthesis from CO_2 is inhibited by KCN. The labelling experiments were therefore also conducted in the presence of this compound. Unexpectedly, the degree of labelling of glucose formed in the presence of labelled acetate was not much changed. However, the labelling patterns were altered; they diverged from "normal" even more in the presence, than in the absence, of KCN. The finding that KCN appears to affect both label rearrangement reactions of acetyl-CoA and the synthesis of this compound from CO_2 therefore supports the idea that these processes are integrated.

A labelling study of this kind cannot completely elucidate a new metabolic pathway. Still this study has revealed several properties of the metabolism of acetate and CO_2 in Cf. aurantiacus. Altogether the results support a new metabolic cycle for acetyl-CoA synthesis from CO_2 in this organism. Such a cycle is schematically presented in Fig. 2. In this cycle acetyl-CoA is an intermediate in its own synthesis from CO_2. Formation of CO_2 from acetate (by exchange?) may also be catalyzed by this cycle. Our data indicate that acetyl-CoA is metabolized at a higher rate by the cycle

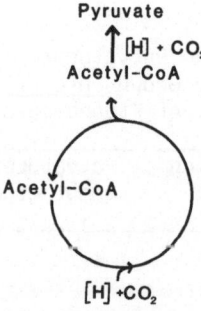

Fig. 2. Proposed pathway for pyruvate formation from CO_2 in Chloroflexus aurantiacus. Details of the pathway are not known.

than by pyruvate synthase. However, further work and other methods are required to investigate the details of the new metabolic pathway. This pathway is of special interest since autotrophic CO_2 fixation is essential to all life on earth and because Cf. aurantiacus is only distantly related to other autotrophs (Woese et al., 1985).

REFERENCES

Cohen, S. M., Ogawa, S., and Schulman, R. G., 1979a, ^{13}C NMR studies of gluconeogenesis in rat liver cells: utilization of labeled glycerol cells from euthyroid and hyperthyroid rats, Proc. Natl. Acad. Sci. USA, 76:1603.

Cohen, S. M., Schulman, R. G., and McLaughlin, X., 1979b, Effects of ethanol on alanine metabolism in perfused mouse liver studied by ^{13}C NMR, Proc. Natl. Acad. Sci. USA, 76:4804.

Fuchs, G., and Stupperich, E., 1985, Evolution of autotrophic CO_2 fixation, in: "The Evolution of Procaryotes," K. H. Schleifer and E. Stackebrandt, eds., Academic Press, London.

Holo, H., and Sirevåg, S., 1986, Autotrophic growth and CO_2 fixation by Chloroflexus aurantiacus, Arch. Microbiol., 145:173.

Holo, H., and Grace, D., 1987, Polyglucose synthesis in Chloroflexus aurantiacus studied by ^{13}C-NMR, Arch. Microbiol., 148:292.

Kondratieva, E. N., Keppen, O. I., and Krasilnikova, E. N., 1985, Physiology and metabolism of Chloroflexus aurantiacus in different growth conditions, V Symp. Photosynthetic Procaryotes, Grindelwald, Switzerland (Abstr. 215).

Lowry, O. H., Rosenbrough, N. J., Farr, A. L., and Randall, R. J., 1951, Protein measurements with the Folin phenol reagent, J. Biol. Chem., 193:265. Peters, R. A., Wakelin, R. W., Rivett, D. E. A., and Thomas, L. C., 1953, Fluoroacetate poisoning: comparison of synthetic fluorocitric acid with the enzymically synthetized fluorotricarboxylic acid, Nature, 171:1111.

Quayle, J. R., and Ferenci, T., 1978, Evolutionary aspects of autotrophy, Microbiol. Rev., 42:251.

Sirevåg, R., 1975, Photoassimilation of acetate and metabolism of carbohydrate in Chlorobium thiosulfatophilum, Arch. Microbiol., 104:105.

Woese, C. R., Stackebrandt, E., Macke, T. J., and Fox, G. E., 1985, A phylogenetic definition of the major eubacterial taxa, Syst. Appl. Microbiol., 6:143.

KINETIC AND REGULATORY PROPERTIES OF CITRATE SYNTHASE FROM THE THERMOPHILIC

GREEN GLIDING BACTERIUM <u>CHLOROFLEXUS AURANTIACUS</u>

D.J. Kelly

Department of Microbiology
University of Sheffield, Western Bank
Sheffield, S10 2TN, U.K.

INTRODUCTION

<u>Chloroflexus aurantiacus</u> is a photosynthetic member of a group of
microbes which represent an extremely deep branching in the eubacterial line
of descent (Gibson et al., 1985; Oyaizu et al., 1987). The other genera in
this grouping, the green nonsulphur bacteria, are <u>Herpetosiphon</u> and
<u>Thermomicrobium</u>, with which <u>Chloroflexus</u> shares little apparent phenotypic
resemblance. Although <u>Chloroflexus</u> contains BChl <u>a</u> and <u>c</u>, the latter
localized in chlorosomes attached to the cytoplasmic membrane, 16S rRNA
sequencing studies have shown no phylogenetic relatedness to the green
sulphur bacteria (Gibson et al., 1985). Indeed, in terms of reaction centre
photochemistry (Blankenship et al., 1983), some aspects of electron
transport (Bruce et al., 1982) and the wide range of carbon sources utilized
for photoheterotrophic or chemoheterotrophic growth (Madigan et al., 1974),
the metabolism of <u>Chloroflexus</u> most resembles that of the purple nonsulphur
bacteria. Unlike the Rhodospirillaceae, however, the Calvin cycle is
apparently not used to fix CO_2 into cell material under autotrophic growth
conditions (Holo and Sirevåg, 1986) and a novel mechanism has been
postulated. In addition, the thermophilic character of <u>Chloroflexus</u> (optimum
growth at 55°C) is seemingly rare amongst phototrophic bacteria.

In many groups of microbes there exists a correlation between taxonomic
position and the regulation and/or properties of certain key enzymes of
metabolism. This is particularly apparent in the case of some tricarboxylic
acid cycle enzymes, for example citrate synthase (Weitzman, 1981). In
Gram-negative bacteria citrate synthase (CS) is subject to allosteric
inhibition by NADH, has a native M_r of about 250,000 ("large" type) and is
probably a tetramer (Weitzman and Danson, 1976). Gram-positive bacteria have
a "small" citrate synthase (M_r 100,000) which is not subject to allosteric
inhibition by NADH.

To date, very few thermophiles have been examined with the specific aim
of determining molecular sizes and regulatory properties of their citrate
synthases. Thus, the extent to which growth at high temperatures influences
these enzymic parameters is not known. As the thermophilic, photosynthetic
phenotype is important in phylogenetic terms (Woese, 1987) it was of
interest to characterize the <u>Chloroflexus</u> CS and compare it with that from
mesophilic and non-photosynthetic bacteria.

MATERIALS AND METHODS

Organism and Growth Conditions

Cf. aurantiacus strain J-10-fl was obtained from R. C. Fuller
(University of Massachusetts). The bacteria were grown in the mineral salts
medium described by Madigan et al. (1974) supplemented with 0.08% (w/v)
glycylglycine, 0.1% (w/v) yeast extract and 0.25% (w/v) casamino acids. The
pH was adjusted to 8.2 after the addition of 0.3 mg l^{-1} $FeCl_3$ and 1.0 ml
l^{-1} of the trace element solution described by Weaver et al. (1975). All
cultures were grown at 55°C for 5-6 days at a distance of 30 cm from a 100 W
incandenscent bulb.

Harvesting and Preparation of Cell-Free Extracts

Cultures were harvested by centrifugation (8,000 x g, 30°C, 20 min)
washed once in 50 mM Tris-HCl buffer pH 8.2 and resuspended in the same
buffer. Cells were disrupted by ultrasound and debris was removed by
centrifugation (13,000 x g, 25°C, 10 min). The supernatant was carefully
removed and maintained at room temperature for assays performed the same
day. For partial purification of citrate synthase, chlorosomes and cell
membranes were first removed from the crude extract by ultracentrifugation
(150,000 x g, 30°C, 3 h).

Enzyme Assays

All measurements were performed at 50°C. Citrate synthase was assayed
according to Reeves et al. (1971) using the chromogen dithio(bis)
2-nitrobenzoate (DTNB). The reaction mixture (3 ml) contained Tris-HCl
buffer pH 8.2, 200 mM; DTNB, 200 µM; acetyl CoA, 40 µM and oxaloacetate, 500
µM. The reaction was started by the addition of enzyme (see RESULTS) and the
absorbance increase at 412 nm recorded against a reference cuvette lacking a
acetyl CoA. Deacylase activity was measured after omission of oxaloacetate.

Malate dehydrogenase (MDH) was assayed in 200 mM Tris-HCl pH 8.2, 500
µM oxaloacetate and 150 µM NADH in a reaction volume of 3 ml. The reference
lacked NADH and the reaction was started with enzyme. The following

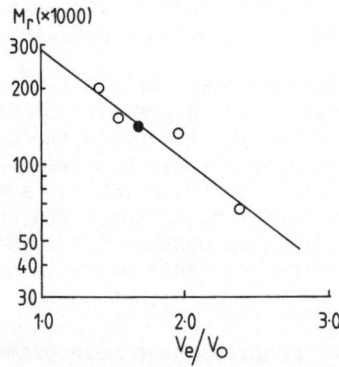

Fig. 1. Determination of the molecular size of citrate synthase. A Cf.
aurantiacus cell-free extract was applied to an Ultrogen AcA34
column and the ratio of elution volume (V_e) to avoid volume (V_o) for
citrate synthase (o) compared to that for β-amylase (M_r 200,000),
alcohol dehydrogenase (M_r 150,000) the dimer of bovine serum albumin
(M_r 133,000) and the monomer of bovine serum albumin (M_r 67,000).
The V_e/V_o value for CS gave an approximate M_r of 140,000.

extinction coefficients were used; NADH, 6.22 mM⁻¹ cm⁻¹, DTNB, 13.6 mM⁻¹ cm⁻¹.

Size Determination of Citrate Synthase by Gel Filtration

Chloroflexus cell-free extracts (from which the bulk of the chlorosomes had been removed) were concentrated 10 fold by aquacide treatment and 0.5 ml (about 125 mg of protein) was applied to a pre-calibrated, 25 x 1 cm column of AcA34 Ultragel (LKB). Proteins were eluted at a flow rate of 0.42 ml min⁻¹ and 2.5 ml fractions collected. The separation was performed at room temperature.

Ion-Exchange and Affinity Chromatography

Ion exchange chromatography was performed using DEAE sepharose CL-6B pre-equilibrated at room temperature with 50 mM Tris-HCl buffer pH 8.2. Protein was eluted using an NaCl gradient from 0-1.0 M. Active fractions were pooled, dialyzed, concentrated with aquacide and applied to a column of blue sepharose CL-6B. The column was successively eluted with 2 ml of 1 mM acetyl CoA and 2 ml 1 mM NADH, washed with 5 ml Tris buffer before a final elution with 2 ml of 0.5 M NaCl. The final eluate was dialyzed overnight against 10 mM Tris-HCl pH 8.2.

RESULTS AND DISCUSSION

Of the several available assay methods for citrate synthase, that using DTNB as the chromogenic reagent to detect the free CoA produced during the reaction is one of the most sensitive and convenient.

However, maximal activities of CS were only obtained in crude cell-free extracts of Chloroflexus when the reaction was started with extract rather than with the substrate oxaloacetate. This suggested the possibility of inhibition of the enzyme by the chromogen DTNB. As no deacylase activity was detectable in such extracts, standard assay conditions were chosen in which all the components were mixed, pre-incubated at 50°C and the reaction started with enzyme.

Gel filtration of clarified extracts of Cf. aurantiacus was performed to determine the size class of citrate synthase (Fig. 1). Activity was eluted from the column as a single symmetrical peak with an elution volume corresponding to an approximate molecular weight of 140,000. This is about the size expected of a "small" CS enzyme. A characteristic of this type of CS is a relatively low K_m for acetyl CoA (below 50 µM; Weitzman and Danson, 1976) combined with a pronounced sensitivity to inhibition by ATP. In contrast, the "large" Gram-negative type of CS shows only a weak inhibition with ATP and lower affinity for acetyl CoA. These effects are thought to be due to competition between ATP and acetyl CoA for binding sites on the enzyme. Consistent with the "small" size class of the Chloroflexus enzyme, determination of the apparent K_m for acetyl CoA in crude extracts gave a value of 25 µM (Fig. 2). The effect of ATP on enzyme activity is shown in Table 1. Significant inhibition occurred over the physiological range of ATP concentrations (1-5 mM).

The most important regulatory property that distinguishes the two size classes of CS is, however, their response to inhibition by reduced nicotinamide nucleotides, particularly NADH. In many microbes this can be studied in crude extracts if malate dehydrogenase and "NADH oxidase" activities are sufficiently low enough so that the added NADH is removed at an insignificant rate. Although NADH oxidation per se was not detectable in Chloroflexus crude cell-free extracts, malate dehydrogenase specific

Table 1. Inhibition of <u>Chloroflexus aurantiacus</u> Citrate
Synthase by ATP. Activities represent averages
of the number of determinations shown in
parentheses. Assays were carried out at 50°C,
using crude cell-free extracts. The acetyl CoA
concentration was 40 µM in each case.

[ATP], mM	CS Activity (nmol/min/mg protein)	% Inhibition
0	79 (3)	0
1	62 (2)	21.5
2	52 (1)	34.2
5	42.5 (2)	46.2

activities were extremely high (3-4 µmol/min/mg protein). Inhibition studies
with NADH were therefore performed using partially purified CS.

Fig. 3 shows the chromatographic behaviour of CS and MDH activities
after elution from an ion-exchange column. Both activities eluted over the
same range of salt concentrations. Nevertheless, the ratio of MDH:CS
activity in the pooled peak fractions after dialysis and aquacide treatment
was reduced substantially from that seen in crude extracts (Table 2). To
reduce this ratio further, dye-linked affinity chromatography was used (Fig.
4). Enzymes with a dinucleotide fold are known to bind to triazine
dye-linked sepharoses and they can often be eluted specifically using the
appropriate substrate (Dean and Watson, 1979). Nevertheless, although both
CS and MDH activities did bind to the dye-linked column used here, CS
activity could not be eluted specifically using acetyl-CoA. Elution with 1
mM NADH, however, removed most of the bound MDH from the column (Fig. 4).
Citrate synthase was then recovered using 0.5 M NaCl as a non-specific
eluent. This procedure resulted in final MDH:CS activity ratios nearing
unity (Table 2). The response of the partially purified CS to NADH is shown

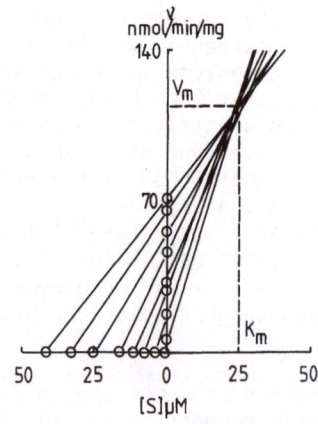

Fig. 2. Determination of the K_m for acetyl CoA. Initial rates (V) of CS
activity in the presence of a range of acetyl CoA concentrations
([S]) were compared by the direct linear plot method. The median
intersection point gave a K_m value of 25 µM and V_{max} of 113
nmol/min/mg protein. All assays were carried out at 50°C.

Table 2. Removal of Malate Dehydrogenase Activity by Affinity Chromatography

Step	Enzyme Activity (nmol/min/100 µL)		MDH:CS Ratio
	MDH	CS	
Crude extract after 150,000 x g spin	4943	94	52.6:1
Pooled ion-exchange peak fractions after aquacide treatment	603	33	18.3:1
0.5 M NaCl eluate from blue sepharose CL-6B column	24.1	17.6	1.36:1

in Table 3. No inhibition of activity was observed at up to 2 mM NADH, which includes the range of likely intracellular concentrations. It should be noted that the residual malate dehydrogenase activity present would only be sufficient to reduce the NADH concentration by about 1% per minute during the course of the assay.

Another known CS effector molecule, 2-oxoglutarate, also had little inhibitory effect on the enzyme (Table 3). This observation is in accord with the demonstration of a complete TCA cycle in this microbe (Holo and Sirevåg, 1986). since 2-oxoglutarate inhibits CS only from bacteria which possess an incomplete TCA cycle (Weitzman, 1981).

Despite the fact that <u>Cf. aurantiacus</u> stains Gram-negatively (Pierson and Castenholz, 1974) the properties of its citrate synthase are very similar to those of Gram-positive bacteria. Interestingly, the

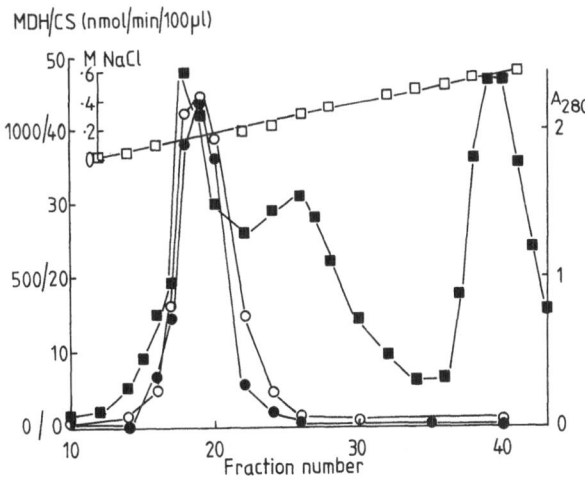

Fig. 3. Elution profile of citrate synthase and malate dehydrogenase activities after ion-exchange chromatography. Enzyme activity was eluted from a DEAE sepharose CL-6B column with a NaCl gradient (□-□). Fractions (1 ml) were assayed for absorbance at 280 nm (□-□), malate dehydrogenase activity (o-o) and citrate synthase activity (o-o).

Table 3. Effects of NADH and 2-Oxoglutarate on Partially
Purified Citrate Synthase from <u>Chloroflexus
aurantiacus</u>. Assays were performed on the
dialyzed 0.5 M NaCl eluate from the blue
sepharose CL-6B column.

Inhibitor	Enzyme Activity (nmol/min/100 μl)	% Inhibition
No additions	11.0	0
1 mM NADH	11.0	0
2 mM NADH	11.0	0
2 mM 2-oxoglutarate	10.0	9

non-photosynthetic but thermophilic Gram-negative genus <u>Thermus</u> is also
characterized by the possession of an NADH insensitive "small" type of
citrate synthase (Weitzman, 1978). It has been proposed that the lack of
allosteric regulation of CS by NADH in such species is related to high
growth temperatures which would not be conducive to the existence of the
"large" type of citrate synthase (Weitzman, 1978; Brock, 1967). However
<u>Chloroflexus</u> is not extremely thermophilic and there is evidence for
allosteric control as a general regulatory phenomenon in several other
thermophiles (Williams, 1975).

It is also possible that the "small" type of citrate synthase evolved
early and would therefore be widespread amongst species representing the
deepest branchings of the eubacterial lineage. Both <u>Chloroflexus</u> and <u>Thermus</u>
are representatives of two such branchings and the deepest of all
eubacterial lineages is represented by another thermophile, <u>Thermotoga</u>

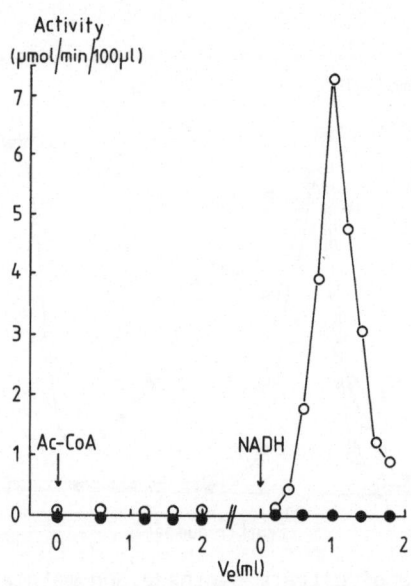

Fig. 4. Affinity chromatography on blue sepharose CL-6B. Peak fractions from
ion-exchange chromatography were applied to a small blue sepharose
column which was successively eluted with acetyl CoA (1 mM) and NADH
(1 mM). Fractions of 0.2 ml were collected and assayed for citrate
synthase (o-o) and malate dehydrogenase (o-o).

<u>maritima</u> (Achenbach-Richter et al., 1987) the citrate synthase of which has
yet to be examined.

It seems most likely, however, that <u>Chloroflexus</u> does in fact represent
an unusual group of Gram-positive photosynthetic bacteria. In a recent
study, Jürgens et al. (1987) have come to the same conclusion, based on
their analyses of the cell wall constituents; no lipopolysaccharide was
detectable and ornithine proved to be the sole diamino acid in the
peptidoglycan. This result explains the isolated phylogenetic position of
<u>Chloroflexus</u> compared to other photosynthetic bacteria; the only other
Gram-positive group known at the present time is the unrelated BChl g
containing heliobacteria. The discovery of further examples of Gram-positive
phototrophs should thus prove to be extremely important.

SUMMARY

Some properties of the citrate synthase from <u>Chloroflexus aurantiacus</u>
have been examined in crude cell-free extracts and partially purified
preparations. The enzyme had an approximate native molecular size of
140,000, was not inhibited by NADH or 2-oxoglutarate but was inhibited by
ATP (about 50% at 5 mM). The K_m for acetyl CoA at pH 8.2 in the presence of
0.5 mM oxalocetate was determined to be 25 µM.

These properties are characteristic of the "small" size class of
citrate synthases normally associated with Gram-positive eubacteria, despite
the fact that <u>Chloroflexus</u> stains Gram negatively. Possible reasons for this
anomaly are considered.

REFERENCES

Achenbach-Richter, L., Gupta, R., Stetter, K. O., and Woese, C. R., 1987,
        Were the original eubacteria thermophiles?, <u>System. Appl. Microbiol.</u>,
        9:34.
Blankenship, R. E., Feick, R., Bruce, B. D., Kirmaier, C., Holton, D., and
        Fuller,R. C., 1983, Primary photochemistry in the facultative green
        photosynthetic bacterium <u>Chloroflexus aurantiacus</u>, <u>J. Cell Biochem.</u>,
        22:251.
Brock, T. D., 1967, Life at high temperatures, <u>Science</u>, 158:1012.
Bruce, B., Fuller, R. C., and Blankenship, R. E., 1982, Primary
        photochemistry in the facultative aerobic green photosynthetic
        bacterium <u>Chloroflexus aurantiacus</u>, <u>Proc. Natl. Acad. Sci. U.S.A.</u>,
        79:6532.
Dean, P. D. G., and Watson, D. H., 1979, Protein purificaiton using
        immobilised triazine dyes, <u>J. Chromatography</u>, 165:301.
Gibson, J., Stackebrandt, E., and Woese, C. R., 1985, The phylogeny of the
        green photosynthetic bacteria: lack of a close relationship between
        <u>Chlorobium</u> and <u>Chloroflexus</u>, <u>System. Appl. Microbiol.</u>, 6:152.
Holo, H., and Sirevåg, R., 1986, Autotrophic growth and CO_2 fixation of
        <u>Chloroflexus aurantiacus</u>, <u>Arch. Microbiol.</u>, 145:173.
Jürgens, U. J., Meissner, J., Fischer, U., König, W. A., and Weckesser, J.,
        1987, Ornithine as a constituent of the peptidoglycan of <u>Chloroflexus
        aurantiacus</u>, diaminopimelic acid in that of <u>Chlorobium vibrioforme</u> f.
        thiosulfatophilum, <u>Arch. Microbiol.</u>, 148:72.
Madigan, M., Peterson, S. R., and Brock, T. D., 1974, Nutritional studies on
        <u>Chloroflexus</u>, a filamentous photosynthetic gliding bacterium, <u>Arch.
        Microbiol.</u>, 100:97.
Oyaizu, H., Debrunner-Vossbrinck, B., Mandelco, L., Studier, J. A., and
        Woese, C. R., 1987, The green non-sulfur bacteria: A deep branching
        in the eubacterial line of descent, <u>System. Appl. Microbiol.</u>, 9:47.

Pierson, B. K., and Castenholz, R. W., 1974, A phototrophic gliding
        filamentous bacterium of hot springs, Chloroflexus aurantiacus, gen.
        and sp. nov., Arch. Microbiol., 100:5.
Reeves, H. C., Rabin, R., Wegener, W. S., and Ajl, S. J., 1971, Assay of
        enzymes of the tricarboxylic acid and glyoxylate cycles, Methods in
        Microbiol., 6A:425.
Weaver, P. F., Wall, J. D., and Gest, H., 1975, Characterization of
        Rhodopseudomonas capsulata, Arch. Microbiol., 105:207.
Weitzman, P. D. J., 1978, Anomalous citrate synthase from Thermus aquaticus,
        J. Gen. Microbiol., 106:383.
Weitzman, P. D. J., 1981, Unity and diversity in some bacterial citric acid
        cycle enzymes, Adv. Microb. Physiol., 22:185.
Weitzman, P. D. J., and Danson, M. J., 1976, Citrate synthase, Curr. Top.
        Cell. Reg., 10:161.
Williams, R. A. D., 1975, Caldoactive and thermophilic bacteria and their
        thermostable proteins, Sci. Prog. (Oxford), 62:373.
Woese, C. R., 1987, Bacterial evolution, Microbiol. Revs., 51:221.

HYDROGENASES OF GREEN BACTERIA

I.N. Gogotov

Institute of Soil Science and Photosynthesis
USSR Academy of Sciences
Pushchino, Moscow Region, 142292, USSR

INTRODUCTION

Green sulfur bacteria can use reduced sulfur compounds (hydrogen sulfide, thiosulfate, S°) or H_2 as electron donors providing for CO_2 photoassimilation (Kondratieva and Gogotov, 1981). Molecular hydrogen uptake occurs with the participation of hydrogenase, the activity of which is found in a number of green bacteria such as Chlorobium, Pelodictyon and Chloroflexus (Kondratieva and Gogotov, 1981, 1983). Recently great attention has been paid to the study of structure and functioning mechanism of these unique biocatalysts as well as to the possibility of their application in native or immobilized states as catalysts of hydrogenation-dehydrogenation reactions. However, in contrast to hydrogenases from chemotrophic and purple bacteria (Kondratieva and Gogotov, 1981, 1983; Colbeau et al., 1983; Serebryakova et al., 1984), the data on the properties of these enzymes from the green bacteria are not abundant.

The aim of the present paper has been to isolate the hydrogenase from the cells of a nitrogen fixing green sulfur bacterium Chlorobium limicola f. thiosulfatophilum and to study the properties of this enzyme.

MATERIALS AND METHODS

The hydrogenase was isolated from Cb. limicola f. thiosulfatophilum strain L cells grown in Larsen's medium (Larsen, 1952) with 0.1% NH_4Cl, 0.1% Na_2S and 0.04% $Na_2S_2O_3$ in the light (2,000 lux; 30°C). Cells from the culture growth exponential phase were used for isolating the hydrogenase and electron carriers.

Enzyme activity assays were carried out using the reactions of reduction of various electron acceptors in the presence of H_2 (Serebryakova and Gogotov, 1981) and H_2 evolution from reduced methyl viologen (Gogotov et al., 1974). Hydrogenase activity using $^3H_2$-H_2O exchange reaction was measured by the increase of radioactivity in a liquid phase of the reaction vessel containing 0.2 mCi of gaseous tritium. The activity of the enzyme was expressed in μmol H_2 min^{-1} mg^{-1} of protein.

Table 1. Purification of <u>Cb. limicola</u> f. <u>thiosulfatophilum</u> Hydrogenase

Sample	Total Protein, mg	Total Activity, units*	Purification fold	Yield %
Acetone extract	238.0	57.25	1.0	100
Ammonium sulfate precipitate I (30-60% saturation)	63.0	60.40	4.8	100
Ammonium sulfate precipitate II (40-60% saturation)	15.6	54.61	20.0	96
Phenyl-sepharose CL-4B	3.8	27.18	36.0	50
Toyopearl HW-55	0.56	10.00	98.0	17

*Units: μmol H_2 oxidized min^{-1} in the presence of benzyl viologen (pH 8.0).

Hydrogenase Purification

The hydrogenase was usually purified from 20 g of cells, the acetone powder obtained being subjected to hydrogenase solubilization using an acetone extraction method (Serebryakova and Gogotov, 1981). The resultant preparation was fractionated with ammonium sulfate (Table 1). Further hydrogenase purification was carried out by chromatography on a column (1.5 x 20 cm) of CL-4B phenyl-sepharose. The enzyme was eluted with 25 mM Tris-HCl buffer (pH 8.0) containing 0.01% Triton X-100. The hydrogenase preparation obtained was further purified by gel filtration on a column (4 x 60 cm) with Toyopearl HW-55, saturated with 0.1 M NaCl in 25 mM Tris-HCl buffer (pH 8.0).

The fractions with hydrogenase activity were combined and concentrated by ultrafiltration on a PM-30 membrane. Thus, a highly purified hydrogenase preparation with a specific activity of about 20 μmol H_2 min^{-1} mg^{-1} of protein was obtained (Table 1). Electrophoretically homogeneous hydrogenase was obtained by preparative electrophoresis on a 7.5% polyacrylamide gel (PAAG) accompanied by partial enzyme inactivation.

Methods for Assaying Hydrogenase Properties

Hydrogenase purity was assayed by analytical electrophoresis in PAAG at pH 8.9. Enzyme M_r was determined by gel filtration on a Sephadex G-100 calibrated column in the presence of sodium dodecyl sulfate (SDS) (Weber and Osborn, 1969). The isoelectric point of the hydrogenase preparations was estimated on the microscale by vertical isoelectric focusing in PAAG columns within the pH range of 3-10 and 4-6 (Serebryakova et al., 1984). The hydrogenase redox potential (E_0') was calculated on the basis of the dependence of the enzyme activity on E_h of the reaction mixture (Zorin et al., 1984).

Isolation of electron carriers was carried out from a soluble fraction of the <u>Cb. limicola</u> f. <u>thiosulfatophilum</u> cells (Serebryakova et al., 1981), using 100 g of cells resuspended in 300 ml 25 mM Tris-HCl buffer (pH 7.2). The initial extract was subjected to chromatography on a column (4.5 x 10 cm) with DEAE-cellulose DE_{52}. Cyt <u>c</u>-555 was in the fraction not bound to DEAE-cellulose. Chromatography on a column with CM-cellulose CM_{32} and gel filtration on a column with Sephadex G-50 were applied for its further purification.

Ferredoxin and rubredoxin absorbed on to DEAE-cellulose were eluted with 0.5 M NaCl in the initial buffer. After desalting, the preparation was again subjected to chromatography on a column of DEAE-cellulose. The proteins were eluted from the column with a linear concentration gradient of NaCl (0.1-0.5 M) in 25 mM Tris-HCl buffer. The fractions containing rubredoxin and ferredoxin were combined separately, concentrated and then purified by gel filtration on a G-50 Sephadex column (4 x 150 cm). The preparations of electron carriers obtained were concentrated by dialysis against 30% polyethylene glycol solution (M_r = 40,000) for 6-8 h and used in the reactions with hydrogenase.

Cyt $\underline{c}_{sss}$, rubredoxin and ferredoxin were identified by their characteristic absorption spectra and molecular weights obtained by gel filtration on a Sephadex G-50 calibrated column. According to the data obtained, M_r values for Cyt $\underline{c}_{sss}$, rubredoxin and ferredoxin were about 10,000, 8,000 and 6,000 Da, respectively.

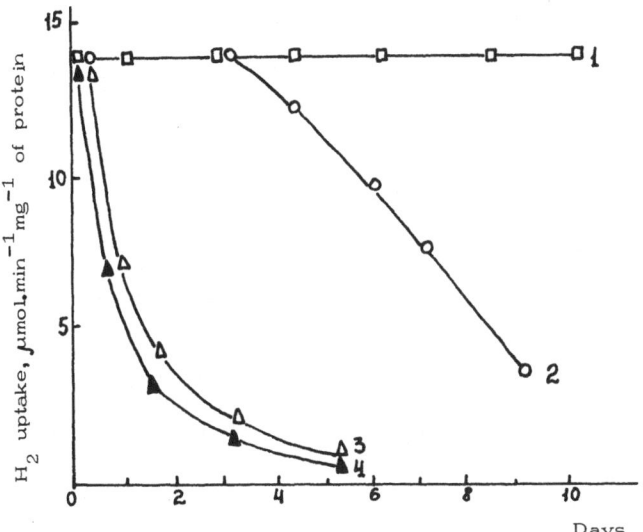

Fig. 1. Effect of storage conditions on <u>Cb. limicola</u> f. <u>thiosulfatophilum</u> hydrogenase stability.
1 - H_2, Ar or air (4°C); 2 - H_2, 20°C; 3 - Ar, 20°C; 4 - air, 20°C. Acceptor - BV^{2+}.

Electron transfer reduction reactions were carried out in a 1 cm anaerobic Thunberg cuvette (20°C). Their spectra and reduction kinetics were registered on a Specord UV VIS spectrophotometer (GDR). Reduction rates were calculated using the following extinction coefficients: 25 mM^{-1} cm^{-1} for Cyt $\underline{c}_{sss}$ at 555 nm, 8.85 mM^{-1} cm^{-1} for ferredoxin at 390 nm (Buchanan et al., 1969; Meyer et al., 1971; Steinmetz and Fischer, 1981).

Protein was determined by the method of Bradford (1976) using bovine serum albumin as standard.

RESULTS AND DISCUSSION

Hydrogenase Localization

During disruption of <u>Cb. limicola</u> f. <u>thiosulfatophilum</u> cells by sonication the major part of the hydrogenase (~ 60%) was found in the soluble fraction as in the case of the purple nonsulfur bacterium <u>Rhodospirillum rubrum</u> (Kondratieva and Gogotov, 1981) and the purple sulfur bacteria <u>Thiocapsa roseopersicina</u> (Kondratieva and Gogotov, 1981) and <u>Chromatium vinosum</u> (Feigenblum and Krasna, 1970). When the cells were treated with lysozyme (Bell et al., 1974) the major part of the hydrogenase activity was found in the spheroplast fraction. However, in contrast to <u>Rhodobacter capsulatus</u> (Serebryakova and Gogotov, 1981), addition of 1-5% Triton X-100 to the reaction mixture with the <u>Cb. limicola</u> f. <u>thiosulfatophilum</u> cells did not affect their hydrogenase activity.

It is known that the oxidized and reduced forms of MV and BV are able to penetrate the cytoplasmatic membrane of purple bacteria to different extents (Kovacs et al., 1983). This work has indicated that the active site of the hydrogenases from a number of purple bacteria is on the outward surface of the membranes. Analogous investigations carried out with <u>Cb. limicola</u> f. <u>thiosulfatophilum</u> showed that in contrast to the purple bacteria the active site of the hydrogenase in this bacterium is on the cytoplasmatic side.

Stability

Low stability during storage is characteristic of preparations of the membrane-bound hydrogenases (Kondratieva and Gogotov, 1981, 1983; Serebryakova and Gogotov, 1981). The half-life of the enzyme from <u>Cb. limicola</u> f. <u>thiosulfatophilum</u> was about 7.5 days upon storage under H_2 at 20°C (Fig. 1). Storage of preparations under Ar or air led to a quicker inactivation whereas H_2 had a stabilizing effect at room temperature. The hydrogenase stability was independent of the gas phase content at 4°C (Fig. 1). Lability to decreased temperatures was exhibited by the <u>Rb. capsulatus</u> hydrogenase (Serebryakova and Gogotov, 1981; Serebryakova et al., 1984). However, in contrast to this bacterium, the preparations of the <u>Cb. limicola</u> f. <u>thiosulfatophilum</u> hydrogenase can be stored for a long time in liquid nitrogen essentially without loss in activity (Serebryakova and Rodionov, 1985).

Sucrose and glycerol had a stabilizing effect on the <u>Rb. capsulatus</u> hydrogenase (Serebryakova and Gogotov, 1981; Serebryakova et al., 1984) but not on the enzymes from <u>Cb. limicola</u> f. <u>thiosulfatophilum</u> and <u>T. roseopersicina</u> (Zorin and Gogotov, 1982). Stability to a high ionic strength is also characteristic of both <u>Cb. limicola</u> f. <u>thiosulfatophilum</u> and <u>T. roseopersicina</u> hydrogenases. It allowed us to use fractionation with ammonium sulfate to purify these enzymes (Table 1). The hydrogenases from <u>Cb. limicola</u> f. <u>thiosulfatophilum</u> and from the purple bacteria (Table 2) are rather stable at high temperatures. Considerable enzyme inactivation occurred only above 70°C, the highest enzyme activity being at 65°C.

Physico-Chemical Properties

The <u>Cb. limicola</u> f. <u>thiosulfatophilum</u> hydrogenase absorption spectrum is characteristic of iron-sulfur proteins and has absorption maxima at 280 and 400 nm. The molecular weight determined by gel filtration on a Sephadex G-100 column was about 65,000 (Table 2). Approximately the same value (66,500) was obtained by electrophoresis of the hydrogenase preparation in PAAG with SDS.

Table 2. Properties of Hydrogenases from Purple and Green Bacteria

Properties	Rb. capsulatus (Gogotov, 1984)	T. roseopersicina (Kondratieva and Gogotov, 1981, 1984)	Cb. limicola f. thiosulfato-philum
Localization	Membranes	Membranes, cytoplasm	Membranes
E, mM^{-1} cm^{-1}	n.d. n.d.	115 29	n.d. n.d.
M_r	73,000	68,000	66,000
Subunits protein	1 x 73,000	1 x 25,000 1 x 47,000	1 x 66,500
Ni/mol	1	1	+
Fe/mol	4	4	4
S^{2-}/mol	4	4	4
sH/mol	n.d.	10	n.d.
pI	4.25	4.15 and 4.20	5.42
EPR-signal	HIPIP-type	HIPIP-type	n.d.
(ox.state)	$g = 2.019$	$g_I = 2.018$ $g_{II} = 2.004$	
E_o' (pH 7), mV	-340	-230	-305
$T_{1/2}$, h: Air, 20°C	20	1440	24
H_2, 20°C	120***	n.d.	168
Thermostability, °C	70	80	70
Natural electron Acceptors (donors)	Cyt $\underline{c}'$ and $\underline{b}_{560}$	Cyt $\underline{c}'_3$, $\underline{c}_3$, $\underline{c}_{552}$ HIPIP (Cyt $\underline{c}_3$)	rubredoxin Cyt $\underline{c}_{555}$
Activity, μmol H_2 Evolved (consumed) min^{-1} mg^{-1} of protein	1.3 (30)*	60(46)* 612(1,000)**	2(20)*

* at 30°C, ** at 70°C, *** in the presence of 25% glycerol; n.d. = not determined.

Preincubation of hydrogenase preparations in PAAG with SDS did not lead to dissociation of molecules into subunits. Consequently, the Cb. limicola f. thiosulfatophilum hydrogenase is a monomer with $M_r \sim 66,000$. Thus, the isolated hydrogenase is similar to those from R. rubrum (Kondratieva and Gogotov, 1981), Rb. capsulatus (Serebryakova et al., 1984) and C. vinosum (Klibanov et al., 1980) in molecular weight and subunit composition. The Cb. limicola f. thiosulfatophilum hydrogenase, like those from purple bacteria, is a weakly acidic protein with pI = 5.42. The hydrogenase contains Fe,S (Table 2) and probably nickel since Ni-limited growth of cells dramatically decreases their hydrogenase activity.

Catalytic Properties

The Cb. limicola f. thiosulfatophilum hydrogenase catalyzes the reactions of H_2 oxidation and hydrogen evolution from the reduced MV and $^3H_2$-H_2O isotopic exchange. However, like the Rb. capsulatus hydrogenase the enzyme from this bacterium catalyzes only H_2 uptake at a high rate. H_2 evolution rate for this enzyme was only 0.5-10% of its H_2-consuming activity (Table 2). This property is characteristic of unidirectional hydrogenases, obtained from a number of microorganisms (Kondratieva and Gogotov, 1981). A unidirectional action of these hydrogenases is due to the dependence of their activity on E_h of the medium (Zorin et al., 1984). The enzyme from Cb. limicola f. thiosulfatophilum exhibits its maximum activity at E_h > -300 mV

Table 3. Reduction of Some Compounds by the Cb. limicola f.
thiosulfatophilum Hydrogenase under H₂

Compound	E_o', mV	Concentration, mM	Reduction rate $\mu mol\ min^{-1}\ mg^{-1}$ hydrogenase
Methyl viologen	-440	5.0	0.40
Benzyl viologen	-360	5.0	12.70
Safranine T	-290	0.2	0.12
Phenosafranine	-252	0.2	0.52
FAD	-214	0.5	0
FMN	-208	0.5	0
Indigodesulfonate	-125	0.2	0.10
Methylene blue	+11	0.2	2.00
Phenazine methosulfate	+80	0.2	1.80

favourable only for H₂ uptake (Serebryakova and Rodionov, 1985). E_h decrease
to -400 mV leads to a 15-fold decrease in the enzyme activity and at the
potential necessary for H₂ evolution the Cb. limicola f. thiosulfatophilum
hydrogenase exhibits low activity. A similar dependence of the enzyme
activity on medium E_h is exhibited by the Rb. capsulatus hydrogenase (Zorin
et al., 1984) participating in in vivo in recyclization of H₂ evolved by
nitrogenase (Kondratieva and Gogotov, 1983).

Hydrogenase activity is considerably affected by the pH value of the
reaction medium (Serebryakova and Rodionov, 1985). At pH 8.0, optimum for
the catalysis of H₂ uptake, practically no hydrogen evolution from reduced
MV occurs, but at pH 5.0, optimal for hydrogen evolution, the hydrogenase is
capable of catalyzing the back reaction. In this case hydrogen uptake and
evolution rates are commensurate, but low. The hydrogenase catalytic
activity in ³H₂-H₂O exchange reaction was increased with a decrease of the
medium pH. An analogous dependence of the enzyme activity on the medium pH
is also characteristic of the unidirectional hydrogenase from Rb. capsulatus
(Serebryakova et al., 1984).

The Cb. limicola f. thiosulfatophilum hydrogenase is capable of
reducing a wide spectrum of artificial electron acceptors (Table 3) of which
benzyl viologen is the most specific (K_m(BV) = 2 mM). This distinguishes
the enzyme from the hydrogenase of thermophilic green bacterium Chloroflexus
aurantiacus Ok-70-fl. H₂ uptake catalyzed by the Cf. aurantiacus hydrogenase
occurs at a higher rate in the presence of FAD, FMN and PMS (Drutschmann and
Klemme, 1985). Activation energy of the H₂ uptake reaction catalyzed by the
Cb. limicola f. thiosulfatophilum hydrogenase in the presence of BV was 24
kJ M⁻¹.

High affinity to molecular hydrogen is characteristic of unidirectional
hydrogenase (Kondratieva and Gogotov, 1983; Serebryakova et al., 1984). For
the enzyme from Cb. limicola f. thiosulfatophilum K_m(H₂) is less than 10 µM.

Cytochrome C₅₅₅ isolated from Cb. limicola f. thiosulfatophilum but not
ferredoxin which has more negative redox potential than the hydrogenase, can
function as a natural electron acceptor for both the Cb. limicola f.
thiosulfatophilum hydrogenase and the enzymes from other phototrophic
bacteria (Gogotov, 1984; Serebryakova et al., 1984). Besides, under an
atmosphere of H₂ the hydrogenase also reduced rubredoxin isolated from this
bacterium (Table 4). The reaction of rubredoxin occurred at a considerably
higher rate and with less lag-period than the Cyt C₅₅₅ reduction.

Table 4. Reduction of Natural Electron Carriers by the Cb. limicola f.
thiosulfatophilum Hydrogenase

Electron carriers	Concentration, mM	E, mM^{-1} cm^{-1}	Reduction rate, nmol min^{-1} mg^{-1} protein
Ferredoxin	0.40	20.0 (390 nm)	0
Rubredoxin	0.45	8.85 (490 nm)	480.0
Cytochrome C_{555}	0.44	25.0 (555 nm)	96.0

Synthesis and Activity Regulation

There are not many data on the regulation of green bacterial
hydrogenase synthesis and activity. Synthesis of the hydrogenase from the
thermophilic facultatively phototrophic green bacterium Cf. aurantiacus
Ok-70-fl is repressed by S^{2-} (5.7 mM) and stimulated specifically by Ni^{2+} as
well as by molecular hydrogen (Drutschmann and Klemme, 1985). According to
the data obtained by us the synthesis of the hydrogenase from Cb. limicola
f. thiosulfatophilum is also dependent on the content of Ni^{2+} and electron
donor in the growth medium. The highest hydrogenase activity was exhibited
by cells grown under electron donor (H_2S) limitation. During Ni^{2+}-limited
growth the hydrogenase activity was notably decreased (3-5 fold).

Thus, the hydrogenase obtained by us from the cells of the green sulfur
bacterium Cb. limicola f. thiosulfatophilum is similar to those from Rb.
capsulatus and other nitrogen-fixing bacteria in a number of the studied
properties (Table 2). The main function of the Cb. limicola f.
thiosulfatophilum hydrogenase and the enzymes from purple bacteria is
probably to enable growth and CO_2 assimilation with molecular hydrogen and
to recycle H_2 produced in the light by nitrogenase action.

REFERENCES

Bell, G. B., Le Gall, J., and Peck, H. D., 1974, Evidence for the
periplasmic location of hydrogenase in Desulfovibrio gigas, J.
Bacteriol., 120:994.
Bradford, M., 1976, A rapid and sensitive method for the quantitation of
microgram quantities of protein utilising the principle protein-dye
binding, Anal. Biochem., 72:248.
Buchanen, B. B., Matsubara, H., and Evans, M. C., 1969, Ferredoxin from the
photosynthetic bacterium, Chlorobium thiosulfatophilum. A link to
ferredoxins from nonphotosynthetic bacteria, Biochim. Biophys. Acta,
189:46.
Colbeau, A., Chabert, J., and Vignais, P. M., 1983, Purification, molecular
properties and localization in the membrane of the hydrogenase of
Rhodopseudomonas capsulata, Biochim. Biophys. Acta, 748:116.
Drutschmann, M., and Klemme, J.-H., 1985, Sulfide-repressed, membrane-bound
hydrogenase in the thermophilic facultative phototroph, Chloroflexus
aurantiacus, FEMS Microbiol. Lett., 28:231.
Feigenblum, E., and Krasna, A. J., 1970, Solubilization and properties of
the hydrogenase of Chromatium, Biochim. Biophys. Acta, 198:157.
Gogotov, I. N., 1984, Hydrogenase of purple bacteria: properties and
regulation of synthesis, Arch. Microbiol., 140:86.
Gogotov, I. N., Mitkina, T. V., and Glinskiy, V. P., 1974, Effect of
ammonium on hydrogen evolution and nitrogen fixation in
Rhodopseudomonas palustris, Mikrobiologiya, 43:586.

Klibanov, A. H., Kaplan, N. O., and Kamen, M. D., 1980. Thermal stabilities of membrane-bound, solubilized, and artificially immobilized hydrogenase from <u>Chromatium vinosum</u>, <u>Arch. Biochem. Biophys.</u>, 199:545.

Kondratieva, E. N., and Gogotov, I. N., 1981, Molecular hydrogen in microbial metabolism. Moscow: Nauka, 343 p.

Kondratieva, E. N., and Gogotov, I. N., 1983, Production of molecular hydrogen in microorganisms, <u>in</u>: "Advances in Biochemistry Engineering/Biotechnology, vol. 28," A. Fiechter, ed., Springer Verlag, Berlin.

Kovacs, K. L., Baginka, Cs., and Serebryakova, L. T., 1983, Distribution and orientation of hydrogenase in various photosynthetic bacteria, <u>Curr. Microbiol.</u>, 9:215.

Larsen, H. J., 1952, On the culture and general physiology of the green sulfur bacteria, <u>J. Bacteriol.</u>, 64:187.

Meyer, T. E., Sharp, J. J., and Bartsch, R. G., 1971, Isolation and properties of rubredoxin from the photosynthetic green sulfur bacteria, <u>Biochim. Biophys. Acta</u>, 234:266.

Serebryakova, L. T., and Gogotov, I. N., 1981, Stability of solubilized and membrane-bound hydrogenase of <u>Rhodopseudomonas capsulata B10</u>, <u>Prikladnaja Biokhimiya i Mikrobiologiya</u>, 17:555.

Serebryakova, L. T., and Rodionov, V. V., 1985, Properties of hydrogenase from green bacteria, <u>Symposium of Molecular Biology of Hydrogenases</u>, Szeged, 23.

Serebryakova, L. T., Zorin, N. A., and Gogotov, I. N., 1984, Purification and properties of the hydrogenase from <u>Rhodopseudomonas capsulata</u>, <u>Biokhimiya</u>, 49:1456.

Steinmetz, M. A., and Fischer, U., 1981, Cytochromes of non-thiosulfate utilizing green sulfur bacterium <u>Chlorobium limicola</u>, <u>Arch. Microbiol.</u>, 130:31.

Weber, K., and Osborn, M., 1969, The reliability of molecular weight determinations by dodecyl sulfate polyacrilamide gel electrophoresis, <u>J. Biol. Chem.</u>, 244:4406.

Zorin, N. A., and Gogotov, I. N., 1982, Stability of hydrogenase of purple sulfur bacterium <u>Thiocapsa roseopersicina</u>, <u>Biokhimiya</u>, 47:827.

Zorin, N. A., Serebryakova, L. T., and Gogotov, I. N., 1984, Effect of the redox potential on the activity of purple bacteria hydrogenases, <u>Biokhimiya</u>, 49:1316.

AMMONIA ASSIMULATION AND AMINO ACID METABOLISM IN <u>CHLOROFLEXUS AURANTIACUS</u>

J.-H. Klemme, G. Laakmann-Ditges and J. Mertschuweit

Institut für Mikrobiologie der Universität Bonn
Meckenheimer Allee 168
D-5300 Bonn 1, F. R. G.

SUMMARY

The thermophilic phototrophic prokaryote, <u>Chloroflexus aurantiacus</u> Ok-70-fl, grows only poorly in defined media lacking amino acids. The organism is unable to fix N_2 (Heda and Madigan, 1986) but contains a Ni^{2+}-stimulated membrane-bound hydrogenase of the uptake type (Drutschmann and Klemme, 1985). Of the main enzymes of ammonia assimilation (glutamate dehydrogenase, EC 1.41.2; alanine dehydrogenase, EC 1.41.1; glutamine synthetase, EC 6.3.1.2, and glutamate synthase, EC 1.41.13), only glutamine synthetase was detected in cell-free extracts of cells grown in complex or defined media.

In contrast to the enzymes in other bacteria, the glutamine synthetase of <u>Cf. aurantiacus</u> was neither repressed nor inactivated by ammonia (Kaulen and Klemme, 1983). Glutamate, alanine and isoleucine were the main constituents of the intracellular free amino acid pool in cells grown photosynthetically in minimal medium. <u>Cf. aurantiacus</u> Ok-70-fl was shown to contain two L-threonine dehydratases (Laakmann-Ditges and Klemme, 1986). Interestingly, the cellular pool of isoleucine attained a level (1 mM range) at which, in the absence of any activation, the activity of the Ile-sensitive "biosynthetic" L-threonine dehydratase would have been completely inhibited.

The two L-threonine dehydratases were purified to electrophoretic homogeneity by a procedure involving anion exchange and hydrophobic interaction chromatography. Only one of the two dehydratases was shown to be sensitive to inhibition by L-isoleucine, L-leucine and L-valine. The inhibition by isoleucine was partly relieved by L-valine.

Although the absorption spectrum of each enzyme had a very high A_{280}/A_{420}-ratio (13.2) suggesting the absence of pyridoxal-5'-phosphate (PLP), both enzymes could be converted to an inactive form by dialysis against phenylhydrazine and subsequently be reconstituted to a fully active form with PLP. The dissociation constants of the enzyme-PLP-complexes were found to be 0.05 and 0.045 μM, for the Ile-sensitive and insensitive enzyme, respectively. At 20°C the molecular weight of each enzyme was 112 ± 6 kDa (estimated with a FPLC gel filtration technique). Both dehydratases contained only one type of subunit (55 ± 3 kDa) judged by SDS gel-electrophoresis. The aggregation state of the Ile-sensitive enzyme was controlled by temperature: At 20-25°C the enzyme existed as a dimer (112 kDa), at temperatures above 45°C as a tetramer (mol. wt. of about 220 kDa). The Ile-sensitive enzyme was mainly active with L-threonine (K_M(Thr) = 1.3 mM, K_M(Ser) = 21 mM), whereas the other dehydratase displayed comparable activities with L-serine (K_M = 10 mM) and L-threonine (K_M = 20 mM).

Since the latter enzyme was shown to be subject to a glucose catabolite repression in continuous culture, it may be classified as a "biodegradative" hydroxy amino acid dehydratase.

REFERENCES

Drutschmann, M., and Klemme, J.-H., 1985, Sulfide-repressed, membrane-bound hydrogenase in the thermophilic facultative phototroph, Chloroflexus aurantiacus, FEMS Microbiol. Lett., 28:231.

Heda, G. D., and Madigan, M. T., 1986, Utilization of amino acids and lack of diazotrophy in the thermophilic anoxygenic phototroph Chloroflexus aurantiacus, J. Gen. Microbiol., 132:2469.

Kaulen, H., and Klemme, J.-H., 1983, No evidence of covalent modification of glutamine synthetase in the thermophilic phototrophic bacterium Chloroflexus aurantiacus, FEMS Microbiol. Lett., 20:75.

Laakmann-Ditges, G., and Klemme, J.-H., 1986, Occurrence of two L-threonine (L-serine) dehydratases in the thermophile Chloroflexus aurantiacus, Arch. Microbiol., 144:219.

NITROGEN METABOLISM AND N$_2$ FIXATION IN PHOTOTROPHIC GREEN BACTERIA

G.D. Heda[a] and M.T. Madigan (Corresponding Author)

Department of Microbiology, Southern Illinois University
Carbondale, IL 62901, U.S.A.
[a] Present Address: Thrombosis Research Center
    Temple University School of Medicine, 3400 N. Broad Street
    Philadelphia, PA 19140, U.S.A.

INTRODUCTION

The ability of phototrophic purple bacteria to fix N$_2$ was discovered during studies of light-dependent H$_2$ production by Rhodospirillum rubrum (Gest and Kamen, 1949; Kamen and Gest, 1949). Subsequent studies of N$_2$ fixation in nonsulfur purple bacteria have shown this important metabolic process to be nearly universally distributed across the group (Madigan et al., 1984). Shortly after the discovery of N$_2$ fixation in R. rubrum, strains of Chromatium and Chlorobium were also shown to be capable of fixing molecular nitrogen (Lindstrom et al., 1950). Until the 1980's all subsequent work on aspects of nitrogen fixation in green bacteria employed the syntrophic mixed culture "Chloropseudomonas" (Zakhvateva et al., 1970; Zakhvateva and Kondrateva, 1971; Evans and Smith, 1971; Evans et al., 1971; Smith et al., 1971). These studies, of course, cannot be considered definitive because of the likelihood that the heterotrophic component of the "Chloropseudomonas" culture was itself a N$_2$-fixer (many sulfate-reducing bacteria are known to fix N$_2$ (Postgate and Kent, 1985).

Interest in the nitrogen-fixing properties of green bacteria has recently focused on the control and expression of nitrogenase in Chlorobium (Keppen et al., 1985; Heda and Madigan, 1986a; Rodionov et al., 1986). In each study ammonia was shown to rapidly inactivate nitrogenase in a fashion similar to the well-known ammonia "switch-off" effect in nonsulfur purple bacteria (Hallenbeck, 1987). In our work with phototrophic green bacteria we have undertaken a study of nitrogen metabolism and N$_2$-fixation in several representatives of this group. In this paper we present the results of this study and give some molecular details of the nitrogen-fixing process and of nitrogenase in the green bacterium Chlorobium.

MATERIALS AND METHODS

Bacterial Strains

Chlorobium limicola f. thiosulfatophilum strains NCIB 8327 and Tassajara were obtained from Reidun Sirevåg, University of Oslo, Norway.

Pelodictyon luteolum strain 2530 was obtained from Norbert Pfennig, Universität Konstanz, Federal Republic of Germany. Chloroflexus aurantiacus strains J-10-fl and OK-70-fl were taken from our culture collection.

Culture Media and Growth Conditions

Both Chlorobium strains were grown in the modified Pfennig's medium described in Heda and Madigan (1986a). Pelodictyon was grown in the same medium containing 2 mM (instead of 4 mM) sulfide and 1% NaCl. Chloroflexus strains were grown as described by Heda and Madigan (1986b). To test for the ability to use alternative nitrogen sources, ammonia was omitted and the medium supplemented with a nitrogen source at the concentrations specified in Table 1. Growth on N_2 was obtained as described in Heda and Madigan (1986a).

Cells were grown photosynthetically in completely filled 17 mm screw cap tubes or in 100 ml volumes in stirred 250 ml side-arm Erlenmeyer flasks containing a headspace of $N_2:CO_2$ (99:1). Cultures of Chlorobium were grown at 1500 lux incandescent illumination and 30°C; Pelodictyon was grown at 700 lux and 24°C. Chloroflexus was grown at 1500 lux and 47°C. Cell growth was measured turbidimetrically in a Klett-Summerson photometer fitted with a No. 66 (red) filter.

Nitrogenase

Nitrogenase activity (acetylene reduction) in intact cells was determined as described in Heda and Madigan (1986a). The light intensity during assay was 1500-2000 lux in all cases except for Pelodictyon where light intensities were about 800 lux. In vitro nitrogenase activity was assayed in cell extracts prepared as previously described (Heda and Madigan, 1986a). The components of the reaction mixture are listed in a footnote to Table 3.

Partial Purification of Chlorobium 8327 Nitrogenase

Crude cell extracts prepared under nitrogen (Heda and Madigan, 1986a) were centrifuged at 40,000 x g for 90 min in Oak Ridge-type centrifuge tubes sealed under nitrogen with sleeve-type serum stoppers. The supernatant was removed with a syringe and passed over a 2.5 x 30 cm anaerobic DEAE Sephacel column equilibrated with 100 mM MOPS buffer (pH 7.5) containing 2 mM $Ns_2S_2O_4$ and 1 mM dithiothreitol. The column was eluted with a step gradient of NaCl (0.1, 0.2, 0.4 M, equilibrated with anaerobic buffer) and fractions collected and concentrated for electrophoresis.

SDS-Polyacrylamide gel electrophoresis was performed according to Laemmli (1970). Samples were denatured by boiling for 5 min and electrophoresed on 10% polyacrylamide slab gels containing 0.1% SDS at a constant current of 30 mA/slab (25°C). Gels were stained with Coomassie Blue and destained in 10% acetic acid:40% methanol in water.

Immunodiffusion

Ouchterlony double diffusion (Stollar and Levine, 1963) was performed in 1% agarose containing 0.15 M NaCl and 0.02% sodium azide in 10 mM potassium phosphate, pH 7.5. Molten agar was pipetted onto cleaned microscope slides and hardened at 4°C before cutting wells. Prior to the application of antiserum (center well) cell extracts or partially purified nitrogenase fractions were added to outer wells as described in the legend to Fig. 5. Gels were stained with Coomassie Blue and destained (as described above) before photographing.

Table 1. Growth of Green Bacteria on Various Compounds as Sole Nitrogen Source[1]

Nitrogen Source[2]	Chlorobium		Chloroflexus		Heliobacterium chlorum[4]	Pelodictyon luteolum
	8327	Tassajara	OK-70-fl	J-10-fl		
Ammonia	3	3	2	2	2	2
N$_2$	3	3	0	0	2	2
Alanine	0	NT[5]	3	2	0	NT
Arginine	NT	NT	3	0	NT	NT
Aspartate	2	0	3	3	0	NT
Asparagine	2	0	3	0	NT	NT
Cysteine	0	0	2	2	NT	NT
Glutamate	2	0	3	3	0	NT
Glutamine	3	3	3	3	2	NT
Glycine	0	NT	3	2	NT	1
Histidine	0	0	2	0	NT	NT
Leucine	0	3	0	1	NT	NT
Lysine	0	1	2	0	NT	NT
Methionine	0	0	1	0	NT	NT
Nitrate	0	2	0	0	0	NT
Proline	3	0	0	0	NT	NT
Serine	1	NT	3	3	NT	NT
Threonine	0	NT	3	0	NT	NT
Tryptophan	0	2	2	0	NT	NT
Urea	3	3	0	0	0	NT
Valine	3	2	2	1	NT	NT

[1]All strains grown photosynthetically under the conditions specified in the Materials and Methods.
[2]Nitrogenous compounds added as follows: 8 mM (Chlorobium and Pelodictyon), 10 mM glutamine, 3 mM in all cases.
[3]Growth response as follows: 3 = > 300 Klett Units (KU); 2 = 150-300 KU; 1 = 50-150 KU; = = no growth. 100 Klett Units was equal to approximately 200 µg of bacterial cell protein/ml (Chlorobium) and 100 µg protein/ml (Chloroflexus). Growth measured after the second passage in the same medium.
[4]From Beer-Romero (1986).
[5]NT, Not Tested.

RESULTS AND DISCUSSION

Nitrogen Nutrition of Green Bacteria

    Nitrogen nutritional studies were initiated with several species of green bacteria to assess the ability of these organisms to use alternative nitrogen sources. Included in this survey were two well characterized strains of Chlorobium (strains 8327 and Tassajara) and Cf. aurantiacus (strains J-10-fl and OK-70-fl); a more limited survey was carried out with P. luteolum strain 2530 and Heliobacterium chlorum (data from Beer-Romero, 1986). Each strain was transferred to media containing a single nitrogenous compound as nitrogen source, and the results of these experiments are shown in Table 1. All species of green bacteria grew with ammonia as nitrogen source; ammonia appears to be a universal nitrogen source for phototrophic bacteria (Madigan and Gest, 1982). All species of green bacteria also grew to reasonably high cell densities on the amino acid glutamine as nitrogen source. As for ammonia, glutamine appears to be universally used by phototrophic bacteria; all nonsulfur purple bacteria as well as the purple sulfur bacterium <u>Chromatium</u> use glutamine as a nitrogen source (see Heda and Madigan 1986a for supporting references).

Chlorobium strains 8327 and Tassajara were quite versatile in utilizing amino acids other than glutamine as nitrogen sources (Table 1); the ability of strain 8327 to incorporate amino acids has previously been noted (Kelly, 1974). Strain Tassajara grew well on leucine, valine, and tryptophan and was the only strain of any green bacterium tested capable of growth on nitrate as sole nitrogen source (Table 1). Strain 8327 used aspartate, asparagine, and glutamate, and grew especially well on proline and valine as nitrogen sources. Both <u>Chlorobium</u> species used urea as nitrogen source. Of particular significance for studies of N_2 fixation in <u>Chlorobium</u> was the observation that strain 8327 grew on glutamate as sole nitrogen source; glutamate-grown cells were maximally derepressed for nitrogenase, and this facilitated studies of nitrogen fixation in this organism (see below). <u>Chloroflexus</u> strains OK-70-fl and J-10-fl were nutritionally also quite versatile. Both strains grew to maximal densities on glutamate and aspartate, and in fact, grew to higher densities on these and several other amino acids than on ammonia; this probably reflects the utilization of these substrates as carbon sources as well (Table 1). Cysteine, glycine and serine also served as excellent nitrogen sources for <u>Chloroflexus</u>, but urea was not used (Table 1). <u>H. chlorum</u>, on the other hand, was relatively restricted in its nitrogen metabolism. Of the amino acids rested only glutamine supported growth (Table 1).

Chlorobium (both strains), <u>P. luteolum</u> and <u>H. chlorum</u> grew with molecular nitrogen (N_2) as sole nitrogen source; however, attempts to grow <u>Chloroflexus</u> on N_2 were unsuccessful (Table 1). In additional growth experiments with <u>Chloroflexus</u> strain J-10-fl, cells were grown on limiting (0.3 mM) NH_4^+ in flasks containing either N_2 or Ar as gas phase. As shown in Fig. 1, no evidence of growth on N_2 could be obtained; cells grew on ammonia until it was exhausted, and the final cell densities were nearly the same

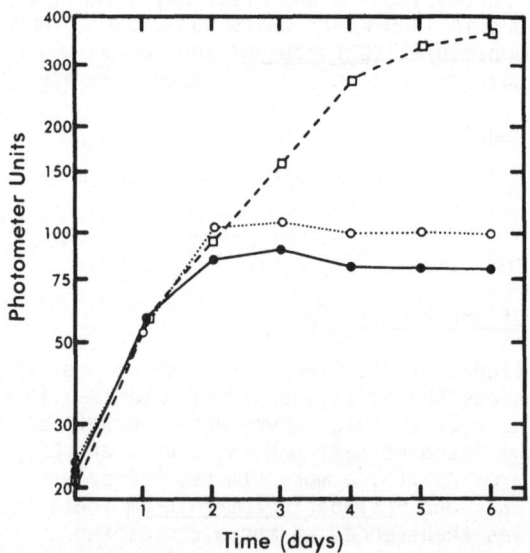

Time (days)

Fig. 1. Growth of <u>Chloroflexus aurantiacus</u> strain J-10-fl on limiting and excess NH_4^+. Cells were grown photosynthetically in 250 ml Erlenmeyer flasks containing 100 ml of medium D supplemented with 0.3 mM NH_4^+ (open circles, N_2 headspace; closed circles, Ar headspace), or 3 mM NH_4^+ (squares, Ar headspace). (From Heda and Madigan, 1986b).

Table 2. Acetylene Reduction Rates in Intact Cells of Various Green Bacteria[1]

Species/Strain	Growth Conditions	Specific Nitrogenase Activity[2]
Chlorobium thiosulfatophilum		
8327	N_2	1700
	Aspartate (8 mM)	2900
	Glutamate (10 mM)	8000
	Ammonia (0.5 mM)	600
	Ammonia (10 mM)	0
Tassajara	N_2	1300
Chloroflexus aurantiacus		
OK-70-fl	Glutamate (3 mM)	< 0.001
	Aspartate (3 mM)	< 0.001
	Ammonia (0.3 mM)	< 0.001
J-10-fl	Glutamate (3 mM)	< 0.001
Heliobacterium chlorum[3]		
Gest/Favinger	N_2	1050
Heliospirillum gestii[3]		
Ormerod	N_2	2450
Pelodictyon luteolum		
2530	N_2	1200

[1]Assay conditions as described in Heda and Madigan (1986a) and in the METHODS section of this paper. Cells grown photosynthetically on the indicated nitrogen compound as sole source of nitrogen.
[2]Nmoles of ethylene per hour per milligram cell protein.
[3]From Beer-Romero (1986).

under Ar or N_2. By contrast, a control flask containing nonlimiting ammonia grew to much higher cell densities (Fig. 1). Similar results were obtained with strain OK-70-fl and two other Chloroflexus isolates (strains Y-400-fl and 396-1) obtained from Yellowstone Hot Springs (data not shown).

Growth of phototrophic purple bacteria on amino acids as nitrogen source generally leads to copious H_2 production by nitrogenase (Madigan and Gest, 1982; Meyer et al., 1978). Curiously, however, H_2 production by amino acid-grown cultures of either Chlorobium or Chloroflexus was not observed. We attribute the absence of photohydrogen production by Chloroflexus to its inability to fix N_2 (see Fig. 1 and Table 2). However, the lack of significant H_2 production by amino acid grown cultures of Chlorobium is puzzling. We suspect that H_2 is actually produced during growth of Chlorobium on amino acids as N-source but that uptake hydrogenase(s) reincorporate the H_2 as a source of reductant before it has a chance to accumulate (Chlorobium is known to contain an active hydrogenase, Kovacs et al., 1983). The alternative hypothesis, that is that Chlorobium nitrogenase does not reduce protons, seems untenable in light of recent findings (Simpson and Burris, 1984). Further work on this aspect of N_2 fixation in Chlorobium would be desirable.

Comparative Nitrogenase Activities

Cell suspensions of N_2-grown Chlorobium, P. luteolum, H. chlorum and Heliospirillum gestii fixed nitrogen at high rates (Table 2); rates of N_2 fixation (acetylene reduction) of between 1000 and 2000 nmoles ethylene produced per hour per milligram cell protein were obtained with all species tested (Table 2). As expected, excess ammonia repressed nitrogenase synthesis in Chlorobium (Table 2) as it does in all wild type N₂-fixing bacteria (Postgate, 1982). Although nitrogenase derepression occurred when strain 8327 was grown on any amino acid (Heda and Madigan, 1986a) highest nitrogenase activities were obtained with glutamate-grown cells (Table 2); average acetylene reduction rates of about 8000 nmoles/min/mg cell protein wre consistently observed with this strain. This rate is comparable to rates obtained with some of the best N₂-fixing nonsulfur purple bacteria, such as Rhodobacter capsulatus (Madigan et al., 1984). We therefore conclude that growth on glutamate maximally derepresses nitrogenase in Chlorobium 8327. Growth of this strain on glutamate probably represents a nutritional state of N-starvation similar to that previously described for nonsulfur purple bacteria grown on yeast extract or amino acids as nitrogen source (Arp and Zumft, 1983; Alef and Kleiner, 1982).

Both N₂-grown H. chlorum and Hsp. gestii reduced acetylene at high rates, Hsp. gestii showing the highest rate of acetylene reduction of any N₂-grown green bacterium tested (Table 2). These data are consistent with the fact that the original enrichment strategy used in the isolation of heliobacteria selected for anoxygenic phototrophs able to fix N₂ (Gest and Favinger, 1983; Beer-Romero, 1986). Although Heliobacillus was not tested in this study, it is also capable of fixing N₂ (Beer-Romero and Gest, 1987).

Acetylene reduction was not observed in cell suspensions of Cf. aurantiacus grown under any nutritional conditions (Table 2). In the studies of Heda and Madigan (1986b), a variety of supplements were made to cultures of Chloroflexus to insure that nutritional conditions were adequate for expression of nitrogenase, if indeed the organism was capable of fixing N₂; none of these nutritional supplements resulted in cell suspensions capable of reducing acetylene. The inability of Chloroflexus to grow on N₂ (Fig. 1) or reduce acetylene under a variety of growth conditions consistent with nitrogenase derepression in purple bacteria (Table 2), leads to the conclusion that Chloroflexus is phenotypically NIF⁻. This is surprising considering that virtually all anoxygenic phototrophs fix N₂ (Madigan et al., 1984; Yoch, 1978). We have no ready explanation for why Chloroflexus does not fix N₂, however the fact that Chloroflexus is a thermophile may be of significance in this connection. In general, N₂ fixaito by thermophilic microorganisms is rare (Postgate, 1982). This hypothesis would also explain why the thermophilic purple sulfur bacterium Chromatium tepidum, which resembles N₂-fixing mesophilic Chromatium species in most respects except for its high optimum growth temperature (-50°C), is also incapable of fixing N₂ (Madigan, 1986). Perhaps diazotrophy is simply incompatible with thermophily in anoxygenic phototrophs.

Ammonia "Switch-Off" of Nitrogenase Activity in Green Bacteria

The addition of ammonia to N₂-fixing cultures of phototrophic purple bacteria catalyzes the rapid inactivation of nitrogenase (Johansson et al., 1983; Hallenbeck, 1987). A similar response occurs in green bacteria. For example, in Chlorobium, cultures of strain 8327 grown on N₂ lost more than 70% of nitrogenase activity within 20 min of adding 1 mM NH₄⁺ to the culture (Fig. 2). Ammonia "switch-off" was equally dramatic in Chlorobium strain Tassajara and could also be demonstrated in N₂-fixing cultures of P. luteolum, H. chlorum and Hsp. gestii (data not shown). Since reversible ammonia "switch-off" has been shown to operate in Chromatium (Gotto and Yoch, 1985), most nonsulfur purple bacteria (Sweet and Burris, 1981; Alef and Kleiner, 1982), Chlorobium (Keppen et al., 1985; Heda and Madigan, 1986a; Rodionov et al., 1986) and other green bacteria (this paper), it is obviously an important mechanism of nitrogenase control in phototrophic bacteria. Presumably ammonia "switch-off" results in ATP savings to the organism from the energy demands of N₂ fixation (Postgate, 1982). Many, but

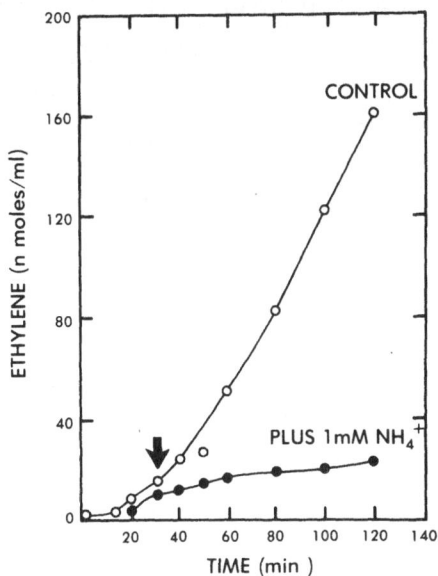

Fig. 2. Ammonia "switch-off" of nitrogenase activity in intact cells of
N₂-grown _Chlorobium_ strain 8327. Ammonia (1 mM) was added at the
point indicated by the arrow. Protein content of cell suspension
equaled 0.7 mg/ml. (From Heda and Madigan, 1986a).

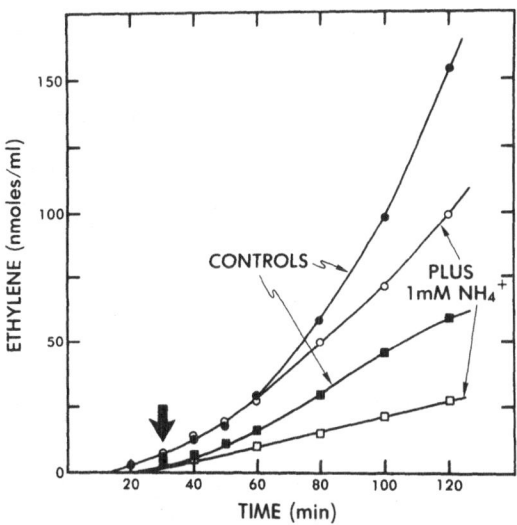

Fig. 3. Ammonia "switch-off" of nitrogenase activity in intact cells of
glutamate and limiting ammonia grown _Chlorobium_ strain 8327.
Ammonia (1 mM) was added at the point indicated by the arrow. Cells
were grown in media previously sparged with argon containing either
10 mM glutamate (circles) or 0.5 mM NH₄Cl (squares). Protein
content of suspensions equaled 0.16 mg/ml (glutamate cells) or 0.2
mg/ml (ammonia cells). (From Heda and Madigan, 1986a).

Table 3. Nitrogenase Activity in Cell Extracts of
Chlorobium Strain 8327[1]

Assay Conditions	Ethylene produced (nmol/20 min)
Complete[2]	104
minus extract	0
minus Mg^{2+}	0
minus dithionite	0
plus Chlorobium membranes[3]	80
plus salt wash of Chlorobium membranes[4]	10

[1]Cells of strain 8327 grown photosynthetically on 10 mM glutamate as sole nitrogen source.
[2]Complete assay mixture (total volume 0.5 ml) contained: 5 mM ATP; 30 mM creatine phosphate; 9 units creatine phosphokinase; 10 mM $MgCl_2$; 100 mM MOPS buffer (pH 7.5); 20 mM $Na_2S_2O_4$, cell extract (2.5 mg protein from 48 K x g supernatant).
[3]1.5 mg protein of Chlorobium membrane fragments.
[4]100 µl of a 0.5 M NaCl wash of Chlorobium membranes.

not all heterotrophic N_2-fixers respond to ammonia in the same way (Postgate, 1982).

Ammonia "switch-off" of nitrogenase activity was much less pronounced in cells of Chlorobium 8327 grown on either glutamate or limiting ammonia as sole nitrogen source (Fig. 3). This phenomenon is observed in R. rubrum (Sweet and Burris, 1981) and other nonsulfur purple bacteria (Alef and Kleiner, 1982). For unknown reasons the enzymic mechanism triggering nitrogenase "switch-off" in phototrophic bacteria is much more active in N_2-grown cells than in nitrogen-starved cells (Sweet and Burris, 1981). This is particularly perplexing when one considers that nitrogenase isolated from nitrogen-starved cells of R. rubrum (which should be in an active form) is inactive (Ludden and Burris, 1976, see next section). From studies of "switch-off" in R. rubrum it has been concluded that several environmental variables (light, oxygen, etc.) can also "switch-off" mechanism is the same system that inactivates nitrogenase in response to these other effectors (Sweet and Burris, 1981).

Properties of Chlorobium Nitrogenase: In Vitro Activity

In vitro acetylene reduction in extracts of glutamate-grown Chlorobium strain 8327 was dependent on ATP, an ATP generating system, a low potential reductant (dithionite), and Mg^{2+} (Table 3); these substrates are required for in vitro activity of all nitrogenases (Postgate, 1982). However, acetylene reduction in crude, membrane-free cell extracts of strain 8327 was only a small fraction of the activity obtained in intact cells. Nitrogenase isolated from glutamate-grown cells of R. rubrum exists in an inactive form and only becomes activated after interaction with a Mn^{2+}-dependent, membrane-bound "activating enzyme" that catalyzes a covalent modification of the Fe protein of nitrogenase, effectively activating the entire nitrogenase preparation (Ludden and Burris, 1976; Saari et al., 1984). Reasoning that Chlorobium nitrogenase may be partially inactivated during preparation of cell extracts, we performed experiments designed to test for the presence of a Chlorobium nitrogenase "activating enzyme". However, as shown in Table 3, neither the addition of a Chlorobium membrane preparation nor a salt wash of Chlorobium membranes stimulated nitrogenase activity in extracts of glutamate-grown strain 8327 (the salt wash actually inhibited activity). This suggests that nitrogenase in extracts of glutamate-grown _Chlorobium_ cells is already in an active form. Further support for this conclusion has come from experiments showing the linear kinetics of _Chlorobium_ nitrogenase activity _in vitro_ (Heda and Madigan, 1986a) and from the fact that _Chlorobium_ can be grown on N_2 in the absence of Mn^{2+} (Keppen et al., 1985), whereas _R. rubrum_ cannot (Yoch, 1979). Thus, although a nitrogenase activating system must be operative in _Chlorobium_ because N_2-grown cells are subject to rapid ammonia "switch-off" (see Fig. 2 and 3), the "switch-on" system is presumably Mn^{2+}-independent and probably only operates in N_2-grown cells.

Properties of Chlorobium Nitrogenase: Structural and Immunological Relationships

In order to assess the structural properties of nitrogenase from _Chlorobium_, partial purification of the enzyme from glutamate-grown strain 8327 was performed. The purification strategy employed was that previously found successful in purifying nitrogenase from _R. rubrum_ (Ludden and Burris, 1978). Crude extracts of glutamate-grown _Chlorobium_ strain 8327 were applied to an ion exchange column and fractions eluted with NaCl. By comparing the elution profiles of extracts of cells grown on excess ammonia (where nitrogenase synthesis is repressed, see Table 2) with those grown on glutamate, physical evidence for a two component nitrogenase system was obtained (Fig. 4). As shown in Fig. 4 major "ammonia-repressible" polypeptides (proteins present in glutamate-grown cells and absent from ammonia-grown cells) were present only in the 0.2 M NaCl eluate. The major

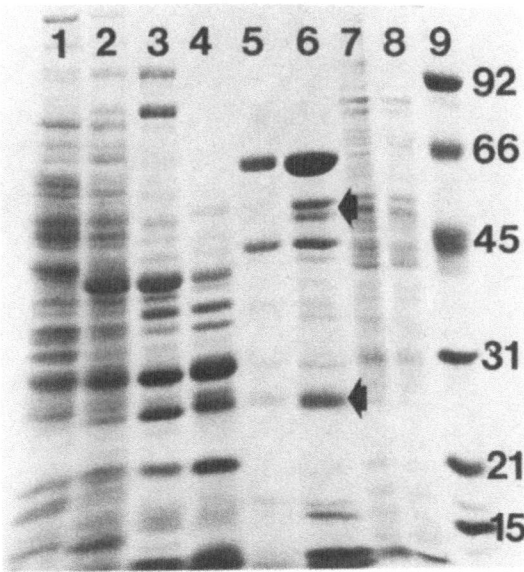

Fig. 4. SDS-polyacrylamide gel electrophoresis of partially purified nitrogenase from _Chlorobium_ strain 8327. Lanes 1, 3, 5, and 7, cells grown on excess ammonia; lanes 2, 4, 6, and 8, cells grown on glutamate as nitrogen source. Lane 9, molecular weight standards (in thousands). Lanes 1 and 2, crude extracts; lanes 3 and 4, 0.1 M NaCl eluate from DEAE Sephacel column; lanes 5 and 6 0.2 M NaCl eluate; lanes 7 and 8, 0.4 M NaCl eluate. Arrows indicate probable position of _Chlorobium_ Mo-Fe protein (upper arrow) and Fe protein (lower arrow) of nitrogenase.

ammonia-repressible polypeptides had molecular weights of about 55, 53, and 27 kDa (Fig. 4). We interpret these bands to be the α and β polypeptides of the Mo-Fe protein and the single polypeptide of the Fe protein of Chlorobium nitrogenase, respectively. Although rates of acetylene reduction by the 0.2 M NaCl fraction of Chlorobium extracts were quite low (data not shown), the presence of 3 major ammonia repressible bands corresponding closely in molecular weight to nitrogenase polypeptides from a variety of other N$_2$-fixing bacteria (Postgate, 1982; Hallenbeck, 1987), is strong evidence that Chlorobium nitrogenase is also a two component enzyme consisting of three distinct polypeptides.

To test for immunological similarities between R. rubrum and Chlorobium nitrogenases, immunodiffusion experiments were carried out using antisera prepared against R. rubrum nitrogenase. Surprisingly, crude extracts or partially purified nitrogenase fractions from glutamate-grown Chlorobium strain 8327 did not form a precipitin line with antisera to R. rubrum nitrogenase (Fig. 5). Extracts of glutamate-grown R. rubrum used as a positive control did react, however, forming easily visible precipitin lines (Fig. 5). In order to ensure that equivalent proportions of antigen and antisera were present, different amounts of Chlorobium extract or partially purifi-d fractions were added to separate wells, but in no case was a precipitin band observed (Fig. 5). We tentatively conclude that Chlorobium nitrogenase is not immunologically related to the enzyme from R. rubrum, even though antisera to R. rubrum nitrogenase does form precipitin lines with Rhodobacter capsulatus and Rhodobacter sphaeroides nitrogenases (D. Yoch, personal communication).

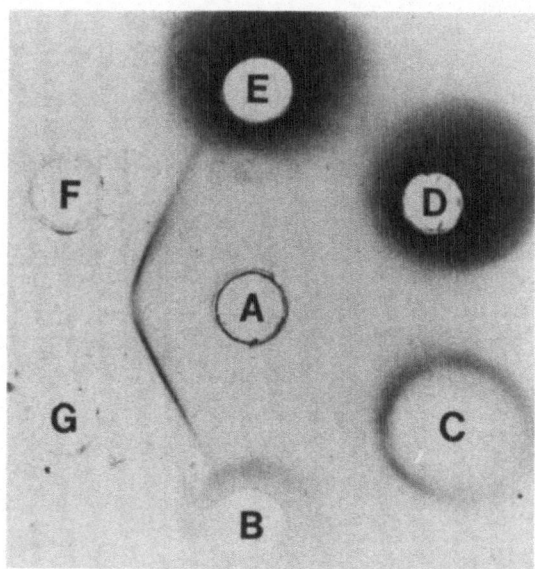

Fig. 5. Immunodiffusion analyses of Chlorobium and Rhodospirillum rubrum
nitrogenase. Center well (A) contained antisera to R. rubrum Mo-Fe
protein and Fe protein. Outer wells contained: B, C, 65 μg protein
of Chlorobium 8327 partially purified nitrogenase; D, 180 μg protein
from crude extracts of glutamate-grown Chlorobium 8327; E, 360 μg
protein from crude extracts of glutamate-grown Chlorobium 8327; F,
110 μg protein from crude extracts of glutamate-grown R. rubrum
1.1.1; G, 220 μg protein from crude extracts of glutamate-grown R.
rubrum 1.1.1.

184

Concluding Remarks

This paper has shown that phototrophic green bacteria share several features in common with purple bacteria in respect to nitrogen metabolism and nitrogen fixation. Representatives of the genera Chlorobium, Pelodictyon, Heliobacterium, Heliobacillus and Heliospirillum are all N_2-fixing organisms; Cf. aurantiacus is not a N_2-fixer. The ability of ammonia to "switch-off" nitrogenase activity in green bacteria implies that enzymic mechanisms exist which function to inactivate previously active nitrogenase. The general pattern of ammonia "switch-off" thus appears very similar in purple and green bacteria. However, because a Mn^{2+}-dependent nitrogenase activating system does not appear to operate in Chlorobium, it is premature to conclude that precisely the same biochemical events trigger nitrogenase "switch-on" (activation) in both groups of phototrophs.

Not surprisingly, the general structure of nitrogenase from Chlorobium appears similar to that of other N_2-fixers, a two component system consisting of three distinct polypeptides. However, because of its unique phylogenetic position (Gibson et al., 1985), further work to obtain homogeneous preparations of Chlorobium nitrogenase components would be of both fundamental and evolutionary interest. It would be particularly interesting to determine whether Chlorobium Fe protein can supply electrons to R. rubrum Mo-Fe protein (as Chromatium Fe protein can, Gotto and Yoch, 1985), and whether the lack of immunological cross reactivity between Chlorobium and R. rubrum is due to any significant structural differences in the two proteins.

ACKNOWLEDGEMENTS

We thank Dr. Paul Ludden for providing antisera to R. rubrum nitrogenase components and Peggy Beer-Romero for sharing unpublished results on N_2 fixation in heliobacteria. This work was supported by grants from the United States Department of Agriculture (83-CRCR-1-1308) and National Science Foundation (PCM 8505492).

REFERENCES

Alef, K., and Kleiner, D., 1982, Regulatory aspects of inorganic nitrogen metabolism in Rhodospirillaceae, Arch. Microbiol., 133:239.

Arp, D. J., and Zumft, W. G., 1983, Overproduction of nitrogenase by nitrogen-limited cultures of Rhodopseudomonas palustris, J. Bacteriol., 153:1322.

Beer-Romero, P., 1986, Comparative studies on Heliobacterium chlorum, Heliospirillum gestii and Heliobacillus mobilis, M.A. thesis, Indiana University, Department of Biology, Bloomington, Indiana.

Beer-Romero, P., and Gest, H., 1987, Heliobacillus mobilis, a peritrichously flagellated anoxyphototroph containing bacterilchlorophyll g, FEMS Microbiol. Lett., 41:109.

Evans, M. C. W., and Smith, R. V., 1971, Nitrogen fixation by the green photosynthetic bacterium Chloropseudomonas ethylicum, J. Gen. Microbiol., 65:95.

Evans, M. C. W., Telfer, A., Cammack, R., and Smith, R. V., 1971, EPR studies of nitrogenase: ATP dependent oxidation of fraction 1 protein by cyanide, FEBS Lett., 15:317.

Gest, H., and Favinger, J. L., 1983, Heliobacterium chlorum, an anoxygenic brownish-green photosynthetic bacterium containing a "new" form of bacteriochlorophyll, Arch. Microbiol., 136:11.

Gest, H., and Kamen, M. D., 1949, Photoproduction of molecular hydrogen by Rhodospirillum rubrum, Science, 109:558.

Gibson, J., Ludwig, W., Stackebrandt, E., andWoese, C. R., 1985, The phylogeny of the green photosynthetic bacteria: absence of a close relationship between Chlorobium and Chloroflexus, Syst. Appl. Microbiol., 6:152.

Gotto, J. W., and Yoch, D. C., 1985, Regulation of nitrogenase activity by covalent modification in Chromatium vinosum, Arch. Microbiol., 141:40.

Hallenbeck, P. C., 1987, Molecular aspects of nitrogen fixation by photosynthetic prokaryotes, CRC Crit. Revs. Microbiol., 14:1.

Heda, G. D., and Madigan, M.T., 1986a, Aspects of nitrogen fixation in Chlorobium, Arch. Microbiol., 143:330.

Heda, G. D., and Madigan, M. T., 1986b, Utilization of amino acids and lack of diazotrophy in the thermophilic anoxygenic phototroph Chloroflexus aurantiacus, J. Gen. Microbiol., 132:2469.

Johansson, Bo. C., Nordlund, S., and Baltscheffsky, H., 1983, Nitrogen fixation and ammonia assimilation, in: "The Phototrophic Bacteria: Anaerobic Life in the Light," J. G. Ormerod, ed., University of California Press.

Kamen, M. D., and Gest, H., 1949, Evidence for a nitrogenase system in the photosynthetic bacterium Rhodospirillum rubrum, Science, 109:560.

Kelly, D. P., 1974, Growth and metabolism of the obligate photolithotroph Chlorobium thiosulfatophilum in the presence of added organic nutrients, Arch. Microbiol., 100:163.

Keppen, D. I., Lebedeva, N. V., Petukhov, S. A., and Rodionov, Yu. V., 1985, The activity of nitrogenase in the green sulfur bacterium Chlorobium limicola forma thiosulfatophilum, Microbology (English translation of Mikrobiologiya) 54:28.

Kovacs, K. L., Bagyinka, Sc., and Serebriakova, L. T., 1983, Distribution and orientation of hydrogenase in various photosynthetic bacteria, Curr. Microbiol., 9:215.

Laemmli, U. K., 1970, Cleavage of structural proteins during the assembly of the head of bacteriophage T4, Nature (London), 227:680.

Lindstrom, E. S., Tove, S. R., and Wilson, P. W., 1950, Nitrogen fixation by the green and purple sulfur bacteria, Science, 112:197.

Ludden, P. W., and Burris, R. H., 1976, Activating factor for the iron protein of nitrogenase from Rhodospirillum rubrum, Science, 194:424.

Ludden, P. W., and Burris, R. H., 1978, Purification and properties of nitrogenase from Rhodospirillum rubrum, and evidence for phosphate, ribose, and an adenine-like unit covalently bound to the iron protein, Biochem. J., 175:251.

Madigan, M. T., 1986, Chromatium tepidum, sp.n., a thermophilic photosynthetic bacterium of the family Chromatiaceae, Intern. J. Syst. Bacteriol., 36:222.

Madigan, M. T., Cox, S. S., and Stegeman, R. A., 1984, Nitrogen fixation and nitrogenase activities in members of the family Rhodospirillaceae, J. Bacteriol., 157:73.

Madigan, M. T., and Gest, H., 1982, Biological dinitrogen fixation by photosynthetic bacteria, in: "CRC Handbook Series of Biosolar Resources, Volume I, part I: Basic Principles," A. Mitsui and C. C. Black, eds., CRC Press, Boca Raton, Florida.

Meyer, J., Kelley, B. C., and Vignais, P. M., 1978, Nitrogen fixation and hydrogen metabolism in photosynthetic bacteria, Biochimie, 60:245.

Postgate, J. R., 1982, The Fundamentals of Nitrogen Fixation, Cambridge University Press, Cambridge, England.

Postgate, J. R., and Kent, H. M., 1985, Diazotrophy within Desulfovibrio, J. Gen. Microbiol., 131:2119.

Rodinov, Yu. V., Lebedeva, N. V., and Kondrateva, E. N., 1986, Ammonia inhibition of introgenase activity in purple and green bacteria, Arch. Microbiol., 143:345.

Saari, L. L., Triplett, E. W., Ludden, P. W., 1984, Purification and properties of the activating enzyme for iron protein of nitrogenase from the photosynthetic bacterium <u>Rhodospirillum rubrum</u>, <u>J. Biol.</u> <u>Chem.</u>, 259:15502.

Simpson, F. B., and Burris, R. H., 1984, A nitrogen pressure of 50 atmospheres does not prevent evolution of hydrogen by nitrogenase, <u>Science</u>, 224:1095.

Smith, R. V., Telfer, A., and Evans, M. C. W., 1971, Complementary functioning of nitrogenase components from a blue-green alga and a photosynthetic bacterium, <u>J. Bacteriol.</u>, 107:574.

Stollar, D., and Levine, L., 1963, Two dimensional immunodiffusion, <u>Meth. Enzymol.</u>, 6:848.

Sweet, W. J., and Burris, R. H., 1981, Inhibition of nitrogenase activity by NH_4^+ in <u>Rhodospirillum rubrum</u>, <u>J. Bacteriol.</u>, 145:824.

Yoch, D. C., 1979, Manganese, an essential trace element for N_2 fixation by <u>Rhodospirillum rubrum</u> and <u>Rhodopseudomonas capsulata</u>: role in nitrogenase regulation, <u>J. Bacteriol.</u>, 140:987.

Yoch, D. C., 1978, Nitrogen fixation and hydrogen metabolism by photosynthetic bacteria, <u>in</u>: "The Photosynthetic Bacteria," R. K. Clayton and W. R. Sistrom, eds., Plenum Press, New York.

Zakhvateva, N. V., and Kondrateva, E. N., 1971, Fixation of molecular nitrogen by photosynthesizing bacteria in relation to presence of ligth and ATP and character of exogenous substrate, <u>Dokl. Akad. Nauk SSSR</u>, 196:72.

Zakhvateva, N. V., Malofeeva, I. V., and Kondrateva, E. N., 1970, Nitrogen fixation capacity of photosynthesizing bacteria, <u>Microbiology</u> (English translation of <u>Mikrobiologiya</u>), 39:661.

METABOLISM OF THIOSULFATE IN <u>CHLOROBIUM</u>

H.G. Trüper, C. Lorenz, M. Schedel and M. Steinmetz

Institut für Mikrobiologie
Rheinische Friedrich-Wilhelms-Universität
Meckenheimer Allee 168, D-5300 Bonn 1, F. R. G.

INTRODUCTION

Thirty-five years ago, Helge Larsen (1952) described the first isolation of a thiosulfate-utilizing <u>Chlorobium</u> which he named <u>Chlorobium thiosulfatophilum</u>. It differed from <u>Chlorobium limicola</u> (Nadson, 1906) just by this one property.

Comparative taxonomic studies with a larger number of strains led Pfennig and Trüper (1971) to transfer the thiosulfate-utilizing forms of <u>Chlorobium</u> to subspecific rank. Thus, today, in the Chlorobiaceae, the utilization of thiosulfate as a photosynthetic electron donor is restricted to the subspecies or "formae" "<u>thiosulfatophilum</u>" of <u>Cb. limicola</u> and <u>Cb. vibrioforme</u>.

The capacity to utilize thiosulfate as photosynthetic electron donor facilitates the cultivation of these forms as compared with those that depend on sulfide. As a consequence, there exist many more detailed investigations of these forms since Larsen first isolated them.

The two subspecies "<u>thiosulfatophilum</u>" show a number of phenomena related to thiosulfate that are not shared by the other Chlorobiaceae: 1) thiosulfate utilization as single photosynthetic electron donor, 2) thiosulfate formation as a free intermediate during phototrophic sulfide oxidation, 3) disproportionation of elemental sulfur in the light in the absence of CO_2, 4) possession of Cyt <u>c</u>-551, 5) possession of high levels of thiosulfate sulfur transferase (rhodanese).

In the following, unpublished experimental results will be reported relating to points 2, 3 and 5 of this list, and the utilization of thiosulfate will be discussed within the present state of knowledge of sulfur metabolism in Chlorobiaceae.

MATERIALS AND METHODS

Bacteria and Cultivation

All strains used during this study were from the German Culture Collection (DSM, cf. Table 2). The bacteria were grown according to Pfennig and Trüper (1981) or Trüper and Imhoff (1981).

Preparation of Elemental Sulfur

Hydrophilic elemental sulfur was prepared according to Janek (1933).

Analytical Determination of Sulfur Compounds

The following methods were used: Sulfide (Pachmayr, 1960; Trüper and Schlegel, 1964), elemental sulfur (Bartlett and Skoog, 1954), thiosulfate (Urban, 1961), sulfite (Grant, 1947), polythionates (Kelly et al., 1969), sulfate (Dodgson, 1961).

Stirred Cell (Cuvette) Experiments

The sulfur disproportionation experiments were carried out in a stirred cuvette according to Trüper (1964). For H_2S removal bubbling with oxygen-free argon was used. Sulfide was trapped in zinc acetate solution. The experiments with growing cell suspensions were performed in smaller flat cells of different sizes (10 and 100 mL), equipped with temperature jacket and magnetic stirring, built to allow optimal illumination even at high suspension densities. The headspace contained an oxygen-free 75% N_2/25% CO_2 mixture. The bacteria were harvested by centrifugation after 24 h growth, and resuspended in the following medium: $CaCl_2 \cdot 2H_2O$, 0.25 mM; NH_4Cl, 6 mM; $MgCl_2 \cdot 6H_2O$, 2 mM; KCl, 5 mM; KH_2PO_4, 7.5 mM; $NaHCO_3$, 15 mM; Na ascorbate 2 mM; pH 7.0; protein concentration was 0.5 or 3.0 mg/mL. The sulfur compounds under study were added to the respective vessels at zero time.

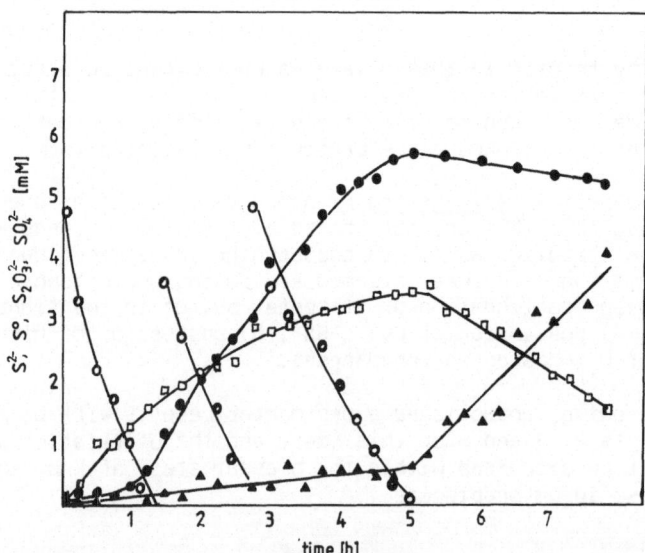

Fig. 1. Photooxidation of sulfide by growing cells of Cb. limicola f. thiosulfatophilum (strain DSM 249). (o) sulfide, (□) thiosulfate, (●) elemental sulfur, (▲) sulfate. Stirred cell experiment. For details see METHODS. 0.5 mg protein/mL. To avoid toxic concentrations, sulfide was added three times in lower concentrations.

Protein Determinations

Whole cell protein was determined according to Stickland (1951) as modified by Schmidt et al. (1963). Protein in cell-free extracts was assayed after Lowry et al. (1951).

Preparation of Cell-Free Extracts

Harvested pellets of cell material were resuspended in the respective buffer at 1 g wet weight per 1-3 ml and a few mg of deoxyribonuclease added. Cells were broken by sonification (1 min/ml at 4°C; Branson Sonifyer Cell Disruptor B15). Coarse materials were removed by centrifugation (30,000 rpm, 30 min, Sorvall Superspeed RC5) and the supernatant then ultracentrifuged for 5 h at 100,000 x g at 4°C (Beckman, L5-65). The dark greenish-brown supernatant obtained contained the soluble proteins.

Enzyme Determinations

Enzymes were tested as follows: Thiosulfate sulfur transferase (rhodanese) according to Smith and Fascelles (1966), thiosulfate reductase according to Hashwa (1972), thiosulfate acceptor oxidoreductase according to Lyric and Suzuki (1970), reverse sulfite reductase as described by Schedel and Trüper (1979).

Chemicals and biochemicals were purchased as reported earlier (Schedel et al., 1979; Steinmetz and Fischer, 1982).

RESULTS

Formation of Thiosulfate During Sulfide Oxidation

Since the experiments of Larsen (1953) and Shaposhnikov et al. (1958) it has been generally believed that in Cb. limicola during phototrophic growth with sulfide the oxidation of sulfide to elemental sulfur and of elemental sulfur to sulfate are strictly sequential processes.

In a short term sulfide oxidation experiment we found in Cb. limicola f. thiosulfatophilum that not only elemental sulfur but also thiosulfate appears in the medium (Fig. 1). Thiosulfate was in fact the predominant intermediate in the first phase, whereas later elemental sulfur became predominant. The specific rate of thiosulfate formation at the beginning of the experiment was 52 nmol $S_2O_3^{2-}$ min^{-1} mg^{-1} protein. The experiment proved, however, that in batch culture the oxidation of sulfide and the appearance of sulfate are strictly sequential, because the concentration of sulfate began to increase (at the expense of thiosulfate and elemental sulfur) only after complete utilization of sulfide.

When thiosulfate was given as the only electron donor under the same experimental conditions, Cb. limicola f. thiosulfatophilum did not form any detectable sulfur intermediate (i.e., elemental sulfur or tetrathionate), but only sulfate (Fig. 2). Also when elemental sulfur alone was given as electron donor, only sulfate was found and no thiosulfate or sulfide were detected (Fig. 3). Table 1 shows the oxidation rates for sulfide, elemental sulfur and thiosulfate derived from the experiments in Fig. 1-3.

Disproportionation of Elemental Sulfur

In 1974 Paschinger et al. reported the photochemical disproportionation of sulfur into sulfide and sulfate by Cb. limicola f. thiosulfatophilum. We have reinvestigated this phenomenon by measuring the amounts of sulfur

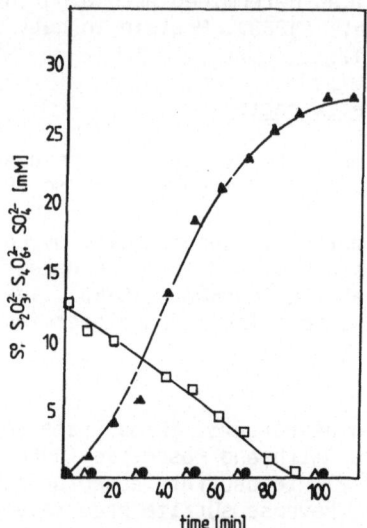

Fig. 2. Photooxidation of thiosulfate by growing cells of <u>Cb. limicola</u> f. <u>thiosulfatophilum</u> (strain DSM 249). (□) thiosulfate, (▲) sulfate, (●) elemental sulfur, (Δ) tetrathionate. Stirred cell experiment; for details see METHODS. 3 mg protein/ml.

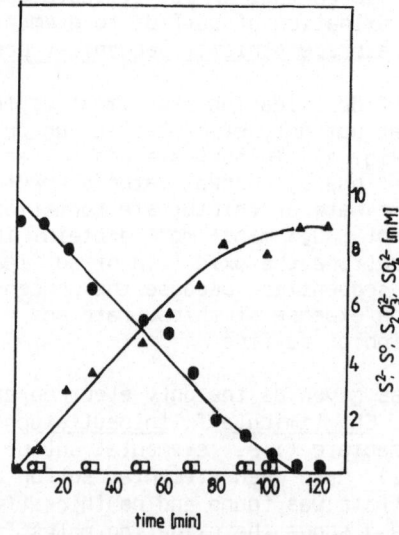

Fig. 3. Photooxidatio of elemental sulfur by growing cells of <u>Cb. limicola</u> f. <u>thiosulfatophilum</u> (strain DSM 249). (●) elemental sulfur, (Δ) sulfate, (□) thiosulfate, (o) sulfide. Strired cell experiment; for details see METHODS. 3 mg protein/ml.

consumed and of possible sulfide, sulfite, sulfate, thiosulfate and polythionates formed in the light in the absence of CO_2.

In partial confirmation of the results of Paschinger et al. (1974) we found:
1) the only readily detectable immediate reaction products are sulfide and thiosulfate but not, however, sulfate and polythionates. Paschinger et al. (1974) had reported sulfide, sulfate and varying amounts of unidentified sulfur compounds. They did not analyse immediately for oxidized sulfur compounds and therefore did not find a satisfactory stoichiometry. We found values for sulfur, sulfide and thiosulfate (Fig. 4) that allow the following stoichiometry to be written:

$$4S° + 3H_2O \quad --) \quad 2H_2S + H_2S_2O_3.$$

Using formaldehyde trapping we found minute amounts of sulfite (e.g., 0.05 μmol SO^{2-} per 1.3 μmol H_2S). Thus it seems likely that the thiosulfate is formed from $S°$ and SO_3^{2-} in an extremely rapid reaction. A comparison of the stoichiometry equation above with van Niel's (1931) equation for sulfur dependent bacterial photosynthesis,

$$2S° +3CO_2 + 5H_2O \quad ---) \quad 2H_2SO_4 + 3 CH_2O,$$

shows that quantitatively the disproportionation of sulfur may represent a replacement of CO_2 by $S°$ as electron acceptor.
2) The reaction occurs only in the light. We found that light saturation for the process is the same as that for CO_2 fixation (Lippert, 1967), namely about 600 Lux.

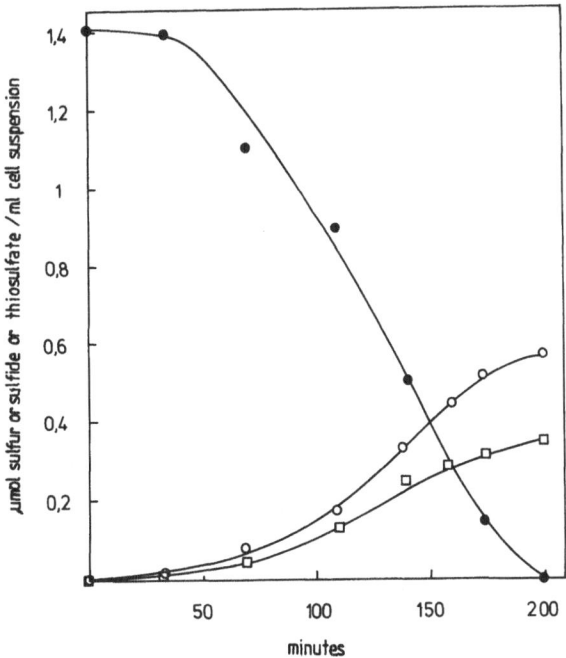

Fig. 4. Disproportionation of elemental sulfur by <u>Cb. limicola</u> f. <u>thiosulfatophilum</u> (strain DSM 249). Optical density of cell suspension at 666 nm: 1.6. (●) elemental sulfur, (o) sulfide blown out, (□) thiosulfate.

Table 1. Oxidation Rates for Reduced Sulfur Compounds by Whole Cells of Cb. limicola f. thiosulfatophilum (strain DSM 249). The values were calculated from the experiments shown in Figs. 1-3.

Sulfur Compound	Oxidation Rate (nmol/min · mg protein)
Sulfide	98
Elemental sulfur	28
Thiosulfate	43

3) The reaction occurs only in the absence of CO_2. Upon addition of bicarbonate, H_2S production stops immediately.
4) The reaction occurs only as long as the H_2S formed is removed by bubbling with an inert gas (N_2 or argon) (Fig. 5).
5) The reaction occurs only in thiosulfate-utilizing subspecies of Chlorobium. Table 2 shows that non-thiosulfate-utilizing members of the Chlorobiaceae as well as members of the purple sulfur bacteria are not capable of sulfur disproportionation.

Enzymes Involved in Thiosulfate Metabolism

Several enzymes considered to be involved in thiosulfate metabolism have been described in the literature:
1) Thiosulfate sulfur transferase (rhodanese) was found to occur in crude extracts of Cb. limicola f. thiosulfatophilum (Yoch and Lindstrom, 1971). Steinmetz and Fischer (1985) have purified two thiosulfate sulfur transferases from Cb. vibrioforme f. thiosulfatophilum, that differed from each other in molecular weights, isoelectric points and K_m-values for thiosulfate and cyanide. We did not find measurable activities in Cb. limicola.
2) Thiosulfate reductase, i.e., an enzymatic activity which reductively splits thiosulfate to sulfide and sulfite, has been discussed for phototrophic bacteria (Hashwa, 1972).

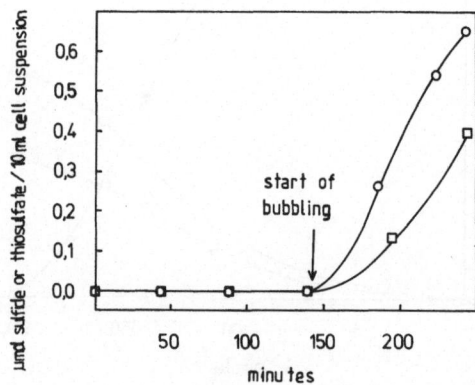

Fig. 5. Dependence of sulfur disproportionation by Cb. limicola f. thiosulfatophilum (strain DSM 249) on removal of H_2S. Optical density of cell suspension at 666 nm: 0.9. (o) sulfide, (□) thiosulfate.

Table 2. Occurrence of Sulfur Disporportionation in Representative Species
of Phototrophic Sulfur Bacteria

Species	Strain	Utilizes Thiosulf.	Sulfur Dispro-portionation
Cb. limicola	DSM 245	-	-
Cb. limicola f. thiosulfatophilum	DSM 249	+	+
	DSM 257	+	+
Cb. vibrioforme f. thiosulfatophilum	DSM 255	+	+
	DSM 263	+	+
Cb. phaeobacteroides	DSM 268	-	-
Pc. aestuarii	DSM 272	-	-
Chromatium vinosum	D=DSM 180	+	-
Ectothiorhodospira mobilis	DSM 237	+	-

3) Thiosulfate: acceptor oxidoreductase, the enzyme combining two molecules
of thiosulfate to tetrathionate with a yield of two electrons, has been
shown to occur in both of the thiosulfate-utilizing Chlorobium subspecies
(Mathewson et al., 1968; Kusai and Yamanaka, 1973a; Steinmetz and Fischer,
1982). Kusai and Yamanaka (1973b) purified the enzyme and found that it
catalysed the reduction of Chlorobium Cyt c-551. The latter then can
transfer electrons to Chlorobium Cyt c-555, which was found to stimulate
Cyt c-551 reduction.

As tetrathionate has never been shown to occur during utilization of
thiosulfate by whole cells of the 2 subspecies the question arises as to
what extent the 3 thiosulfate-utilizing enzymes listed above are really
involved in the in vivo process. We have done studies with Cb. vibrioforme
f. thiosulfatophilum with this in mind and found a distribution of these
enzymes as given in Table 3. The results clearly show that thiosulfate
sulfur transferase is by far the predominant enzyme. The two other rather
low activities, particularly that of thiosulfate reductase, are unlikely to
participate at least in the main pathway of dissimilatory thiosulfate
oxidation.

Thus the splitting of thiosulfate is undoubtedly the primary step in
this pathway. The distribution of the 2 enzymes is not significantly
influenced by the sulfur source in the growth medium (Table 4).

Table 3. Thiosulfate-Transforming Enzymes in Cell-Free Extracts
of Cb. vibrioforme f. thiosulfatophilum (strain DSM
263). Cells were grown in a medium containing sulfide
and thiosulfate.

Enzyme	Specific Activity (nmol $Na_2S_2O_3$/ min mg protein)	Activity (%)
Thiosulfate sulfur transferase	41	95
Thiosulfate acceptor oxidoreductase	2	4.6
Thiosulfate reductase	0.17	0.4

Table 4. Influence of Sulfur Source during Growth upon Activities of
Thiosulfate-Transforming Enzymes in Cb. vibrioforme f.
thiosulfatophilum (strain DSM 263). Specific activities given in
nmol $Na_2S_2O_3$/min mg protein

Sulfur Source in Growth Medium	Thiosulfate Sulfur Transferase		Thiosulfate: Acc. Oxidoreductase	
	Specific Activity	Percentage	Spec. Activity	Percentage
$Na_2S_2O_3$ + Na_2S	41	94	2.7	6
$Na_2S_2O_3$	65	95	3.4	5
Na_2S	34	93	2.7	7

The oxidation of the sulfane moiety to sulfite is assumed to be
catalysed by a reverse sulfite reductase. We measured such an activity in
cell-free extracts of Cb. limicola f. thiosulfatophilum (strain DSM 279)
with 10 nmol H_2/min mg protein. We have not been able, however, to detect a
siroheme as the typical prosthetic group of that enzyme, when we followed
the procedures that had proven successful for Thiobacillus denitrificans and
Chromatium vinosum (Schedel and Trüper, 1979; Schedel et al., 1979).

DISCUSSION

The results presented above do not support a dominating role of
thiosulfate: acceptor oxidoreductase in the photometabolism of
thiosulfate-utilizing chlorobia. As a consequence, the participation of Cyt
c-551 in thiosulfate oxidation as postulated by Kusai and Yamanaka (1973a,
b) becomes unlikely. Therefore the search for the in vivo role of this
cytochrome must continue.

The dominating role in thiosulfate utilization is played by thiosulfate
sulfur transferase (rhodanese). This enzyme is usually tested by an
unphysiological reaction, i.e., transfer of the sulfane moiety from
thiosulfate to cyanide, forming sulfite and rhodanide. Westley (1973)
summarized the reactive capacity of thiosulfate sulfur transferase as
follows: The enzyme is a rather versatile sulfane sulfur transferase;
sulfane donors may be thiosulfate, polysulfides (H-S_n-H), organic
trisulfides (R-S-S-S-R), organic thiosulfonates (R-SO_2-S^-) and organic
persulfides (R-S-S^-); sulfane acceptors may be cyanide, sulfite, organic
sulfinates (R-SO_2^-), thiols (R-SH), dithiols (e.g., lipoate), borohydride and
hydrosulfite (dithionite), of which only the reaction with cyanide is
clearly irreversible. Thiosulfate (acceptor oxido-)reductase activity is
seen as another property of thiosulfate sulfur transferase in the presence
of dithiols (Westley, 1973).

The further oxidation of sulfite to sulfate in Chlorobiaceae is
catalyzed by the well established adenylylsulfate reductase plus ATP
sulfurylase (or ADP sulfurylase) pathway (Kirchhoff and Trüper, 1974; Khanna
and Nicholas, 1983; Bias and Trüper, 1987). The strictly sequential
occurrence of sulfide oxidation and sulfate production in batch cultures
points toward a strong inhibitory action of sulfide upon further oxidation
of sulfite. Sulfite oxidation to sulfate is also blocked under
disproportionating conditions, most probably owing to high sulfide levels in
the cell suspension. In spite of the removal of H_2S by gassing, the sulfide
concentration in the cellular periplasm must be high as a result of
elemental sulfur reduction.

The presence of high activities of thiosulfate sulfur transferase (Table 3), actually on the same level as thiosulfate utilization measured in whole cells (Table 1) allows the fast removal of sulfite formed by reverse sulfite reductase during growth. Under conditions of disproportionation the enzyme transfers sulfane sulfur from polysulfides (formed during sulfide oxidation) to sulfite yielding thiosulfate. Polysulfides give rise to hydrophilic elemental sulfur (S°).

When sulfide is exhausted in batch culture (Fig. 1), further oxidation of sulfite becomes possible and thiosulfate is split into sulfite and polysulfides again. Sulfite is continuously removed by oxidation to adenylylsulfate. Polysulfides are oxidized to sulfite by reverse sulfite reductase. As a measurable consequence sulfate appears in the medium.

An interesting difference between the two thiosulfate-utilizing subspecies of Cb. limicola and Cb. vibrioforme is, that (in batch culture) the latter forms elemental sulfur during thiosulfate oxidation (J. G. Ormerod, personal communication; Steinmetz and Fischer, 1982), whereas the former does not (Fig. 2). An explanation for this difference could be that the two subspecies differ in their in vivo activity of reverse sulfite reductase, as compared with the other enzymes involved in the pathway.

There exists yet another possibility to explain thiosulfate formation from sulfide in the light (Fig. 1). Kusai and Yamanaka (1973b, c) have reported that flavocytochrome c-553 of Cb. limicola f. thiosulfatophilum functions as a sulfide: Cyt c reductase. Fischer (1977) found that this flavocytochrome is heat labile and is, in vitro, rapidly reduced by sulfide, which is oxidized to thiosulfate. It is doubtful, however, that thiosulfate and not hydrophilic elemental sulfur is formed in this system (D. B. Knaff, personal communication). Furthermore, flavocytochrome c occurs not only in the two subspecies "thiosulfatophilum" (Kusai and Yamanaka 1973b, c; Steinmetz and Fischer, 1982), but also in non-thiosulfate utilizing species (Steinmetz and Fischer, 1981). In the thiosulfate-utilizing forms it is a more acidic protein than in the others. As most cytochromes can be reduced by sulfide the role of flavocytochromes c in thiosulfate formation from sulfide needs more detailed investigation.

Sulfur disproportionation may be explained using the following simplified electron flow scheme:

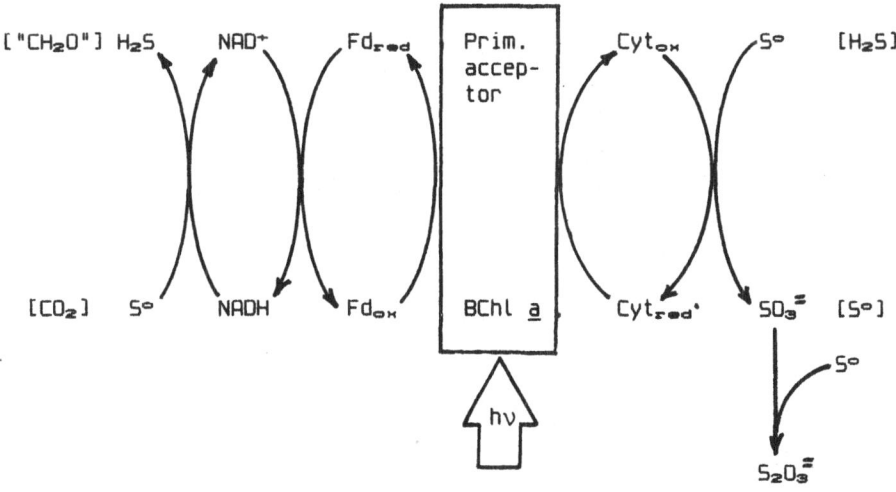

In the light without CO_2 the whole system is reduced. When CO_2 is added, the system runs, oxidizing H_2S and producing elemental sulfur (both in brackets) at the electron input side (right) and reducing CO_2 to carbohydrate ("CH_2O") at the output side (left), i.e., under growth conditions. If, in the absence of CO_2, H_2S is continuously removed, the system runs as long as hydrophilic sulfur (S°) is available and a strong thiosulfate sulfur transferase keeps removing the sulfite as well.

ACKNOWLEDGEMENT

This work was sponsored by grants of the Deutsche Forschungsgemeinschaft.

REFERENCES

Bartlett, P. D., and Skoog, D. A., 1954, Colorimetric determination of elemental sulfur in hydrocarbons, Analyt. Chem., 26:1008.

Bias, U., and Trüper, H. G., 1987, Species specific release of sulfate from adenylyl sulfate by ATP sulfurylase or ADP sulfurylase in the green sulfur bacteria Chloborium limicola and Chlorobium vibrioforme, Arch. Microbiol., 147:406.

Dodgson, J. S., 1961, Determination of inorganic sulphate in studies on the enzymic and non-enzymic hydrolysis of carbohydrate and other sulphate esters, Biochem. J., 78:312.

Fischer, U., 1977, Die Rolle von Cytochromen in Schwefelstoffwechsel phototropher Schwefelbakterien, Doctoral thesis, Univ. Bonn, F.R.G.

Grant, W. M., 1947, Colorimetric determination of sulfur dioxide, Ind. Engg. Chem. Anal. Edit., 19:345.

Hashwa, F., 1972, Die enzymatische Thiosulfatspaltung bei phototrophen Bakterien, Doctoral thesis, Univ. Göttingen, F.R.G.

Janek, A., 1933, Ein neues Verfahren zur Herstellung von Schwefelsolen, Kolloid-Zeitschrift, 64:31.

Kelly, D. P., Chambers, L. A., and Trudinger, P. A., 1969, Cyanolysis and spectrophotometric estimation of trithionate in a mixture with thiosulfate and tetrathionate, Analyt. Chem., 41:898.

Khanna, S., and Nicholas, D. J. D., 1983, Substrate phosphorylation in Chlorobium vibrioforme f. sp. thiosulfatophilum, J. Gen. Microbiol., 129:1365.

Kirchhoff, J., and Trüper, H. G., 1974, Adenylylsulfate reductase of Chlorobium limicola, Arch. Microbiol., 100:115.

Kusai, A., and Yamanaka, T., 1973a, A novel function of cytochrome c (555, Chlorobium thiosulfatophilum) in oxidation of thiosulfate, Biochem. Biophys. Res. Comm., 51:107.

Kusai, A., and Yamanaka, T., 1973b, The oxidation mechanism of thiosulphate and sulphide in Chlorobium thiosulfatophilum. Roles of cytochrome c-551 and cytochrome c-553, Biochim. Biophys. Acta, 325:304.

Kusai, A., and Yamanaka, T., 1973c, Cytochrome c (553, Chlorobium thiosulfatophilum) is a sulfide-cytochrome c reductase, FEBS Lett., 34:235.

Larsen, H., 1952, On the culture and general physiology of the green sulfur bacteria, J. Bacteriol., 64:187.

Larsen, H., 1953, On the microbiology and biochemistry of the photosynthetic green bacteria, Kgl. Norske Vidensk. Selsk. Skr., N1:1.

Lippert, K. D., 1967, Die Verwertung von molekularem Wasserstoff durch Chlorobium thiosulfatophilum. Doctoral thesis, Univ. Göttingen, F.R.G.

Lowry, O. H., Rosebrough, H. J., Farr, A. L., and Randall, R. J., 1951, Protein measurement with the folin phenol reagent, J. Biol. Chem., 193:265.

Lyric, M. R., and Suzuki, I., 1970, Enzymes involved in the metabolism of thiosulfate by Thiobacillus thioparus. III. Properties of thiosulfate oxidizing enzyme and proposed pathway of thiosulfate oxidation, Canad. J. Biochem., 48:355.

Mathewson, J. H., Burger, L. J., and Millstone, H. G., 1968, Cytochrome c-551: thiosulfate oxido-reductase from Chlorobium thiosulfatophilum, Fed. Proc., 27:774.

Nadson, 1906, The morphology of inferior algae. III. Chlorobium limicola Nads., the green chlorophyll bearing microbe (Russian), Bull. Jard. Bot. St. Peterburg, 6:190.

Pachmayr, F., 1960, Vorkommen und Bestimmung von Schwefelverbindungen in Mineralwasser, Doctoral Thesis, Univ. München, F.R.G.

Paschinger, H., Paschinger, J., and Gaffron, H., 1974, Photochemical disproportionation of sulfur into sulfide and sulfate by Chlorobium limicola forma thiosulfatophilum, Arch. Microbiol., 96:341.

Pfennig, N., and Trüper, H. G., 1971, New nomenclatural combinations in the phototrophic sulfur bacteria, Int. J. System. Bacteriol., 21:11.

Pfennig, N., and Trüper, H. G., 1981, Isolation of members of the families Chromatiaceae and Chlorobiaceae, in: "The Prokaryotes," M. P. Starr, H. Stolp, H. G. Trüper, A. Balows, and H. G. Schlegel, eds., Springer-Verlag, Berlin.

Schedel, M., and Trüper, H. G., 1979, Purification of Thiobacillus denitrificans siroheme sulfite reductase and investigation of some molecular and catalytic properties, Biochim. Biophys. Acta, 568:454.

Schedel, M., Vanselow, M., and Trüper, H. G., 1979, Siroheme sulfite reductase isolated from Chromatium vinosum. Purification and investigation of some of its molecular and catalytic properties, Arch. Microbiol., 121:29.

Schmidt, K., Liaaen-Jensen, S., and Schlegel, H. G., 1963, Die Carotinoide der Thiorhodaceae. I. Okenon als Hauptcarotenoid von Chromatium okenii Perty, Arch. Mikrobiol., 46:117.

Shaposhnikov, V. N., Kondratieva, E. N., and Fedorov, V. D., 1958, A contribution to the study of green sulphur bacteria of the Chlorobium genus, Mikrobiologiya (Russian), 27:529.

Smith, A. J., and Lascelles, J., 1966, Thiosulfate metabolism and rhodanese in Chromatium sp. strain D. J. Gen. Microbiol., 42:357.

Steinmetz, M. A., 1984, Cytochrome und Eisenschwefelproteine in Chlorobiaceae-Arten, Doctoral thesis, Univ. Bonn, F.R.G.

Steinmetz, M. A., and Fischer, U., 1981, Cytochromes of the non-thiosulfate-utilizing green sulfur bacterium Chlorobium limicola, Arch. Microbiol., 130:31.

Steinmetz, M. A., and Fischer, U., 1982, Cytochromes of the green sulfur bacterium Chlorobium vibrioforme f. thiosulfatophilum. Purification, characterization and sulfur metabolism, Arch. Microbiol., 131:19.

Steinmetz, M. A., and Fischer, U., 1985, Thiosulfate sulfur transferases (Rhodaneses) of Chlorobium vibrioforme f. thiosulfatophilum, Arch. Microbiol., 142:253.

Stickland, L. H., 1951, The determination of small quantities of bacteria by means of the biuret reaction, J. Gen. Microbiol., 5:698.

Trüper, H. G., 1964, CO_2-Fixierung und Intermediärstoffwechsel bei Chromatium okenii Perty, Arch. Mikrobiol., 49:23.

Trüper, H. G., and Imhoff, J., 1981, The genus Ectothiorhodospira, in: "The Prokaryotes," M. P. Starr, H. Stolp, H. G. Trüper, A. Balows, and H. G. Schlegel, eds., Springer-Verlag, Berlin.

Trüper, H. G., and Schlegel, H. G., 1964, Sulphur metabolism in Thiorhodaceae. I. Quantitative measurements on growing cells of Chromatium okenii, Antonie van Leeuwenhoek J. Microbiol. Serol., 30:225.

Urban, P. J., 1961, Colorimetry of sulphur anions. I. An improved colorimetric method for the determination of thiosulphate, Z. Analyt. Chem., 179:415.

van Niel, C. B., 1931, On the morphology and physiology of the purple and
    green sulphur bacteria, <u>Arch. Mikrobiol</u>., 3:1.
Westley, J., 1973, Rhodanese, <u>in</u>: "Advances in Enzymology, Vol. 39,"
    A. Meister, ed., John Wiley, New York.
Yoch, D. C., and Lindstrom, E. S., 1971, Survey of the photosynthetic
    bacteria for rhodanese (thiosulfate: cyanide sulfur transferase)
    activity, <u>J. Bacteriol</u>., 106:700.

PHYLOGENETIC, EVOLUTIONARY, AND TAXONOMIC ASPECTS OF PHOTOTROPHIC EUBACTERIA

E. Stackebrandt[a], M. Embley[b] and J. Weckesser[c]

[a]Institut für Allgemeine Mikrobiologie
 Christian-Albrechts-Universität, D-2300 Kiel, F. R. G.
[b]Department of Paramedical Sciences
 North East London Polytechnic, London E15 4LZ, U.K.
[c]Institut für Biologie II, Mikrobiologie
 Albert-Ludwigs-Universität, D-7800 Freiburg, F. R. G.

EARLY TAXONOMY OF PHOTOTROPHIC BACTERIA

The first microbiologist to investigate the taxonomy of phototrophic bacteria comprehensively was Winogradsky (1888) who used the term "sulfur bacteria" to combine the sulfur phototrophic bacteria and sulfur chemosynthetic bacteria in one systematic unit. This was later elevated by Migula (1900) to an order Thiobacteria, comprising the families Rhodobacteriaceae and Beggiatoaceae. Molisch (1907) combined the purple sulfur bacteria (Thiorhodaceae) and the purple nonsulfur bacteria (Athiorhodaceae) into an order Rhodobacteria, a move which was later criticised by van Niel (1946) who suggested it had as little justification as grouping the colorless and purple sulfur bacteria. It was Orla-Jensen (1909) who for the first time claimed a phylogenetic validity for the ranks he proposed. In his scheme the order Cephalotrichinae embraced the phototrophic bacteria in a family Rhodobacteriaceae. The most widely accepted classification scheme and the one that was later adopted for "Bergey's Manual of Determinative Bacteriology" was established by Buchanan from 1916 onwards. Here the order Thiobacteriales was described for several families of colorless, purple sulfur and purple nonsulfur bacteria. Thiobacillus, on the other hand, was not included in Thiobacteriales but was classified with the Eubacteriales. In his perceptive paper "The classification and natural relationships of bacteria" van Niel (1946) summarized the events leading to the overclassification of these groups of bacteria and discussed the problems concerning the obvious phylogenetic relationship of the purple bacteria to many colorless sulfur bacteria. Van Niel was aware of the failure of special metabolic properties (i.e., sulfur metabolism), cell morphology or the occurrence of unusual pigment systems to reflect natural relationships. For him there were a number of possible phenotypic characters which could be used to group these bacteria into higher taxa (Fig. 1), each as justified and as arbitrary as the combinations adopted by Buchanan. Van Niel recommended that microbiologists give up all attempts to treat morphologically and/or physiologically similar organisms as "taxonomic entities", or as families or orders in a phylogenetic sense and to recognize that phenotypic characters could only be used for identification. Despite van Niel's suggestions, taxonomists continued to use the Greek-Latin nomenclature to describe higher taxa for the phototrophic bacteria as seen in the 7th and 8th editions of "Bergey's Manual". The 8th edition classified all phototrophic eubacteria into a single order, Rhodospirillales (Pfennig and Trüper, 1974), even though phylogenetic data were available which clearly demonstrated its heterogeneity (Gibson et al., 1979; Fox et al., 1980). Volume 1 of "Bergey's Manual of Systematic Bacteriology" (Murray, 1984) also presents a hierarchic structure of higher ranks which is inconsistent with the phylogenetic data. Thus, for example, the phenotypically defined class Anoxyphotobacteria is invalid in phylogenetic terms, as it embraces more non-phototrophic than phototrophic lineages. Murray was aware of this and he discussed the possibility that the systematics of these organisms might require revision as the phylogenetic data accumulated.

The following discussion summarizes the phylogenetic data which are then used 1. to briefly discuss the major hypotheses concerning certain aspects about the evolution of phototrophic bacteria and 2. compare the phylogenetic groupings with the distribution of a valuable chemotaxonomic marker, lipopolysaccharide. This paper does not discuss the evolution of sulfur metabolism or of the reaction center of the photosynthetic apparatus. These topics have been examined in depth by Trüper (1982) and Olson and Pierson (1987), respectively.

THE PHYLOGENY OF PHOTOSYNTHETIC BACTERIA: A SHORT OVERVIEW

Sequence analyses of conserved and ubiquitously distributed macromolecules have brought microbiologists into a position to objectively test a number of hypotheses regarding the evolution and systematics of phototrophic bacteria. This work started 15 years ago by analysis of Cyt $\underline{c}$ and was successfully extended by investigations on ribosomal ribonucleic acids. At present the phylogenetic positions of almost all of the described species of phototrophic bacteria have been determined and we have a fairly comprehensive picture of the evolution of their several lineages.

During the last 10 years microbiologists have been accustomed to view bacterial phylogenies in terms of dendrograms derived from 16S rRNA catalogs (Fox et al., 1980; Stackebrandt and Woese, 1981; Fox and Stackebrandt, 1987). One of the major findings of this work was that bacteria can be divided into two kingdoms, the archaebacteria and the eubacteria (Woese and Fox, 1977). As yet, phototrophic organisms have been found only in the eubacterial kingdom. Although 16S rRNA oligonucleotide cataloging is a very powerful method for determining the relationship of moderately related

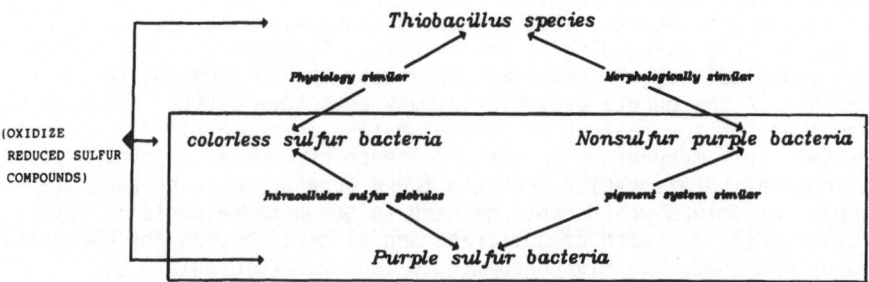

Fig. 1. Morphological and biochemical interrelationships of colorless sulfur, purple sulfur and purple nonsulfur bacteria. This scheme was devised by van Niel (1946) to illustrate the possible ways in which these organisms could be organised. Taxa enclosed in the box represent the order Thiobacteriales (Buchanan, 1916).

divisions, it fails to unravel the order of the most ancient lineages. Thus, members of the major eubacterial divisions share the same low binary similarity values, suggesting that they evolved at approximately the same time during evolution. That this is not the case has been demonstrated by comparison of complete or nearly complete 16S rRNA sequences (Woese, 1987) (Fig. 2). The number and to a certain extent the order of the divisions remain the same as those derived from cataloging. However, as determined by full sequences two eubacterial divisions evolved significantly earlier than the rest. The most ancient division yet detected contains <u>Thermotoga maritima</u>, followed by the divison of the green nonsulfur bacteria. All other divisions, i.e., <u>Deinococcus</u> and relatives, the spirochaetes, the green sulfur bacteria, the planctomycetes, chlamydiae, Gram-positives, the purple bacteria and their non-phototrophic relatives, the <u>Bacteroides-Cytophaga</u> cluster and the cyanobacteria, as well as the single species lineage of <u>Verrucomicrobium</u> (not included in Fig. 2) appear to stem from the same small region of the tree. It is now clear that 5 of the 11 main eubacterial lines of descent harbor phototrophic bacteria. The purple sulfur and nonsulfur bacteria, the cyanobacteria, <u>Heliobacterium chlorum</u> (a representative of the Gram-positive bacteria) and the green sulfur bacteria originate from the one region of the tree, while the green nonsulfur bacteria represent a much more ancient phenotype (Fig. 2).

THE INTRADIVISIONAL STRUCTURE

<u>Chloroflexus and Relatives</u>

The representative of the green nonsulfur bacteria, the filamentous and

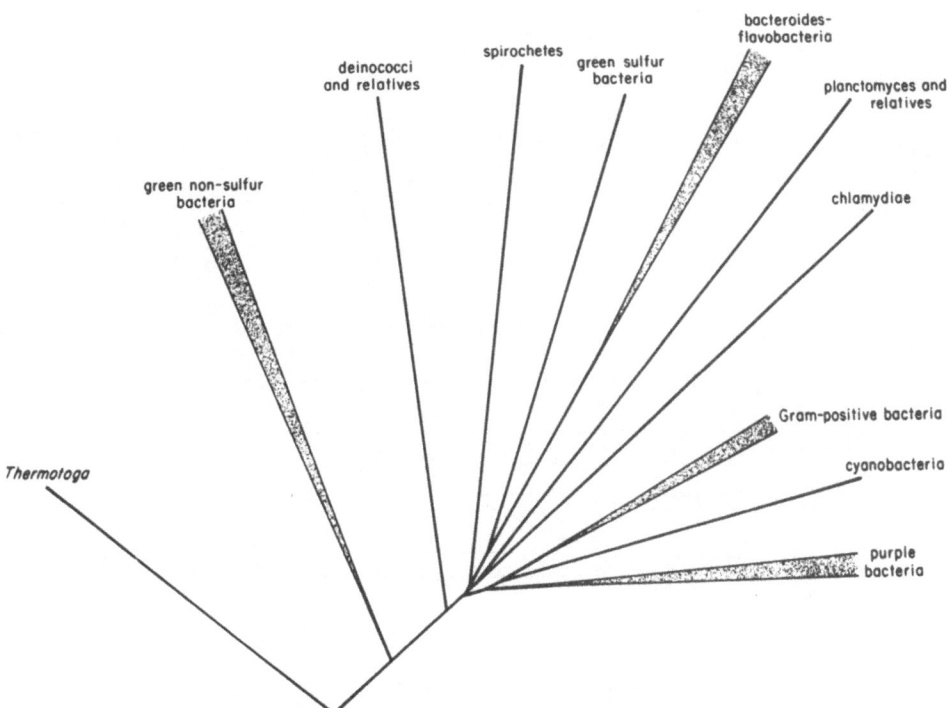

Fig. 2. Phylogenetic relationships of the major lineages (divisions) of the eubacterial kingdom, based upon full 16S rRNA sequences. (From Woese (1987) with permission.)

gliding _Chloroflexus aurantiacus_, shows a moderate relationship to _Herpetosiphon aurantiacum_. Even more distantly related is _Thermomicrobium roseum_ with which _Cf. aurantiacus_ shares unusual long-chain diols (Pond et al., 1986). This is the only common phenotypic character as yet detected for these organisms. Taxa such as _Heliothrix_, _Oscillochloris_ and _Chloronema_, which are believed to be related to _Chloroflexus_, have not been included in phylogenetic studies.

Chlorobium and Relatives

Chloroflexus and the green sulfur bacteria, i.e., _Chlorobium_ and _Chloroherpeton_, possess chlorosomes and resemble each other in light harvesting pigments. They differ significantly in the structure of their photoreaction centers (Feick and Fuller, 1982; Olson and Pierson, 1987) and in the chemistry of the cell wall (Weckesser and Mayer, 1987; see below). Ribosomal RNA sequence analysis shows these organisms to be only distantly related. As determined by 5S rRNA sequences _Prosthecochloris_ is also a member of this division (quoted in Fox and Stackebrandt, 1987).

The Cyanobacteria

The cyanobacteria, _Prochloron_ (Seewaldt and Stackebrandt, 1982), _Prochlorothrix_ (Giovannoni, personal communication) and all of the chloroplasts from algae and higher plants investigated, form a coherent group which as yet contains no non-phototrophic representative. While the cataloging approach showed the chloroplasts as very deep branching organelles (Stackebrandt, 1983) (Fig. 3), comparison of longer 16S rRNA fragments tend to cluster cyanobacteria and chloroplasts closer together.

The Purple Bacteria

The name of this broad division is misleading for the highly diverse phenotypes that are represented (Stackebrandt and Woese, 1984; Woese et al.,

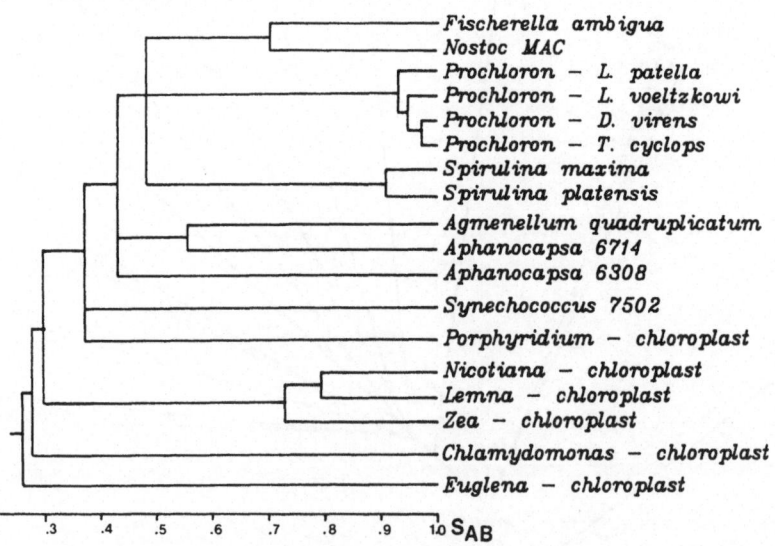

Fig. 3. Relationships of the cyanobacteria and chloroplasts, derived from comparison of 16S rRNA catalogs. The deep branching points of the individual chloroplast rRNAs are due to tachytelic evolution (Bonen et al., 1979; Stackebrandt, 1983).

1984a,b; 1985a; Fowler et al., 1986; Woese, 1987). Most of the taxa traditionally associated with Gram-negative bacteria are members of one of its four subdivisions, abbreviated alpha to delta. Members of the beta and gamma subdivisions appear to share a common ancestry (Lane et al., 1985; Woese, 1987).

The photosynthetic phenotype is by no means predominant within this division. In fact, a photosynthetic member has yet to be detected in the delta subgroup. The most interesting result, which supports earlier findings on Cyt c structure (Dickerson, 1980), is the incoherency of the traditionally defined family Rhodospirillaceae. With the exception of the purple sulfur bacteria (Fig. 6), the genera do not form individual sublines of descent devoid of non-phototrophic bacteria but are found to be highly intermixed with various other phenotypes. As a consequence the traditional genera Rhodopseudomonas and Rhodospirillum have been dissected and additional new genera described (Rhodopila, Rhodobacter, Rhodocyclus) (Imhoff et al., 1984). Some of the relationships of the purple nonsulfur and sulfur bacteria to non-phototrophic relatives of different phenotypes are shown in Figs. 4-6. A more comprehensive picture is provided in recent papers by Woese and colleagues (1984a, b; 1985a). Non-phototrophic bacteria such as Erythrobacter longus and Methylomonas extorquens, which nevertheless contain BChl a, are members of the alpha subdivisions but do not share a close relationship with phototrophic organisms.

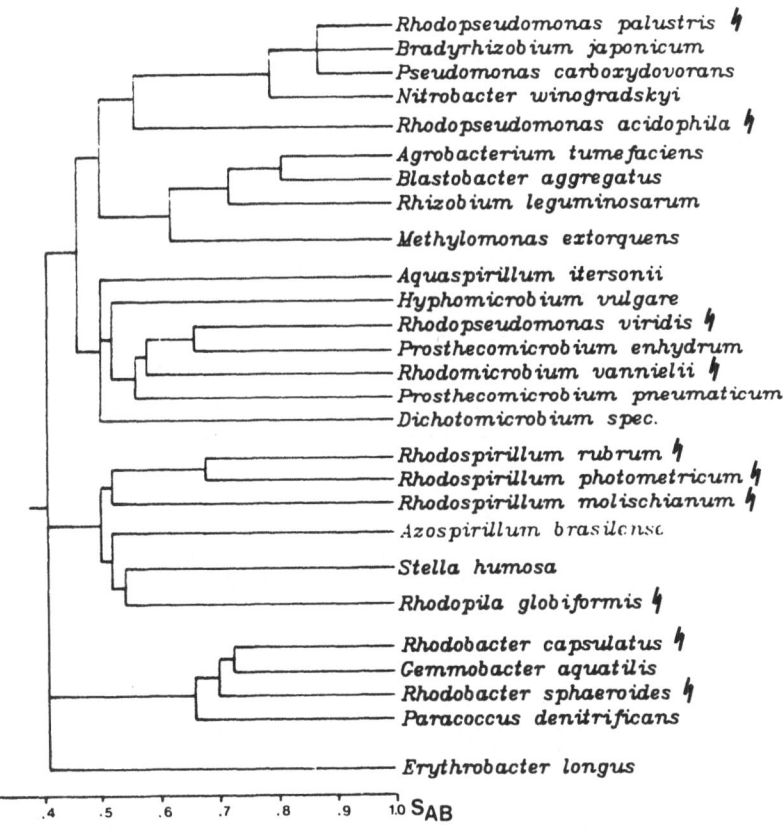

Fig. 4. Relationships of members of the alpha subdivision of purple bacteria and related taxa, derived from comparison of 16S rRNA catalogs (Woese et al., 1984a; Rothe et al., 1987; Stackebrandt, unpublished data). ↯ = phototrophic organisms.

Supporting evidence for the highly complex structure of this division comes from sequence analyses of 5S rRNA (Lane et al., 1985). These data demonstrate the phylogenetic incoherency of thiobacilli which span the full breadth of the purple photosynthetic bacteria. The arguments advanced by van Niel (1946) (see also INTRODUCTION) that sulfur metabolism is not an "indication of relationship" valid for the base description of higher taxa are thus vindicated.

Heliobacterium chlorum and Heliobacillus mobilis

The membership of Heliobacterium chlorum in the division of Gram-positives (Woese et al., 1985b) was completely unexpected as this organism lacks the typical, multilayered Gram-positive cell wall. On the other hand, it has been previously demonstrated (Stackebrandt et al., 1985) that other taxa without a rigid cell wall, i.e., Sporomusa, Megasphaera, Butyrivibrio and Selenomonas, also belong to the phylogenetically defined Gram-positives. H. chlorum differs significantly from other phototrophic bacteria in that it contains BChl g and has a low carotenoid content and a photosynthetic reaction center containing both a Fe-S and a quinone-like molecule as secondary acceptors (Olson and Pierson, 1987). Its exact phylogenetic position is still a moot point. As deduced from 16S rRNA cataloging with more than 150 reference organisms, H. chlorum branches within the Bacillus-Lactobacillus-Streptococcus cluster and the various more ancient clostridial lines (Fig. 7). On the basis of full 16S rRNA sequences and a more limited data base, H. chlorum represents an individual line of descent, equivalent in rank to the subdivisions of actinomycetes and clostridia. The recently isolated Heliobacillus mobilis (Gest, 1987) shares an extremely high level of 16S rRNA sequence homology with H. chlorum.

This summary of currently available data on phototrophic bacteria is unlikely to be the final picture. For example, there is an unconfirmed

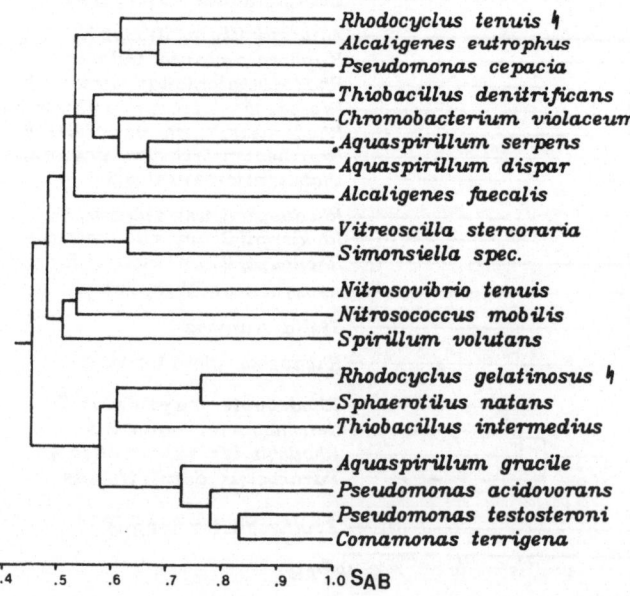

Fig. 5. Relationships of members of the beta subdivision of purple bacteria and related taxa, derived from comparisons of 16S rRNA catalogs (Woese et al., 1984b; Stackebrandt, unpublished data).
⤵ = phototrophic organisms.

report of a <u>Planctomyces</u> strain which contains a novel type of chlorophyll (P. Hirsch, personal communication).

Evolutionary Aspects of Phototrophic Bacteria

Prior to phylogenetic studies employing conserved macromolecules as probes, biochemical traits were used to deduce the course of bacterial evolution. Two major hypotheses have been presented, Margulis's segregation hypothesis (Margulis, 1968) and Broda's conversion hypothesis (Broda, 1978). Both ideas take as their starting point Oparin's assumption (Oparin, 1938) that an anaerobic and heterotrophic, but otherwise undefined prokaryotic type, preceded the evolution of the photosynthetic apparatus. The development of the complex photosynthetic apparatus was believed to be a response to the depletion of nutrients form the primitive oceans by anaerobic and heterotrophic bacteria of the <u>Clostridium</u> type (Oparin, 1938). The view that phototrophic bacteria evolved from a fermenting bacterium having developed an electron transport system was later also expressed by Gest (1983) and Gest and Schopf (1983). The currently available phylogenetic data can neither support nor refute this sequence of events. However, the most ancient eubacterial lineage yet described and the various archaebacterial divisions do not contain organisms which possess a phototrophic system. Despite this Woese (1987) has argued that "given the complexity of the photosynthetic apparatus, it seems unlikely that the process has evolved more than once within the eubacteria and so its origin is deep in the eubacterial tree, possibly at the stage of the common eubacterial ancestor". Even more challenging to the Oparin theory is the recent view that the origin of photosynthesis was probably associated with the origin of life (Woese, 1979; Olson and Pierson, 1987). In other words life started with phototrophs which contained a much simpler photosynthetic apparatus than that found in modern bacteria.

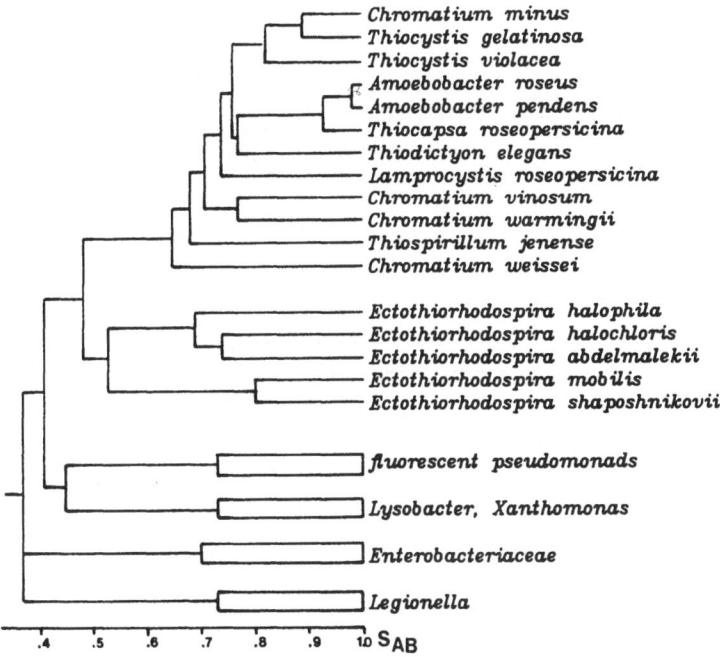

Fig. 6. Relationships of members of the gamma subdivision of purple bacteria and related taxa, derived from comparison of 16S rRNA catalogs (Woese et al., 1985a).

While the exact sequence of events will never be determined, the validity of certain aspects of the segregation and the conversion hypotheses can be examined. According to Margulis all phototrophic bacteria evolved from anaerobic respirers, while in Broda's hypothesis the order of events is reversed. A second major difference concerns the phylogenetic coherency of phototrophic bacteria. The segregation hypothesis places all phototrophic organisms (including the Halobacteria) (Schwemmler, 1979) within a phenotypically homogeneous and distinct subline. In contrast, Broda interprets the same information to give a scheme in which different phototrophic phenotypes give rise to non-phototrophic organisms. The latter hypothesis therefore predicts that phototrophic organisms will be found in many, if not all, major lines of descent.

The phylogenetic branching pattern strongly supports Broda's view although not in all details. As pointed out above, phototrophs do not cluster together to the exclusion of non-phototrophs but are found in several lines of descent. His prediction of a photosynthetic organism within the division of Gram-positives has also been confirmed. In addition, and in contrast to Margulis' view, the available evidence suggests that aerobic bacteria independently evolved from phototrophic ancestors many times in evolution (Figs. 4, 5, 7).

The phylogenetic position of anaerobic respirers, e.g., <u>Desulfuromonas</u> or <u>Desulfovibrio</u> and relatives (Fowler et al., 1986) within the <u>Clostridium</u> subdivision and delta subdivision, respectively, indicate that this bioenergetic process did not evolve before the invention of the photosynthetic apparatus. This apparently supports the conversion hypothesis. If however carbon respiration similar to that present in methanogenic bacteria was already present in the common ancestor of the three major lineages, the respective part of the segregation hypothesis would assume significance.

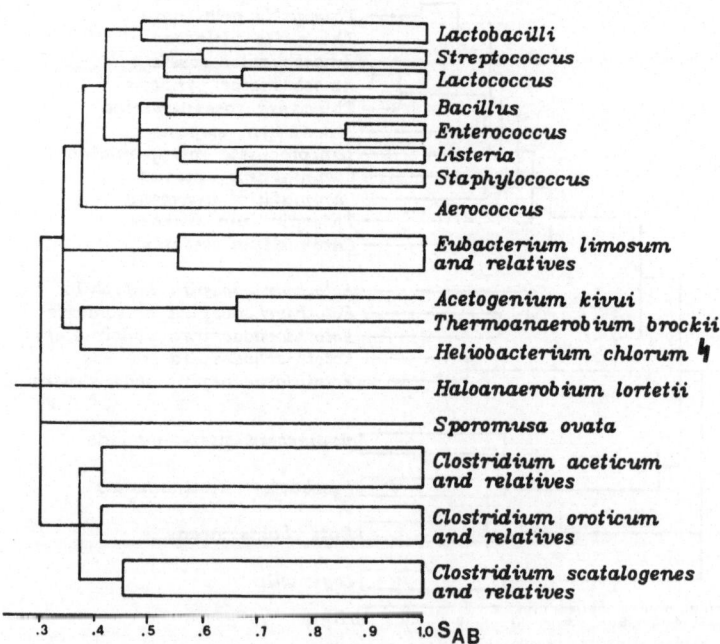

Fig. 7. Phylogenetic position of <u>Heliobacterium chlorum</u> among members of the <u>Clostridium</u> subdivisions of Gram-positive eubacteria, derived from comparison of 16S rRNA catalogs. 🌱 = phototrophic organism.

The use of phylogenetic data to develop a hierarchic system contradicts the traditional classification of the diverse types of phototrophic bacteria into a single order. Interpretation of the data shows that there are at least 7 higher ranks to be described for phototrophic organisms. Possession of a photosynthetic apparatus _per se_ should not be seen as the most important taxonomic criterion upon which to base higher ranks. In this respect photosynthesis carries no more weight than other complex characters such as gliding motility (Reichenbach et al., 1985), spore formation (Stackebrandt et al., 1985) or budding reproduction (Rothe et al., 1987). Instead of constantly searching for unifying epigenetic characters there is a need for features which can be used to recognize the individual status of the new taxa. Although there is only a small amount of data currently available, their distinct phylogenetic positions necessitate the existence of such features. Besides the composition of the photosynthetic apparatus examples include the long chain diols in Chloroflexaceae, the stable variation in chain length of ubiquinones between members of different subdivisions of the purple bacteria (Collins and Jones, 1981; Lane et al., 1985), and the isoprenoid quinone and polyunsaturated fatty acid composition of cyanobacteria.

From a taxonomic point of view, one of the potentially most useful structures of the Gram-negative cell wall is the lipopolysaccharide (LPS) moiety of the outer membrane. This consists of the two conserved domains of the endotoxic lipid A, the core region and the highly variable O-chains (Weckesser et al., 1979; Lüderitz et al., 1982). The LPS composition of several phototrophic organisms and their non-phototrophic relatives have been analyzed. The data not only reflect the phylogenetic grouping of these organisms but can be used for phenotypic circumscription of the phylogenetic groupings in future formal proposals.

Chloroflexaceae and Chlorobiaceae

The disparate phylogeny of the chlorosome-containing green sulfur and green filamentous bacteria is impressively reflected in the composition of their cell walls (Fig. 8). In contrast to meso-diaminopimelic (m-A₂pm), which is the most common diamino-acid found in the peptidoglycan of Gram-negative eubacteria (e.g., _Chlorobium vibrioforme_), _Cf. aurantiacus_ contains L-ornithine (Jürgens et al., 1987). In addition, a large

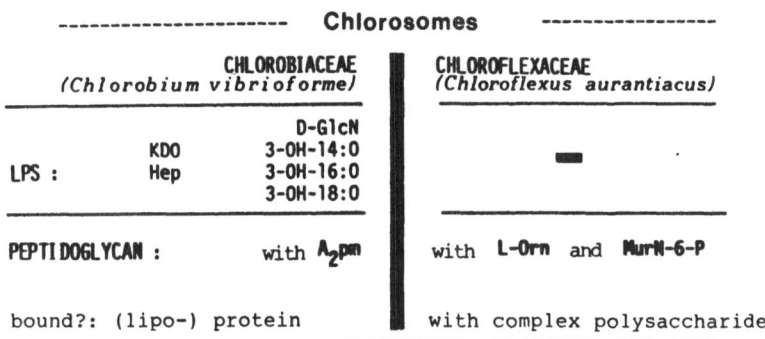

Fig. 8. Chemical composition of the cell wall of Chlorobiaceae and Chloroflexaceae. _Chloroflexus aurantiacus_ possesses a peptidoglycan which is typical for Gram-positive eubacteria. (From Weckesser and Mayer (1987) with permission.)

polysaccharide with a complex composition seems to be bound via MurN-6-P to this peptidoglycan. In this respect Chloroflexus resembles certain spirochetes. A second major difference between members of the two lines of green bacteria is the lack of LPS in Chloroflexus. A distinct outer layer could not be detected for this organism using electron microscopy; in this respect Chloroflexus resembles certain spirochetes. The lipid A of the LPS of Cb. vibrioforme lacks mannose which is commonly found in Chromatium species. The polysaccharide moiety, however, contains the usual core components such as 2-keto-3-deoxyoctonate (KDO) and heptoses (L-glycero-D-manno- and D-glycero-D-glycero-D-mannoheptoses) (Fig. 6).

Purple Bacteria of the Alpha Subdivision

Variations in the structure of the LPS found in the various sublines of purple nonsulfur bacteria and their non-phototrophic relatives of the alpha-subdivision are often in excellent agreement with the phylogenetic structure. Conserved structures of the LPS may be used to describe higher taxa while more variable regions provide data for the circumscription of species and strains.

As depicted in Fig. 4, a subgroup of this subdivision embraces Rhodopseudomonas palustris, Nitrobacter winogradskyi and Rps. viridis as

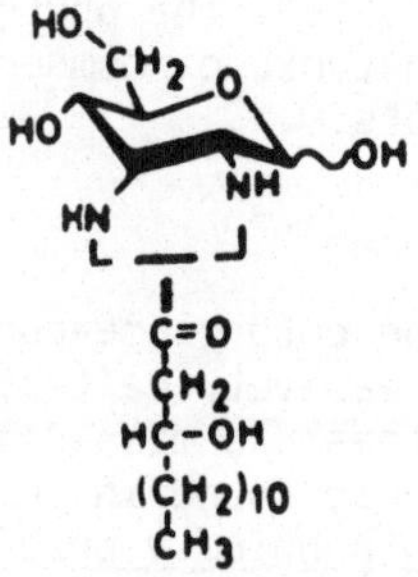

Lipid A$_{DAG}$

(with 2,3-DIAMINO-D-GLUCOSE)

Rhodopseudomonas *viridis*
Rhodopseudomonas sulfoviridis
Rhodopseudomonas palustris

Nitrobacter *winogradskyi*
Nitrobacter hamburgensis

Pseudomonas *diminuta*
Pseudomonas vesicularis

Phenylobacterium *immobile*

Brucella *melitensis*

Thiobacillus *thiooxidans*
Thiobacillus ferrooxidans
Thiobacillus novellus

Fig. 9. Lipid A$_{DAG}$ from Rhodopseudomonas viridis. Lipid A$_{DAG}$ is found predominantly in members of the alpha subdivision of purple bacteria and related taxa.

well as _Pseudomonas diminuta_, _Phenylobacterium immobile_ and related taxa
(Woese et al., 1984a; Stackebrandt and Woese, 1984). The presence of an
unusual lipid A is one of the few uniting phenotypic features which support
the phylogenetic grouping of these species. These lipid As lack glucosamine
but contain instead a 2,3-diamino-2,3-dideoxy-D-glucose (DAG). They all
exhibit an amide-bound 3-OH-14:0 fatty acid. The close relationship between
Rps. palustris and _Nitrobacter_ sp. (Seewaldt et al., 1982) is also expressed
by the presence of L-glycero-D-mannoheptose in the core region, which is
absent in other members of this group.

Fig. 9 lists the species which have already been found to contain the
rare lipid A_{DAG}. It is interesting to note that this feature is not
uniformly distributed in all members of the phylogenetically defined
groupings. This is true for _Rhodomicrobium vannielii_, which clusters with
Rps. viridis, and for _acidophila_, which clusters with _palustris_ and
relatives. Lipid A from the former contains a non-phosphorylated
β-1,6-linked D-glucosamine-disaccharide, carrying a D-mannosepyranosyl
moiety as a polar head group. The lipid A structure of _acidophila_ is less
well characterised. Lipid A_{DAG} has also been found in _Thiobacillus_
thiooxidans and _T. ferrooxidans_ which have been shown to be members of the
beta subdivision by 5S rRNA analysis (Lane et al., 1985). If this finding is
confirmed, lipid A_{DAG} is more widely distributed than originally believed.

A second subgroup of the alpha subdivision harbors _Rhodobacter_
sphaeroides, _Rb. capsulatus_ and a few non-phototrophic relatives which share
an interesting variation of the lipid A structure. Although the _Rb._
sphaeroides structure crossreacts with _Salmonella_ lipid A (they have in
common a phosphorylated β-1,6-linked D-glucosamine-disaccharide as a
backbone), significant differences in structure exist. In addition to
ester-bound 3-OH-10:0, the 3-OH-14:0 amide-bound fatty acid is partially
replaced in _Rb. sphaeroides_ by an unusual keto fatty acid 3-oxo-14:0. The
same variation has been found in other members of this cluster (Fig. 10).

Purple Bacteria of the Beta Subdivision

Despite their phylogenetically separate positions within the beta
subdivision (Fig. 5), _Rhodospirillum tenue_ and _Rps. gelatinosa_ have been
lumped into a single genus, _Rhodocyclus_ (Imhoff et al., 1984). The
phylogenetic position of _Rc. purpureus_, a third species of this genus, has
not been ascertained. Lipophilic behaviour is a common feature of the LPS of
these organisms and this distinguishes them from organisms within the alpha
subdivision. The low degree of of relatedness between _Rc. gelatinosus_ and
Rc. tenuis is also reflected in the chemistry of their LPS. _Rc. gelatinosus_
possesses a lipid A which strongly resembles that of _Salmonella_ and shows
the same high toxicity and pyrogeneity. The difference lies only in the
chain length of the amide-bound fatty acid and in the polar head groups.

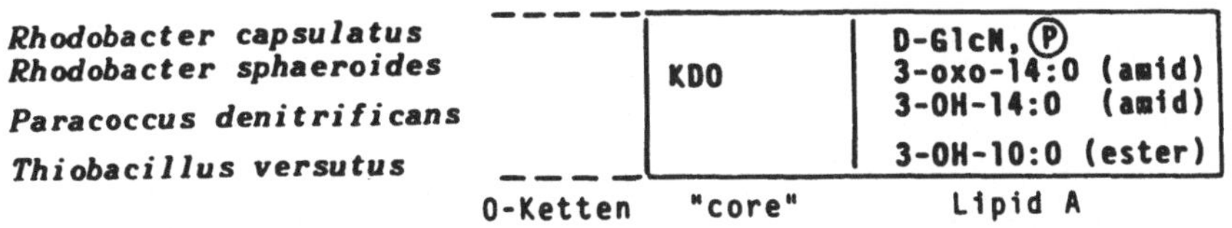

			D-GlcN, (P)
Rhodobacter capsulatus		KDO	3-oxo-14:0 (amid)
Rhodobacter sphaeroides			3-OH-14:0 (amid)
Paracoccus denitrificans			
Thiobacillus versutus			3-OH-10:0 (ester)
	O-Ketten	"core"	Lipid A

Fig. 10. Lipid A with phosphorylated glucosamine-disaccharide and a rare
amide-bound 3-oxo fatty acid (3-oxo-14:0). All these species are
closely related (Weckesser and Mayer, 1987; Lane et al., 1985;
Woese et al., 1984a). (From Weckesser and Mayer (1987) with
permission.)

The LPS from Rc. gelatinosus lacks O-chains. The core region of LPS from this species contains D-glycero-D-mannoheptose and KDO (Fig. 11).

In Rc. tenuis and Rc. purpureus the lipid A shows a structural variation: the β-1,6-linked D-glucosamine-disaccharide is substituted on position C-4 with a third D-glucosamine (Fig. 10). The phosphate moieties at C-1 and C-4' are substituted with D-arabino-furanose and 4-amino-L-arabinose, respectively. The two species are distinguishable by the composition of the polysaccharide moiety in that Rc. purpureus has O-chains very lightly built up from repeated units (Fig. 11).

Purple Bacteria of the Gamma-Subdivision

Only the LPS of four species of Chromatiaceae have been investigated from this subdivision (Weckesser and Mayer, 1987). The close phylogenetic relationship of this family is reflected by this taxonomic probe. Of all organisms so far investigated they are unique in containing D-mannose in their lipid A and D-glycero-D-mannoheptose in the core region. Lipid A from these bacteria is also free of phosphate and contains D-glucosamine as the only amino sugar and 3-OH-14:0 as the only hydroxy fatty acid. The O-chains resemble those of the Enterohacteriaceae (repeating units) to which the Chromatiaceae are more closely related than to the purple nonsulfur bacteria.

CONCLUSIONS

From the very first, the presence of a complex photosynthetic apparatus has led taxonomists to consider phototrophic bacteria as a coherent unit, separate from non-phototrophic prokaryotes. One of the most important findings from phylogenetic studies is that this is not a valid assumption as phototrophic bacteria occur in many branches of the phylogenetic tree. The task is now to circumscribe the new, and often surprising', relationships with readily determined taxonomic characters. Only in this way can be begin

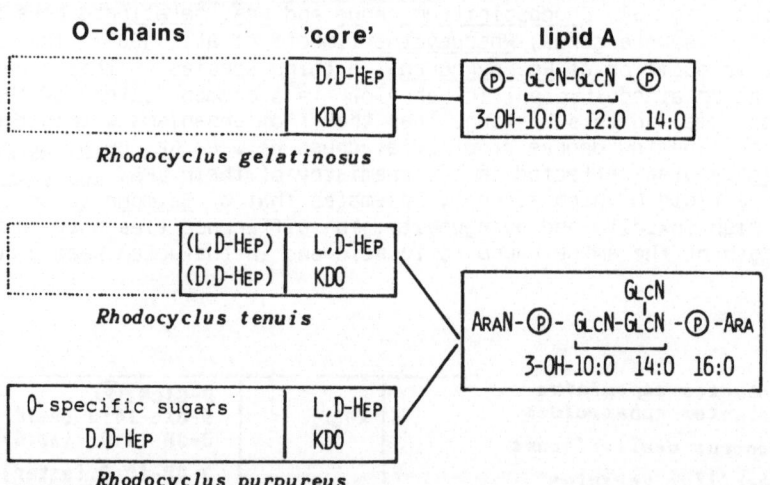

Fig. 11. Closely related species, i.e., Rc. tenuis and Rc. purpureus seem to have identical lipid A and core compositions, differing only in the chemistry of the O-chains. Rc. gelatinosus is only remotely related to these two species which is reflected by significant differences in lipid A, core composition and the lack of O-chains. (From Weckesser and Mayer (1987) with permission.)

to talk about the classification of phototrophic organisms in a meaningful way.

REFERENCES

Bonen, L., Doolittle. W. F., and Fox, G. E., 1979, Cyanobacterial results of 16S ribosomal ribonucleic acid sequence analysis, Can. J. Biochem., 57:879.

Broda, E., 1978, "The Evolution of Bioenergetic Processes", Pergamon Press, Oxford.

Buchanan, R. E., 1916, Studies in the nomenclature and classification of the bacteria I-X, J. Bacteriol., 1:591.

Collins, M. D., and Jones, D., 1981, Distribution of isoprenoid quinone structural types in bacteria and their taxonomic implications, Microbiol. Rev., 45:316.

Dickerson, R. E., 1980, Structural conservatism in proteins over three billion years: cytochrome with a touch of collagen, in: "Biomolecular Structure, Conformation and Evolution, Vol. 1," R. Srinivasan, ed., Pergamon Press, Oxford.

Feick, R. G., and Fuller, R. C., 1984, Topography of the photosynthetic apparatus of Chloroflexus aurantiacus, Biochemistry, 32:3693.

Fowler, V. J., Widdel, F., Pfennig, N., Woese, C. R., and Stackebrandt, E., 1986, Phylogenetic relationships of sulfate- and sulfur-reducing eubacteria, Syst. Appl. Microbiol., 8:32.

Fox, G. E., and Stackebrandt, E., 1987, The application of 16S rRNA cataloguing and 5S rRNA sequencing in bacterial systematics, in: "Methods in Microbiology, Vol. 18," R. Colwell, ed., Academic Press, London (in press).

Fox, G. E, Stackebrandt, E., Hespell, R. B., Gibson, J., Maniloff, J., Dyer, T. A., Wolfe, R. S., Balch, W. E., Taner, R., Magrum, L., Zablen, L. B., Blakemore, R., Gupta, R., Bonen, L., Lewis, B. J., Stahl, D. A., Luehrsen, K. R., Chen, K. N., and Woese, C. R., 1980, The phylogeny of prokaryotes, Science, 209:457.

Gest, H., 1983, Evolutionary roots of anoxygenic photosynthetic energy conversion, in: "The Phototrophic Bacteria," J. G. Ormerod, ed., Blackwell, Oxford.

Gest, H., 1987, Physiological and biochemical characteristics of heliobacteria. Abstract, EMBO Workshop on Green Photosynthetic Bacteria, 19-21.8.87, Nyborg, Denmark.

Gest, H., and Schopf, J. W., 1983, Biochemical evolution of anaerobic energy conversion: the transition from fermentation to anoxygenic photosynthesis, in: "Earth's Earliest Biosphere," J. W. Schopf, ed., Princeton University Press, Princeton, New Jersey.

Gibson, J., Stackebrandt, E., Zablen, L. B., Gupta, R., and Woese, C. R., 1979, A phylogenetic analysis of the purple photosynthetic bacteria, Current Microbiol., 3:59.

Imhoff, J. F., Trüper, H. G., and Pfennig, N., 1984, Rearrangement of the species and genera of the phototrophic "purple nonsulfur bacteria," Int. J. System. Bacteriol., 34:340.

Jürgens, U. J., Meissner, J., Fischer, U., König, W. A., and Weckesser, J., 1987, Ornithine as a constituent of the peptidoglycan of Chloroflexus aurantiacus, diaminopimelic acid in that of Chlorobium vibrioforme f. thiosulfatophilum, Arch. Microbiol., 148:72.

Lane, D. J., Stahl, D. A., Olsen, G. J., Heller, D. J., and Pace, N. R., 1985, Phylogenetic analysis of the genera Thiobacillus and Thiomicrospira by 5S rRNA sequences, J. Bacteriol., 163:75.

Lüderitz, O., Freudenberg, M. A., Galanos, C., Lehmann, V., Rietschel, E. Th., and Shaw, D. H., 1982, Lipopolysaccharides of Gram-negative bacteria, Curr. Top. Membr. Transp., 17:79.

Margulis, L., 1968, Evolutionary criteria in thallophytes: a radical alternative, _Science_, 161:1020.

Migula, W. W., 1900, Spezielle Systematik der Bakterien, G. Fischer, Jena.

Molisch, H., 1907, Die Purpurbakterien nach neuen Untersuchungen, G. Fischer, Jena.

Murray, R. G. E., 1984, The higher taxa, or, a place for everything ...?, _in_: "Bergey's Manual of Systematic Bacteriology," N. R. Krieg and J. G. Holt, eds., Williams and Wilkins, Baltimore.

Olson, J. M., and Pierson, B. K., 1987, Evolution of reaction centers in photosynthetic prokaryotes, _Int. Rev. Cytol._, 108:209.

Oparin, A. I., 1938, The Origin of Life (Trans. by S. Morgulis). The Macmillan Company, New York.

Orla-Jensen, S., 1909, Die Hauptlinien des natürlichen Bakteriensystems, _Zentralbl. Bakteriol., II Abt._, 22.305.

Pfennig, N., and Trüper, H. G., 1974, The phototrophic bacteria, _in_: "Bergey's Manual of Determinative Bacteriology, Vol. 8," R. E. Buchanan and N. E. Gibbons, eds., The Williams and Wilkins Company, Baltimore.

Pond, J. L., Langworthy, T. A., and Holzer, G., 1986, Long-chain diols: a new class of membrane lipids from a thermophilic bacterium, _Science_, 231:1134.

Reichenbach, H., Ludwig, W., and Stackebrandt, E., 1986, Lack of relationship between gliding cyanobacteria and filamentous gliding heterotrophic eubacteria: comparison of 16S rRNA catalogues of _Spirulina_, _Saprospira_, _Vitreoscilla_, _Leucothrix_ and _Herpetosiphon_, _Arch. Microbiol._, 145:391.

Rothe, B., Fischer, A., Hirsch, P., Sittig, M., and Stackebrandt, E., 1987, The phylogenetic position of the budding bacteria _Blastobacter aggregatus_ and _Gemmobacter aquatilis_ gen. nov., spec. nov., _Arch. Microbiol._, 147:92.

Schwemmler, W., 1979, "Mechanismen der Zellevolution," Walter de Gruyter, Berlin.

Seewaldt, E., and Stackebrandt, E., 1982, Partial sequence of 16S ribosomal RNA and the phylogeny of _Prochloron_, _Nature_, 295:618.

Seewaldt, E., Schleifer, K. H., Bock, E., and Stackebrandt, E., 1982, The close phylogenetic relationship of _Nitrobacter_ and _Rhodopseudomonas palustris_, _Arch. Microbiol._, 131:287.

Stackebrandt, E., 1983, A phylogenetic analysis of _Prochloron_, _in_: "Endocytobiology II," H. E. A. Schenk and W. Schwemmler, eds., Walter de Gruyter, Berlin.

Stackebrandt, E., and Woese, C. R., 1981, The evolution of prokaryotes, _in_: "Molecular and Cellular Aspects of Microbial Evolution," M. J. Carlile, R. J. Collins, and B. E. B. Moseley, eds., Cambridge University Press, Cambridge.

Stackebrandt, E., and Woese, C. R., 1984, The phylogeny of prokaryotes, _Microbiol. Sci._, 1:117.

Stackebrandt, E., Pohla, H., Kropenstedt, R., Hippe, H., and Woese, C. R., 1985, 16S rRNA analysis of _Sporomusa_, _Selenomonas_ and _Megasphaera_: on the phylogenetic origin of Gram-positive eubacteria, _Arch. Microbiol._, 143:270.

Trüper, H. G., 1982, Microbial processes in the sulfur cycle through time, _in_: "Mineral Deposits and the Evolution of the Biosphere," H. D. Holland and M. Schidlowski, eds., Dahlem Conferences, Springer, Berlin.

van Niel, C. B., 1946, The classification and natural relationships of bacteria. Cold Spring Harbor Symp. Quant. Biol., 11:285.

Weckesser, J., Drews, G., and Mayer, H., 1979, Lipopolysaccharides of photosynthetic prokaryotes, _Annu. Rev. Microbiol._, 33:215.

Weckesser, J., and Mayer, H., 1987, Lipopolysaccharides of phototrophic bacteria, a contribution to phylogeny and to endotoxin research, _Forum Mikrobiologie_, GIT Verlag, 7/8:242.

Winogradsky, S., 1888, Beiträge zur Morphologie und Physiologie der

Bacterien. Heft 1. Schwefelbacterien, A. Felix, Leipzig.

Woese, C. R., 1979, A proposal concerning the origin of life on the planet earth. _J. Mol. Evolution_, 13:95.

Woese, C. R., 1987, Bacterial evolution, _Microbiol. Rev._, 51:221.

Woese, C. R., and Fox, G. E., 1977, Phylogenetic structure of the prokaryotic domain: the primary kingdoms, _Proc. Natl. Acad. Sci. U.S.A._, 74:5088.

Woese, C. R., Stackebrandt, E., Weisburg, W. G., Paster, B. J., Madigan, M. T., Fowler, V. J., Hahn, C. M., Blanz, P., Gupta, R., Nealson, K. H., and Fox, G. E., 1984a, The phylogeny of the purple bacteria: the alpha subdivision, _Syst. Appl. Microbiol._, 5:315.

Woese, C. R., Weisburg, W. G., Paster, B. J., Hahn, C. M., Tanner, R. S., Krieg, N. R., Koops, H.-P., Harms, H., and Stackebrandt, E., 1984b, The phylogeny of the purple bacteria: the beta subdivision, _Syst. Appl. Microbiol._, 5:327.

Woese, C. R., Weisburg, W. G., Hahn, C. M., Paster, B. J., Zablen, L. B., Lewis, B. J., Macke, T. J., Ludwig, W., and Stackebrandt, E., 1985a, The phylogeny of the purple bacteria: the gamme subdivision, _Syst. Appl. Microbiol._, 6:25.

Woese, C. R., Debrunner-Vossbrinck, B., Oyaizu, H., Stackebrandt, E., and Ludwig, W., 1985b, Gram-positive eubacteria: possible photosynthetic ancestry, _Science_, 229:762.

INFERRING EUBACTERIAL PHYLOGENY FROM 5S RIBOSOMAL RNA STRUCTURE ANALYSIS

H. Van den Eynde, Y. Van de Peer and R. De Wachter

Departement Biochemie, Universiteit Antwerpen (UIA)
Universiteitsplein 1
B-2610 Antwerpen, Belgium

INTRODUCTION

Based on analysis of 16S ribosomal RNA oligonucleotide catalogs of about 400 eubacterial species Woese et al. (1985) made a proposal to divide the eubacterial primary kingdom into 10 major phyla. Two years later with over 500 species characterized and with more than 50 nearly complete sequences, the earlier conclusions were significantly refined and extended (Woese, 1987). The picture obtained is substantially different from the classification system based on morphological and metabolic data in the 8th edition of Bergey's Manual (Buchanan and Gibbons, 1974). Investigation of another universally occurring molecule such as 5S ribosomal RNA ought to help clarify this matter. So far about 160 eubacterial 5S RNA sequences have been published, but the distribution over the 10 major phyla is very uneven as can be seen from Table 1. For example, in the phylum of purple bacteria and relatives almost 80 species have been examined, whereas in the phylum of green sulfur bacteria, along with 4 other major phyla, no representatives whatsoever have been investigated. It is our aim to lessen these gaps, in order to make possible a comparison with the view based on 16S ribosomal RNA. At present, progress has been made for several phyla, while work has started on others (Table 1). In this paper we present a tree derived from the weighted pairwise grouping of over 180 eubacterial 5S RNA sequences including several sequences of formerly undocumented phyla.

METHODS

5S RNA sequences were taken mainly from the latest compilation of Erdmann and Wolters (1986). Additional sequences are from the recent literature and from unpublished results from our laboratory. The sequences were aligned to optimize both primary and secondary structure homology (De Wachter et al., 1982; Erdmann et al., 1985). Dendrograms were constructed by the clustering algorithm described in detail by Huysmans and De Wachter (1986) and by Dams et al. (1987). This algorithm includes the computation of a dissimilarity matrix for all the sequences involved, correction for multiple substitution and back mutation, factorial correction of the dissimilarities for unequal evolutionary rates along different lineages, and clustering by weighted pairwise grouping. The external reference group used for factorial correction comprised cytoplasmic 5S RNAs from 10 red algae.

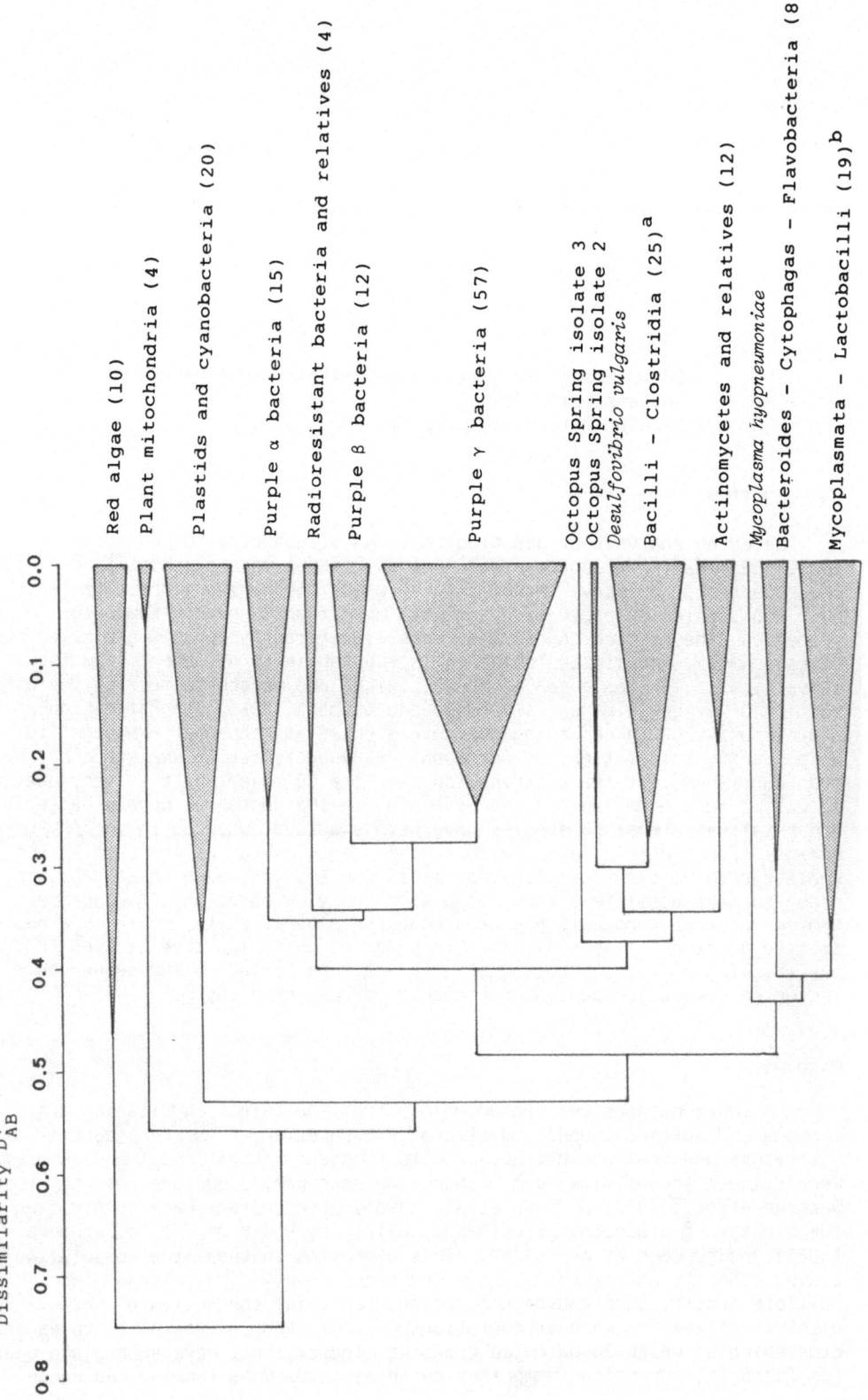

Table 1. Distribution of Published 5S RNA Sequences from 10
Eubacterial Phyla

Phylum[1]	Subdivision (Woese, 1987)	Number of Published 5S RNA Sequences[2]
Purple bacteria	Alpha	15
and relatives	Beta	12
	Gamma[1]	57
	Delta[1]	1
Gram-positive bacteria	High (G+C)	12
	Low (G+C)	41
	Photosynthetic	0
	Gram-negative	0
Cyanobacteria and plastids[1]		18
Radioresistant bacteria	Deinococcus group[1]	0
and relatives	Thermophiles	2
Green non-sulfur bacteria and relatives		1
Green sulfur bacteria		0
Bacteroides, Cytophagas, Flavobacteria and relatives[1]		0
Planctomyces and relatives		0
Chlamydiae		0
Spirochetes and relatives		0

[1]Additional unpublished sequences were available to us for species
belonging to these groups.
[2]The number of published sequences outnumbers the number of
examined species due to sequence heterogeneity reported for
several species. Most of the sequences are compiled in Erdmann and
Wolters (1986). Mitochondrial 5S RNA sequences are omitted in this
table.

Fig. 1. Dendrogram constructed by cluster analysis of 182 5S RNA sequences
from eubacteria. The dendrogram was constructed as described in the
METHODS section from an alignment of 182 eubacterial 5S RNA
sequences available to us at present plus 10 cytoplasmic 5S RNA
sequences from red algae (see text). The latter were used as an
external reference group and allowed a correction for different
evolutionary rates among the eubacterial lineages. D'_{AB} is the
corrected dissimilarity value. Each major cluster is represented as
an isosceles triangle with its top situated at the D'_{AB} value
corresponding to the first divergence within the cluster, and with a
base proportional to the number of species in the cluster. This
number is given in parentheses after the cluster name. The cluster
designated "Actinomycetes and relatives" coincides with the
subdivision "High (G+C)" of Table 1. The clusters "Bacilli -
Clostridia" and "Mycoplasmata - Lactobacilli" contain species with a
low (G+C) content. Clusters that, to some extent, are heterogeneous
are indicated by superscripts [a,b]. The Mycoplasmata - Lactobacilli
cluster includes occasional Clostridia species (Dams et al., 1987).
See text for details about the Bacilli - Clostridia cluster.

RESULTS AND DISCUSSION

Fig. 1 shows the outline of a dendrogram constructed on the basis of 182 eubacterial 5S RNA sequences belonging to 6 different phyla. In addition 10 sequences of cytoplasmic 5S RNAs from red algae were included. The following clusters can be distinguished in our tree: (I) plant mitochondria, (II) plastids and cyanobacteria, (III) purple bacteria with subdivisions alpha, beta and gamma, (IV) radioresistant bacteria and relatives, (V) Bacilli and most Clostridia, (VI) Actinomycetes and relatives, (VII) Bacteroides - Cytophagas - Flavobacteria, (VIII) Mycoplasmata - Lactobacilli Streptococci - Staphylococci. Compared to the tree based on 16S rRNA data (Woese, 1987), the phylum subdivisions and to a lesser extent the major phyla are similar. However, some differences from the 16S RNA picture are apparent, the main one being the polyphyletic nature in the 5S RNA tree of the 16S RNA phylum designated by Woese as "gram-positives and relatives". Thus cluster VIII is more distant from clusters V and VI than from cluster VII. Also, no relationship is seen between plant mitochondria and the alpha subdivision of the purple bacteria, a relationship suggested by Yang et al. (1985) on the basis of 16S RNA data. Concerning the taxonomic position of Myxobacteria, sulfate and sulfur reducers and Bdellovibrios (formerly a separate phylum but now placed with the purple bacteria in subdivision delta (Woese et al., 1985; Woese, 1987)), we do not think it safe to draw any conclusions as yet. The three Myxobacterial 5S RNA sequences included in our tree do not cluster with representatives of the purple bacteria, but rather with the gram-positive Bacilli and Clostridia. Tree topology in this particular area (not shown) is rather unstable, however. <u>Desulfovibrio vulgaris</u>, a member of the subdivision of sulfate and sulfur reducers, clusters with one of two Octopus Spring isolates, which were originally thought to belong to the cluster formed by the radioresistant bacteria and their relatives on the basis of 5S RNA analysis (Stahl et al., 1985). From our tree this is clearly not the case.

The conflicting results which emerge when one compares our tree to others based on 16S RNA or even 5S RNA data do not necessarily have to lead to despair. While 16S RNA data are analyzed by distance matrix analysis and maximum parsimony analysis (Woese, 1987; and references therein), we have analyzed the 5S RNA data by cluster analysis. Furthermore, the 16S RNA tree benefits from signature analysis (Woese, 1987), a way of looking at things that is only gradually gaining acceptance for the analysis of 5S RNA data (e.g., Wolters and Erdmann, 1986). Apart from this, we have to realize that the 5S RNA data-bank is significantly smaller than the one for 16S RNA. As a result 5S RNA tree topology tends to change as more sequences of poorly represented phyla are added. We expect 5S RNA tree topology and the evolutionary picture thus presented to stabilize, as more sequences representing different phyla and phylum subdivisions are added to the tree. This has recently been shown with the alpha, beta and gamma subdivisions of the phylum of purple bacteria and relatives. In all previously published trees from our laboratory (e.g., Huysmans and De Wachter, 1986; Dams et al., 1987), alpha-division purple bacteria were severed from beta-division and gamma-division purple bacteria and were incorporated into the gram-positive cluster. The changes presented in this paper are solely the result of the inclusion of additional sequences representing subdivisions alpha and beta.

ACKNOWLEDGEMENTS

Our research was supported in part by a grant from F.K.F.O.

REFERENCES

Buchanan, R. E., and Gibbons, N. E., eds., 1974, "Bergey's Manual of Determinative Bacteriology," 8th edn., Williams and Wilkins, Baltimore.

Dams, E., Yamada, T., De Baere, R., Huysmans, E., Vandenberghe, A., and De Wachter, R., 1987, Structure of 5S rRNA in Actinomycetes and relatives and evolution of eubacteria, J. Mol. Evol., in press.

De Wachter, R., Chen, M.-W., and Vandenberghe, A., 1982, Conservation of secondary structure in 5S ribosomal RNA: a uniform model for eukaryotic, eubacterial, archaebacterial and organelle sequences is energetically favourable, Biochimie, 64:311.

Erdmann, V. A., Wolters, J., Huysmans, E., and De Wachter, R., 1985, Collection of published 5S, 5.8S and 4.5S ribosomal RNA sequences, Nucl. Acids Res., 13:r105.

Erdmann, V. A., and Wolters, J., 1986, Collection of published 5S, 5.8S and 4.5S ribosomal RNA sequences, Nucl. Acids Res., 14:r1.

Huysmans, E., and De Wachter, R., 1986, The distribution of 5S ribosomal RNA sequences in phenetic hyperspace. Implications for eubacterial, eukaryotic, archaebacterial and early biotic evolution, Endocytob. Cell Res., 3:133.

Stahl, D. A., Lane, D. J., Olsen, G. J., and Pace, N. R., 1985, Characterization of a Yellowstone hot spring microbial community by 5S rRNA sequences, Appl. Environ. Microbiol., 49:1379.

Woese, C. R., Stackebrandt, E., Macke, T. J., and Fox, G. E., 1985, A phylogenetic definition of the major eubacterial phyla, Syst. Appl. Microbiol., 6:143.

Woese, C. R., 1987, Bacterial evolution, Microbiol. Rev., 51:211.

Wolters, J., and Erdmann, V. A., 1986, Cladistic analysis of 5S rRNA and 16S rRNA secondary and primary structure. The evolution of eukaryotes and their relation to archaebacteria, J. Mol. Evol., 24:152.

Yang, D., Oyaizu, Y., Oyaizu, H., Olsen, G. J., and Woese, C. R., 1985, Mitochondrial origins, Proc. Natl. Acad. Sci. U.S.A., 82:4443.

LIPIDS, FATTY ACIDS AND QUINONES IN TAXONOMY AND PHYLOGENY OF ANOXYGENIC PHOTOTROPHIC BACTERIA

J.F. Imhoff

Institut für Mikrobiologie
Meckenheimer Allee 168
D-5300 Bonn, F. R. G.

INTRODUCTION

Taxonomic differentiation of phototrophic purple and green bacteria has been achieved on the basis of a number of morphological properties like form, size, flagellation and internal membrane structures of the cells, pigment composition, DNA base ratio, and also by physiological properties like carbon and nitrogen substrate utilization, ability of aerobic and anaerobic growth in the dark, tolerance of and oxidation products of sulfide and thiosulfate, and others (Pfennig and Trüper, 1974). From these properties anoxygenic phototrophic bacteria have long been recognized as a very heterogeneous group.

Also chemotaxonomic methods have been applied in taxonomy of phototrophic bacteria. The lipid A structure of the lipopolysaccharides shows significant differences among members of the purple nonsulfur bacteria (Weckesser et al., 1974, 1979). These structures have been used to show taxonomic and phylogenetic relatedness of phototrophic bacteria to chemotrophic Gram-negative bacteria (Mayer, 1984). The cellular compositions of polar lipids, fatty acids and quinones appear to be very useful as additional criteria for taxonomic characterization of bacteria, but have been used in the taxonomy of phototrophic bacteria only quite recently (Imhoff et al., 1982, 1984; Imhoff, 1982a, b, 1984a; Hansen and Imhoff, 1985).

Both taxonomy and phylogeny use similarities as criteria. Whereas the former uses all available data, the latter requires data which can be quantified like amino acid and nucleotide sequences of proteins and DNA. Both amino acid sequences of cytochromes (Dickerson, 1980) and rRNA oligonucleotide pattern (e.g., Gibson et al., 1979; Stackebrandt and Woese, 1981) have been used to verify the phylogenetic relationship of anoxygenic phototrophic bacteria. These methods have revealed deep branches among different groups of phototrophic bacteria as well as the close relationship of some phototrophic purple bacteria with certain heterotrophic bacteria. Phototrophic bacteria (including cyanobacteria and Heliobacterium) are found within 5 of 10 recognized divisions among the prokaryotes (Woese et al., 1985; Stackebrandt et al., 1988).

In this contribution taxonomic as well as phylogenetic implications of the composition of lipids, quinones and fatty acids of anoxygenic phototrophic bacteria are discussed.

MATERIALS AND METHODS

     Lipids were extracted with a modification of the method of Bligh and
Dyer (1959) as described previously (Imhoff et al., 1982). Cells were
suspended in 1 M NaCl and extracted with a mixture of chloroform, methanol
and 1M NaCl in the ratio 1/2/0.8. After removal of cell debris by
centrifugation the lipids were brought into the chloroform phase by changing
the ratio of chloroform/methanol/NaCl to 1/1/0.9. The chloroform phase was
concentrated to a small volume and polar lipids were precipitated with ice
cold acetone. Precipitated lipids were removed by centrifugation and washed
once with ice cold acetone. Finally the combined acetone supernatants were
concentrated to a small volume for analysis of quinones. The pellet was
dried with N_2, redissolved in a small volume of chloroform/methanol (4/1)
and used for analysis of polar lipids and fatty acids. Polar lipids were
separated on silica gel thin-layer plates with chloroform/methanol/acetic
acid/water = 85/15/10/3.5 and identified by cochromatography with lipid
standards and with different spray reagents as given earlier (Imhoff et al.,
1982).

     Quinones were separated on silica gel thin-layer plates (Sil G-25
UV_{254}, Machery and Nagel, Düren) with petrol ether (60-80°C)/diethylether =
85/15 (v/v) as described earlier (Imhoff, 1984a). Determination of the
isoprenoid chain length was carried out on reverse-phase thin-layer plates
(Nano-Sil C_{18}-100 UV_{254}, Machery and Nagel, Düren) with acetone/acetonitrile
= 60/40 for ubiquinones and rhodoquinones and with acetone/water = 97/3 for
menaquinones.

     Fatty acids were methylated by addition of 2.5 ml each of methanol and
toluene and 0.1 ml sulfuric acid (conc.) to a dried lipid sample and heating
in closed screw cap tubes at 70°C overnight. Fatty acid methyl esters were
extracted from this reaction mixture into hexane, purified on thin-layer
plates and identified by gas chromatographic analysis by comparison with
standard mixtures of fatty acid methyl esters as references.

RESULTS AND DISCUSSION

Polar Lipids

     A comparative survey of the lipid composition has revealed significant
differences among the three groups of phototrophic purple bacteria (Imhoff
et al., 1982).

     Ectothiorhodospiraceae characteristically contain predominant amounts
of CL and PG, significant amounts also of PC and varying but minor amounts
of PE. The glycolipids, found in Chromatiaceae, purple nonsulfur bacteria
and phototrophic green sulfur bacteria, are absent.

     In Chromatiaceae phospholipids predominate quantitatively with
comparable amounts of CL, PE, and PG as major components, none of which is
predominant. In addition several glycolipids were found. One glycolipid,
which according to chromatographic and staining properties is similar to
MGDG, was found in all tested strains in considerable amounts (Imhoff et
al., 1982). This lipid does not contain galactose, but is a
monoglucosyldiglyceride, as shown for <u>Chromatium vinosum</u> (Steiner et al.,
1969). Two other glycolipids were described as mannosyl,
glucosyl-diglyceride and dimannosyl, glucosyl-diglyceride (Steiner et al.,
1969).

224

Table 1. Polar Lipids of Phototrophic Prokaryotes

	PC	PG	PE	CL	OL	PI	SQDG	DGDG	GLII	MGDG
Cyanobacteria[a, b]	-	+	-	-	nd	-	+	++	-	++
Chlorobiaceae	-	+	+/-	+	-	-	-	-	+•	+
Chloroflexaceae	-	+	-	+	-	+	-	+	-	+
Heliobacterium	-	+	+	+	nd	-	-	-	-	-
Chromatiaceae[c]	-	+	+	+	-	-	?	+[d]	-	+[d]
Ectothiorhodospira	+	+	(+)	+	-	-	-	-	-	-
Rhodospirillum	-*	+	+*	tr	+°	-	-*	-	-	-*
Rhodopseudomonas	+	+	+	+	+°	-	-*	-	-	-*
Rhodobacter	-/+	+	+	-°	+°	-	+/-	-	-	-*
Rhodocyclus	-	+	+	+	nd	-	-	-	-	-

Data from [a]Kates, 1970; [b]Kenyon, 1978; [c]Imhoff et al., 1982, [d]no galactolipids, •galactose and another sugar present, *property of most of the species of this group (exceptions known), °not all species investigated, nd = not determined.

Rhodospirillaceae (purple nonsulfur bacteria) demonstrate a great diversity in their polar lipid composition. Phospholipids are clearly predominant, but small amounts of different glycolipids were found in many species. Most of them contain a variety of major lipids, among which CL, PE, OL, PG, PC and SQDG are the most common. Characteristic lipid patterns are found for the different genera. Species of Rhodopseudomonas have significant amounts of PC, PE, PG, and CL, while PC is absent from all Rhodospirillum species (except R. salinarum) and CL is present in this genus in very small proportions only or absent (Table 1). A glycolipid with similar properties to MGDG or the MGDG-like glycolipid that is present in Chromatiaceae, Chlorobiaceae and Chloroflexus is also found in some species like R. salexigens and Rps. sulfoviridis. Several sulfolipids have been found in various species of the purple nonsulfur bacteria (Imhoff, 1984b). Significant amounts of the plant sulfolipid (SQDG) have been found in some species of Rhodospirillum, Rhodopseudomonas and Rhodobacter. All tested species were able to use reduced sulfur sources as well as sulfate for biosynthesis of these lipids (Imhoff, 1984b).

Both Chlorobiaceae and Chloroflexus contain considerable amounts of glycolipids. The major part of these glycolipids (81-86%) was found to be associated with the chlorosomes of Chlorobium, whereas phospholipids were similarly predominant in the cytoplasmic membrane (Cruden and Stanier, 1970). On a molar basis the chlorosomes have 1-3 moles of sugar per mol of phosphate, whereas the membranes contain only 1 mol of sugar per 5-12 moles of phosphate. Assuming each phospholipid contains only one phosphate and each glycolipid only one sugar molecule (which is not true for CL and di-and triglycosides) about one out of three lipid molecules of the whole cell is a glycolipid. A similar ratio has been found in whole cells of Chloroflexus, where chlorosomes contain phosphate and sugar in nearly equal molarity and membranes have four times the amount of phosphate compared to sugars (Schmidt, 1980).

Several glycolipids are present in Chlorobiaceae (Fig. 1): First Constantopoulos and Bloch (1967) found glycosyl glycerides in phototrophic green bacteria (lipids of "Chloropseudomonas ethylica", later found to be a mixed culture, have been analyzed) and characterized them as monogalactosyl diglyceride (MGDG) and glycolipid II containing galactose, rhamnose and a third unidentified sugar. Cruden and Stanier (1970) substantiated these findings with pure cultures of Cb. limicola. They separated membranes and chlorosomes and found the monogalactosyl diglyceride associated with the chlorosomes and the glycolipid II in the membranes. It has been suggested later that the galactolipid that is specifically associated with the chlorosomes forms a monolayer covering the cytoplasmic boundary of the chlorosome (Staehelin et al., 1980).

Using the diphenylamine spray reagent for glycolipid detection only two glycolipids have been detected (Constantopoulos and Bloch, 1967; Cruden and Stanier, 1970), because this reagent gives no reaction with several other glycolipids of Chlorobiaceae (Bias, 1985). With 1-naphthol, periodate-Schiff, and anisaldehyde spray reagents at least 5 different glycolipids could be found by one-dimensional thin-layer chromatography. Two of these are identical with MGDG and Glycolipid II. Two other glycolipids were described as lipid 1 and lipid 2 by Knudsen et al. (1982). Lipid 1 has a similar R_f-value to PI, but shows different color reactions with 1-naphthol and periodate-Schiff reagents, contains no phosphate, and stains positive with ninhydrin (Bias, 1985; Fig. 1). This is the fluorescent aminolipid characterized by Olson et al. (1983). Lipid 2 moves similarly to DGDG on thin layer plates, but shows different coloration with spray reagents. Although two minor glycolipids with similar chromatographic properties to SQDG and with negative staining behaviour with ninhydrin, molybdate and diphenylamine are present in lipid extracts of Chlorobiaceae, their color reactions with 1-naphthol, periodate-Schiff, and anisaldehyde spray reagents are different from those of SQDG (Bias, 1985). In addition there is no evidence for a sulfolipid in <u>Cb. limicola</u> (strains 6230 and 6430) and

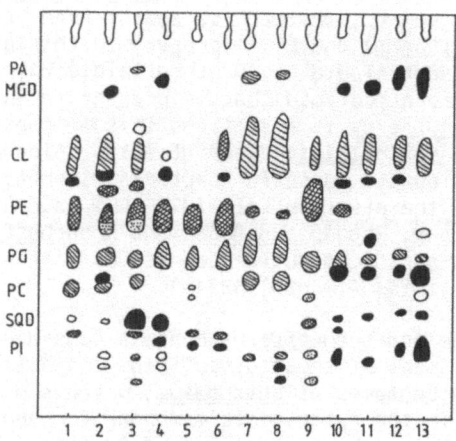

Fig. 1. Thin layer chromatogram of polar lipids from representative species of anoxygenic phototrophic bacteria. Position of lipid standards as indicated: CL cardiolipin, MGD monogalactosyl diglyceride, PA phosphatidic acid, PC phosphatidyl choline, PE phosphatidyl ethanolamine, PG phosphatidyl glycerol, PI phosphatidyl inositol, SQD sulfoquinovosyl diglyceride.
Polar lipids of the following species are shown:
1 <u>Rps. rutila</u>, 2 <u>Rps. sulfoviridis</u>, 3 <u>Rps. marine</u>, 4 <u>R. salexigens</u>, 5 <u>R. molischianum</u>, 6 <u>R. photometricum</u>, 7 <u>E. mobilis</u>, 8 <u>E. halochloris</u>, 9 <u>H. chlorum</u>, 10 <u>Cb. vibrioforme 8327</u>, 11 <u>Cb. vibrioforme 6030</u>, 12 <u>P. luteolum 2530</u>, 13 <u>Cf. aurantiacus</u> OK-70-fl. The signature of the lipids is black for glycolipids, hatched for phospholipids, cross-hatched for amino-phospholipids, and stippled for aminolipids, which are not phospholipids. Lipids showing no signature did not react with any of the spray reagents used but showed up with iodine vapour and sulfuric acid.

Cb. vibrioforme (strain 8327) from experiments with radioactively labelled ^{35}S-cysteine and ^{35}S-thiosulfate (Bias, 1985). It is therefore clear that the lipid with a similar R_f-value to SQDG, which was assumed to be SQDG (Kenyon and Gray, 1974; Knudsen et al., 1982) is not SQDG. Further minor components were detected in two-dimensional thin-layer chromatography only (Bias, 1985).

Three phospholipids present in extracts from Cb. limicola and Cb. vibrioforme (thiosulfate-utilizing strains) were identified as CL, PE, and PG (Knudsen et al., 1982; Bias, 1985; Fig. 1). In contrast to these findings Kenyon and Gray (1974) did not find PE (and PS) in several Chlorobiaceae. As shown in Fig. 1 the lipid composition of Cb. vibrioforme (8327, thiosulfate-utilizing), Cb. vibrioforme (6030) and P. luteolum are similar. PE is present in strain 8327, but absent in the two other strains investigated. An additional glycolipid moving faster than PG was found only in strain 6030.

In Cf. aurantiacus PG but not PE is present (Kenyon and Gray, 1974; Knudsen et al., 1982; Fig. 1). In contrast to the findings of Knudsen et al. (1982) we found considerable amounts of a phospholipid with chromatographic properties and staining behaviour like CL (Fig. 1). Glycolipid II, lipid 1, and lipid 2 of Chlorobiaceae are absent. Two glycolipids containing galactose (and another sugar) are present (Kenyon and Gray, 1974) showing R_f-values and staining behaviour like MGDG and DGDG. The DGDG-like lipid has an R_f-value similar to that of lipid 2 of Chlorobiaceae but is clearly differentiated from this lipid by staining with 1-naphthol. It contains glucose and galactose in equal amounts (Knudsen et al., 1982). Again the presence of SQDG assumed on the basis of R_f-value and negative results in staining with molybdate and ninhydrin (Kenyon and Gray, 1974; Knudsen et al., 1982) is doubtful. The lipid spot does not show the intense red color typical for SQDG with 1-naphthol and 40% sulfuric acid.

Another glycolipid of Cf. aurantiacus has a similar R_f-value to lipid 1 of Chlorobiaceae, but is ninhydrin negative, phosphate positive and shows similar color development on staining with 1-naphthol like PI. Knudsen et al. (1982) found PI in cells grown in the presence or absence of yeast extract. Two further unidentified lipids that do not show positive reaction with ninhydrin, molybdate and 1-naphthol are present (Fig. 1).

The lipid composition of Heliobacterium chlorum is quite simple, completely different from those of Chlorobiaceae and Chloroflexus, and resembles that of Rhodocyclus species. Glycolipids are absent and the phospholipids CL, PE, and PG are the dominating components. Lyso-PE and three minor aminolipids are also present (Fig. 1).

Fatty acids

The fatty acids of phototrophic bacteria are mainly saturated and mono-unsaturated straight chain C-16 and C-18 fatty acids. Within the different groups variation occurs in respect to chain length and degree of saturation as well as in the presence of C-17 and C-19 cyclopropan fatty acids and smaller amounts of short-chain or long-chain fatty acids.

Within all Chlorobiaceae tested high proportions of 14:0 and C-16 fatty acids were found, whereas the content of C-18 fatty acids was very low (Kenyon and Gray, 1974; Knudsen et al., 1982; Table 2). The ratio of saturated to unsaturated fatty acids is high (0.95-1.2) compared to the ratio in most phototrophic purple bacteria.

In Cf. aurantiacus C-14 fatty acids were found in traces only and C-18 fatty acids were dominant. Small amounts of 17:0, 17:1, 18:2, 19:0, 19:1, 20:0 and 20:1 (not shown in Table 2) were found in all strains investigated

Table 2. Major Fatty Acids of Phototrophic Green Bacteria and _Heliobacterium chlorum_

		14:0	14:1	16:0	16:1	17cy	18:0	18:1
Cb. vibrioforme	1930	12	tr	23	52	3	1	2
Cb. phaeobacteroides	2430	16	1	15	64	1	tr	1
Cb. phaeovibriodes	2631	10	tr	29	51	2	tr	2
Cb. limicola	6230*	27	nd	20	37	7	tr	3
Cb. limicola	6330	13	2	17	57	3	tr	-
Cb. limicola	2230*	24	nd	22	43	3	tr	3
P. luteolum	2530	14	1	21	47	11	tr	tr
Cb. vibrioforme	8327*	22	nd	23	43	1	3	1
Cf. aurantiacus	OK-70-fl*	tr	-	8	3	-	10	46
H. chlorum**		1	nd	10	56	-	10	16

Data are from Kenyon and Gray (1974), *Knudsen et al. (1982), and ** our unpublished results.

(Kenyon and Gray, 1974; Knudsen et al., 1982). In addition fatty alcohols and wax esters are present (Knudsen et al., 1982). The higher proportions of C-18 to C-16 fatty acids and the presence of saturated and mono-unsaturated odd-chain fatty acids clearly distinguishes Chloroflexus from the Chlorobiaceae. The ratio of saturated to unsaturated fatty acids is very high in Cf. aurantiacus (3.6-3.8). This and the predominance of long-chain fatty acids is in accordance with its thermophilic nature. To decide whether the observed fatty acid composition is an inherent property of Chloroflexaceae the analysis of mesophilic representatives of this family, as isolated by Gorlenko (1975), is necessary.

The fatty acid composition of H. chlorum is quite different from that of the phototrophic green bacteria. Mainly C-16 and C-18 fatty acids with 16:1 as main component and a high excess of mono-unsaturated fatty acids (Table 2) were found. In this respect Heliobacterium is more similar to some purple nonsulfur bacteria, in particular to the Rhodocyclus species, than to Chlorobiaceae and Chloroflexaceae. The ratio of saturated to unsaturated fatty acids is 0.3 in H. chlorum compared to values of 0.05-0.9 (in most cases below 0.5) in the phototrophic purple bacteria. Values higher than 0.5 were found in the Rhodocyclus species and two halophilic Rhodospirillum species only (Imhoff, unpublished).

In phototrophic purple bacteria C-14 fatty acids are generally present in only small amounts and C-16 and C-18 fatty acids are dominant (Table 3). Although 18:1 is the dominant fatty acid in most groups (except Rhodocyclus), the proportions of 16:0, 16:1, 18:0, and 18:1 are variable and characteristic for species, genera or even families.

Quinones

A number of different quinone structures are known in various phototrophic prokaryotes with variation in the ring structure, substitution of the rings, and differences in isoprenoid chain length, chain saturation and substitution by oxo-groups (Collins and Jones, 1981; Hiraishi and Hoshino, 1984; Hiraishi et al., 1984; Imhoff, 1984a).

In the Rhodospirillaceae species either Q-10 is the sole quinone

Table 3. Major Fatty Acids of Some Phototrophic Purple Bacteria

		14:0	14:1	16:0	16:1	18:0	18:1
Chromatium vinosum	D	1	tr	19	36	1	39
Thiocapsa roseopersicina	6311	tr	tr	22	25.	1	44
Ectothiorhodospira halochloris	9850	-	-	19	tr	8	70
Ectothiorhodospira halophila	51/9	-	-	10	1	7	72
Rhodospirillum rubrum	51	2	-	14	27	1	55
Rhodospirillum molischianum	241	1	-	18	37	1	44
Rhodocyclus tenuis	230	1	-	34	50	1	15
Rhodocyclus gelatinosus	156	5	-	35	41	2	16
Rhodobacter adriaticus	6II	1	-	2	tr	20	72
Rhodopseudomonas marina	985	tr	-	2	tr	14	69
Rhodopseudomonas blastica	138	tr	-	2	3	5	90

component; ubiquinone is present together with MK or with RQ; or Q, RQ, and MK all are present. Menaquinone is never found as the sole quinone component in this group. In Chromatiaceae MK-8 and Q-8 have been found. In Ectothiorhodospira MK and Q are present. The extremely halophilic species contain in addition another component with unknown structure.

In Cb. limicola f. thiosulfatophilum, MK-7 and 1'-oxomenaquinone (chlorobium quinone) have been found (Frydman and Rapaport, 1963; Powls et al., 1968). These two components were also found in Cb. vibrioforme 6030, in the thiosulfate-utilizing Cb. vibrioforme 8327 and in P. luteolum 2530. In Chloroflexus MK is present, but Q, RQ, and chlorobium quinone are lacking (Khoff, unpublished).

During preliminary analysis of the quinones of H. chlorum small amounts of menaquinone have been found, together with another quinone compound of so far unidentified structure. Ubiquinone, rhodoquinone and chlorobium quinone are absent.

Phylogenetic aspects

A quantitative phylogenetic relationship cannot be obtained by comparing lipid, fatty acid, and quinone compositions of bacteria. Some interferences can, however, be drawn with the assumptions that the most primitive structures and those present in different bacterial groups that diverged early are also the most ancient ones.

Among the phospholipids PE, PG and CL apparently were present already quite early in evolution. A simple pattern of lipid composition with these lipids as major components may therefore be regarded as primitive. Such a lipid composition as found n H. chlorum is typical for many Gram-negative bacteria, but a similar pattern is also found in some Gram-positive bacteria, particularly in the genus Bacillus and related bacteria (Goldfine, 1972). From this point of view Heliobacterium, Rhodocyclus and to some degree also Rhodospirillum have a primitive lipid composition. The ability to synthesize methylated derivatives of PE developed later in evolution (Goldfine, 1972). In this respect phototrophic bacteria containing PC, like Rhodopseudomonas, Rhodobacter and Ectothiorhodospira, are more recent developments compared to other phototrophic bacteria.

The phylogenetic interpretation of the fatty acid composition is difficult because it changes with the growth conditions. In general, however, the predominance of short chains and a high degree of saturation appear to be more primitive characters than longer chain and more unsaturated and branched chain fatty acids. In this regard Chlorobiaceae appear the most primitive and also Heliobacterium, Rhodocyclus, Chromatium and Rhodospirillum are more ancient than other groups of phototrophic bacteria, whereas Rhodopseudomonas and Rhodobacter are more recent.

Among the quinone structures probably menaquinone and its biosynthesis arose earlier than ubiquinone and its derivatives. Manequinones and derivatives thereof are found in Gram-positive bacteria and also in a number of "primitive" Gram-negative bacteria as sole quinone component. In this regard Chlorobiaceae, Chloroflexus and also Heliobacterium are more ancient than other anoxygenic phototrophic bacteria. In some purple nonsulfur bacteria like Rhodobacter the ability to synthesize menaquinone has apparently been lost during more recent evolution.

These considerations are congruent with the lines developed by Dickerson (1980) on the basis of the size and sequence homology of soluble c-type cytochrome (c_{ss1} and c_2) in purple nonsulfur bacteria. Also the idea that poorly developed intracytoplasmic membrane systems in phototrophic bacteria are a primitive character points to Heliobacterium, Rhodocyclus, and although on a different line, to Chloroflexus and the Chlorobiaceae as ancient bacteria. (For a discussion of data on 16S rRNA analyses see Stackebrandt et al., 1988.)

ACKNOWLEDGEMENTS

The author wishes to thank Dr. U. Bias, Mrs. J. Mertschuweit, and Dr. M. Steinmetz for cell material from Cf. aurantiacus OK-70-fl, Cb. vibrioforme 8327, Cb. vibrioforme 6030 and P. luteolum 2530 and Dr. U. Bias for valuable discussion. Part of the work presented here was financially supported by the Deutsche Forschungsgemeinschaft.

REFERENCES

Bias, U., 1985, Zur Freisetzung von Sulfat, Verwertung von Cystein und Vorkommen von Sulfolipiden bei Chlorobium, Doctoral thesis, University of Bonn, F.R.G.
Bligh, E. G., and Dyer, W. J., 1959, A rapid method of total lipid extraction and purification, Can. J. Biochem. Physiol., 37:911.
Collins, M. D., and Jones, D., 1981, Distribution of isoprenoid quinone structural types in bacteria and their taxonomic implications, Microbiol. Rev., 45:316.
Constantopoulos, G., and Bloch, K., 1967, Isolation and characterization of glycolipids from some photosynthetic bacteria, J. Bacteriol., 93:1788.
Cruden, D. J., and Stanier, R. Y., 1970, The characterization of chlorobium vesicles and membranes isolated from green bacteria, Arch. Mikrobiol., 72:115.
Dickerson, R. E., 1980, Evolution and gene transfer in purple photosynthetic bacteria, Nature, 203.210.
Frydman, B., and Rapaport, H., 1963, Non-chlorophyllous pigments of Chlorobium thiosulfatophilum in chlorobium-quinone, J. Am. Chem. Soc., 85:823.
Gibson, J., Stackebrandt, E., Zablen, L. B., Gupta, R., and Woese, R. W.,

1979, A phylogenetic analysis of the purple photosynthetic bacteria, Current Microbiol., 3:59.

Goldfine, H., 1972, Comparative aspects of bacterial lipids, Adv. Microbiol. Physiol., 8:1.

Gorlenko, V. M., 1975, Characteristics of filamentous phototrophic bacteria from freshwater lakes, Mikrobiologiya, 44:756.

Hansen, T. A., and Imhoff, J. F., 1985, Rhodobacter veldkampii, a new species of phototrophic purple nonsulfur bacteria, Intern. J. Syst. Bacteriol., 35:115.

Hiraishi, A., Hoshino, Y., and Kitamura, H., 1984, Isoprenoid quinone composition in the classification of Rhodospirillaceae, J. Gen. Appl. Microbiol., 30:197.

Hiraishi, A., and Hoshino, Y., 1984, Distribution of rhodoquinone in Rhodospirillaceae and its taxonomic implications, J. Gen. Appl. Microbiol., 30:435.

Imhoff, J. F., 1982a, Taxonomy and molecular evolution of phototrophic bacteria, Abstr. IV, Intern. Symp. Photosynthetic Prokaryotes, Bombannes-Bordeaux, France.

Imhoff, J. F., 1982b, Taxonomic and phylogenetic implications of lipid and quinone compositions in phototrophic microorganisms, in: "Biochemistry and Metabolism of Plant Lipids," J. F. G. M. Wintermans and P. J. C. Kuiper, eds., Elsevier Biomedical Press, Amsterdam.

Imhoff, J. F., 1984a, Quinones of phototrophic purple bacteria, FEMS Microbiol. Lett., 25:85.

Imhoff, J. F., 1984b, Sulfolipids in phototrophic purple nonsulfur bacteria, in: "Structure, Function and Metabolism of Plant Lipids," P.-A. Siegenthaler, and W. Eichenberger, eds., Elsevier Science Publishers, Amsterdam.

Imhoff, J. F., Kushner, D. J., Kushwaha, S. C., and Kates, M., 1982, Polar lipids in phototrophic bacteria of the Rhodospirillaceae and Chromatiaceae families, J. Bacteriol., 150:1192.

Imhoff, J. F., Trüper, H. G., and Pfennig, N., 1984, Rearrangement of the species and genera of the phototrophic "Purple Nonsulfur Bacteria", Intern. J. Syst. Bacteriol., 34:340.

Kates, M., 1970, Plant phospholipids and glycolipids, Adv. Lipid Res., 8:225.

Kenyon, C. N., 1978, Complex lipids and fatty acids of photosynthetic bacteria, in: "The Photosynthetic Bacteria," R. K. Clayton and W. R. Sistrom, eds., Plenum Publ. Corp., New York.

Kenyon, C. N., and Gray, A. M., 1974, Preliminary analysis of lipids and fatty acids of green bacteria and Chloroflexus, J. Bacteriol., 120:131.

Knudsen, E., Jantzen, E., Bryn, K., Ormerod, J. G., and Sirevåg, R., 1982, Quantitative and structural characteriatics of lipids in Chlorobium and Chloroflexus, Arch. Microbiol., 132:149.

Mayer, H., 1984, Significance of lipopolysaccharide structure for taxonomy and phylogenetical relatedness of Gram-negative bacteria, in: "The Cell Membrane," E. Haber, ed., Plenum Publish. Corp., New York.

Olson, J. M., Shaw, E. K., Gaffney, J. S., and Scandella, C. J., 1983, A fluorescent aminolipid from a green photosynthetic bacterium, Biochem., 22:1819.

Pfennig, N., and Trüper, H. G., 1974, The phototrophic bacteria, in: "Bergey's Manual of Determinative Bacteriology," R. E. Buchanan and N. E. Gibbons, eds., Williams and Wilkins Comp., Baltimore.

Powls, R., Redfearn, E. R., and Tripett, S., 1968, The structure of chlorobiumquinone, Biochem. Biophys. Res. Comm., 33:408.

Schmidt, K., 1980, A comparative study on the composition of chlorosomes (chlorobium vesicles) and cytoplasmic membranes from Chloroflexus aurantiacus strain OK-70-fl and Chlorobium limicola f. thiosulfatophilum strain 6230, Arch. Microbiol., 124:21.

Stackebrandt, E., and Woese, C. R., 1981, The evolution of prokaryotes, in:
    "Molecular and Cellular Aspects of Microbial Evolution," M. I.
    Carlile, I. F. Collins, and B. E. B. Mosely, eds., Univ. Press,
    Cambridge.
Stackebrandt, E., Embley, M., and Weckesser, J., 1988, Phylogenetic,
    evolutionary, and taxonomic aspects of phototrophic eubacteria, This
    volume.
Staehelin, L. A., Golecki, J. R., and Drews, G., 1980, Supramolecular
    organization of chlorosomes and of their membrane attachment sites in
    Chlorobium limicola, Biochim. Biophys. Acta, 589:30.
Steiner, S., Conti, S. F., and Lester, R. L., 1969, Separation and
    identification of the polar lipids of Chromatium strain D, J.
    Bacteriol., 98:10.
Weckesser, J., Drews, G., Mayer, H., and Fromme, I., 1974,
    Lipopolysaccharide aus Rhodospirillaceae, Zusammensetzung und
    taxonomische Relevanz, Zbl. Bakt. Hyg., I. Abt. Orig. A, 228:193.
Weckesser, J., Drews, G., and Mayer, H., 1979, Lipopolysaccharides of
    photosynthetic prokaryotes, Ann. Rev. Microbiol., 33:215.
Woese, C. R., Stackebrandt, E., Macke, T. J., and Fox, G. E., 1985, A
    phylogenetic definition of the major eubacterial taxa, System. Appl.
    Microbiol., 6:143.

A NEW GREEN SULFUR BACTERIUM FROM A FRESHWATER POND

B. Eichler and N. Pfennig

Fakultät für Biologie
Universität Konstanz
F.R.G.

SUMMARY

Several filamentous bacteria belonging to the phototrophic green
bacteria have been described so far. Chloroflexus aurantiacus (Pierson and
Castenholz, 1974) is a gliding, unbranched, multicellular bacterium
occurring in hot springs. It contains BChls a and c and is able to grow
aerobically in the light or dark as well as anaerobically in the light. As
mesophilic representatives of the multicellular filamentous green bacteria
two species, Chloronema giganteum and Cn. spiroideum, were isolated by
Dubinina and Gorlenko (1975). Both species contain gas vacuoles and show
gliding motility. They contain BChl c or d and seem to be organotrophic.
Finally, a unicellular, gliding bacterium, Chloroherpeton thalassium (Gibson
et al., 1984) was isolated from marine habitats. As a member of the green
sulfur bacteria it is obligately phototrophic and contains BChl c and
γ-carotene as the major photosynthetic pigments. Gas vacuoles have been
observed in most strains.

We isolated a new green sulfur bacterium which is similar to Ch.
thalassium but is inhibited by salt concentrations required by the latter
species. The organism is presented in this paper as a new species of the
genus Chloroherpeton, Ch. limnophilum.

Source of Organism

A combined water-plus-sediment sample was taken from a highly eutrophic
freshwater forest pond near Sigmaringen, F.R.G. The pond is 1-2 m deep, and
during summer the sediment and plant residues reach up to the surface.
During this time the water body is completely anoxic. The sample was
incubated in a glass vessel on a laboratory desk in daylight. After some
weeks red layers appeared on the surface of the decaying plant material
indicating growth of phototrophic purple sulfur bacteria.

Isolation and Cultivation

Samples of the reddish material were inoculated into a deep agar
dilution series using bicarbonate-buffered and sulfide-reduced medium
(Pfennig, 1978). After two weeks of incubation green fluffy colonies
developed together with red colonies of a purple sulfur bacterium which was
later identified as Lamprocystis roseopersicina. The green colonies were

further purified by repeated deep agar dilution series. Fluffy colonies, indicating motility of the cells, always appeared. In our cultures the new organism only grew attached to surfaces of agar or $AlPO_4$-precipitates and not in liquid medium. For all further characterizations the medium contained per liter of distilled water (formula modified from Wagener and Pfennig, 1987): KH_2PO_4, 0.15 g; NH_4Cl, 0.03 g; NaCl, 0.7 g; $MgCl_2.6H_2O$, 0.25 g; KCl, 0.2 g; $CaCl_2.2H_2O$, 0.15 g; resazurine, 0.35 mg; repeatedly washed bactoagar, 2 g; $NaHCO_3$, 1.5 g; 0.5 ml of trace element solution SL10 (Widdel et al., 1983) and 0.5 ml of vitamin B_{12}-solution (0.04 g/l) was added. Dithionite was added as reducing agent until resazurine became colorless, then 0.5 mM Na_2S was applied. The pH was adjusted to 6.9-7.1.

Morphology

Single cells of Ch. limnophilum were 0.8 μm wide and 5-20 μm long (Fig. 1). The filaments were unicellular, and no crosswalls could be detected. The cells were able to flex and often bent through 180° in the same manner as those of Ch. thalassium. Slow gliding motility was observed. Gas vacuoles were never detected.

Physiological Properties

Ch. limnophilum is a typical member of the phototrophic green sulfur bacteria. It only grew obligately anaerobically; not even microaerobic conditions were tolerated. Photoautotrophic growth occurred with hydrogen sulfide as electron donor, which was oxidized to sulfate. Elemental sulfur was formed as intermediary product and stored outside the cells. A definite increase in cell density was observed when acetate (5 mM), propionate (1 mM) or pyruvate (5 mM) was added to the medium. Not utilized were fructose, glutamate, butyrate, valerate and succinate.

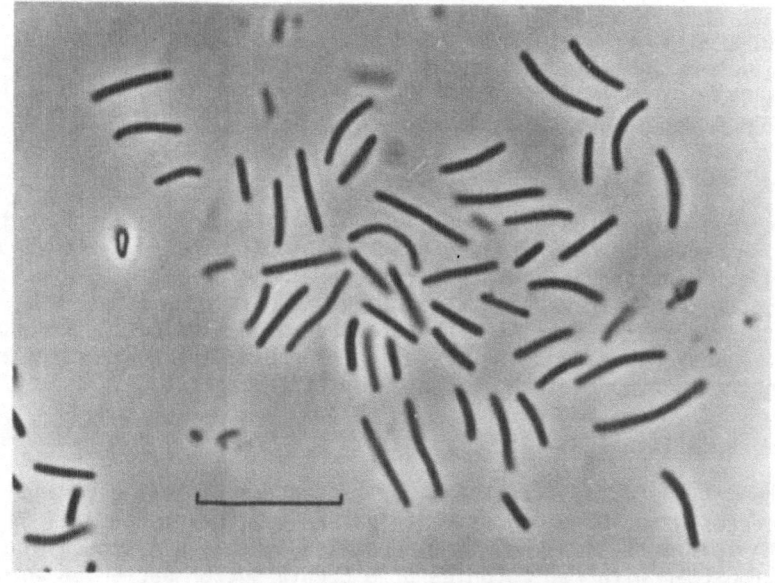

Fig. 1. Phase contrast micrograph of Ch. limnophilus. Bar = 10 μm.

In vivo absorption spectra of green cell suspension of <u>Ch. limnophilum</u> exhibited absorption maxima at 332, 398, 421, 454 and 757 nm, probably due to the presence of BChl <u>c</u> and carotenoids of the chlorobactene series. Methanol extracts showed the presence of small amounts of BChl <u>a</u>.

The organism grew well under low light conditions (100 lx). Light intensities higher than 250 lx considerably inhibited growth. Optimal growth occurred at pH 6.9 and 23°C; growth was inhibited at temperatures higher than 25°C. In contrast to <u>Ch. thalassium</u> the new species grew well in diluted freshwater medium and tolerated salt concentrations only up to 34 mM NaCl (= 0.2%) and 2 mM MgCl$_2$ (= 0.04%). Under optimal growth conditions photoautotrophic growth occurred with a doubling time of about 1 week.

Taxonomy

On the basis of the aforementioned morphological and physiological characteristics the new green sulfur bacterium has to be considered as a member of the genus <u>Chloroherpeton</u>. The new bacterium differs, however, from <u>Ch. thalassium</u> in the following properties:
1. <u>Ch. limnophilum</u> is an obligate freshwater bacterium and does not tolerate even moderate salt concentrations. In contrast, <u>Ch. thalassium</u> is a marine bacterium, unable to grow in freshwater media.
2. The cells of <u>Ch. limnophilum</u> are definitively smaller than those of <u>Ch. thalassium</u>. The cells never develop gas vacuoles. Only a few organic substrates are photoassimilated in the presence of bicarbonate and sulfide: acetate, propionate and pyruvate.

REFERENCES

Dubinina, G. A., and Gorlenko, V. M., 1975, New filamentous phototrophic green bacteria containing gas vacuoles, <u>Mikrobiologiya</u>, 44:511.

Gibson, J., Pfennig, N., and Waterbury, J. B., 1984, <u>Chloroherpeton thalassium</u> gen. nov. et spec. nov., a non-filamentous, flexing and gliding green sulfur bacterium, <u>Arch. Microbiol.</u>, 138:96.

Pfennig, N., 1978, <u>Rhodocyclus purpureus</u> gen. nov. and sp. nov., a ring-shaped, vitamin B$_{12}$-requiring member of the family Rhodospirillaceae, <u>Int. J. Syst. Bacteriol.</u>, 23:283.

Pierson, B. K., and Castenholz, R. W., 1974, A phototrophic gliding filamentous bacterium of hot springs, <u>Chloroflexus aurantiacus</u>, gen. and sp. nov., <u>Arch. Microbiol.</u>, 100:5.

Widdel, F., Kohring, G. W., and Mayer, F., 1983, Studies on dissimilatory sulfate-reducing bacteria that decompose fatty acids. III Characterization of the filamentous gliding <u>Desulfonema limicola</u> gen. nov. sp. nov., and <u>Desulfonema magnum</u> sp. nov., <u>Arch. Microbiol.</u>, 134:286.

Wagener, S., and Pfennig, N., 1987, Monoxenic culture of the anaerobic ciliate <u>Trimyema compressum</u> Lackey, <u>Arch. Microbiol.</u>, in press.

PRELIMINARY CHARACTERIZATION OF A TEMPERATE MARINE MEMBER OF THE

CHLOROFLEXACEAE

E.E. Mack[a] and B.K. Pierson (Corresponding Author)

Biology Department, University of Puget Sound
Tacoma, WA 98416, U.S.A.
[a]Present address: Universität Konstanz, Postfach 5560
D-7750 Konstanz 1, F.R.G.

SUMMARY

Fossil records of the Precambrian era indicate that mat-forming phototrophic prokaryotes were some of the earth's earliest life forms (Schopf and Walter, 1983). Modern analogs of stromatolites (fossils of ancient mat communities) can be seen today in a number of extreme environments: salt marshes, hypersaline lakes and lagoons, and thermal springs (Awramik, 1984). One organism of these communities, the phototrophic prokaryote Chloroflexus, has been shown to be a likely descendent of early phototrophic bacteria on the basis of morphological (Olson and Pierson, 1987) and phylogenetic studies (Stackebrandt, 1985). To date, only a thermophilic species of Chloroflexus, Cf. aurantiacus, has been maintained in pure culture and characterized (Pierson and Castenholz, 1974; Castenholz and Pierson, 1981). One mesophilic fresh-water strain was isolated, but the culture has not been maintained (Gorlenko, 1975). Chloroflexus-like organisms have been observed and identified by morphology in a number of temperate marine and hypersaline environments: in Baja Mexico (Stolz, 1984; D'Amelio, 1987), Shark Bay, Western Australia (unpublished observation, BKP), Solar Lake, Sinai (Cohen, 1984), and Great Sippewissett Salt Marsh, Cape Cod, MA. Chloroflexus has been clearly identified from the Baja and Solar Lake sites by electron microscopy (D'Amelio, 1987; Cohen, 1984). Little is known about its physiology or ecology. An analysis of the morphology, pigmentation, ecology and ultrastructure of the Chloroflexus-like organism of Great Sippewissett Salt Marsh, MA, is reported here.

The Chloroflexus-like organism from Sippewissett was first observed by R. W. Castenholz in 1983 (Castenholz, unpublished observation). In 1984 the organism was observed as one of a 3-membered bacterial community by one of us (BKP). The community consisted of the Chloroflexus-like organism, Chloroherpeton thalassium, and a species of Oscillatoria. The community was observed on the surface of the sandy sediments in a shallow intertidal pool surrounded by grass. The community formed an ephemeral dark green mat during the night and the low light levels of early morning, but the organisms moved deeper into the sediments during the full light of day, leaving the surface sediments covered by a pink layer of purple sulfur bacteria.

In August, 1986, the unstable pond community had changed. The Chloroflexus-like organism was observed occasionally with Ch. thalassium and purple sulfur bacteria, primarily species of Thiocapsa. Very few filaments of Oscillatoria were observed. The pool was filled and drained by a shallow tidal channel. At high tide the pool was approximately 10 cm deep. At low tide there was no standing water, and the sediment was just damp enough to hold the shape of a core. Because of the many summer storms of late July and early August of 1986, fresh sand was continually deposited into the pool. The summer temperature ranged from 25 to 30°C. In winter (January, 1987) the sediment was frozen solid. Under normal conditions in the summer, the surface of the sediment was a beige-pink due to the presence of species of purple sulfur bacteria, primarily Thiocapsa. The Chloroflexus-like organism formed a distinct green layer about 1-2 mm in thickness 4 mm beneath the pink surface layer and directly over a layer of grey sand that was noticeably rich in sulfide.

Because the Chloroflexus-like organism had been observed to migrate under low light levels (Noll and Castenholz, 1983, unpublished), a series of neutral density filters were constructed out of fine nylon mesh and were set over the sediment to create areas of reduced light. The incident sunlight was 700 W/m², and the filters created areas of sediment exposed to 450 W/m², 150 W/m², and 30 W/m². An area of total darkness was created by covering the sediment with foil. Each test area was examined and photographed every 30 min for a period of 3 h. A distinct green layer appeared within 30 min on the surface of the areas exposed to 30 W/m² and total darkness. Microscopic examination of the green layer revealed abundant filaments of the Chloroflexus-like organism. There was no further change during the remainder of the period. No green filaments migrated to the surface under the higher light intensities.

Light microscopy with a Zeiss 16 microscope equipped for epifluorescence revealed that the Sippewissett Chloroflexus-like organism was a gliding, dark green, flexible filament 1.0-1.5 μm in diameter and indeterminate in length. The filaments showed no red fluorescence when excited by blue light. Samples for transmission electron microscopy were fixed in the field in 2.5% glutaraldehyde in sea water. Secondary fixation was in 1% OsO₄ in cacodylate buffer (pH 7.4). Following dehydration in ethanol, the cells were embedded in LADD 112 Epon. The fixation and embedding procedure was provided by E. D'Amelio (personal communication) and was a modification of a recently published method (D'Amelio et al., 1987).

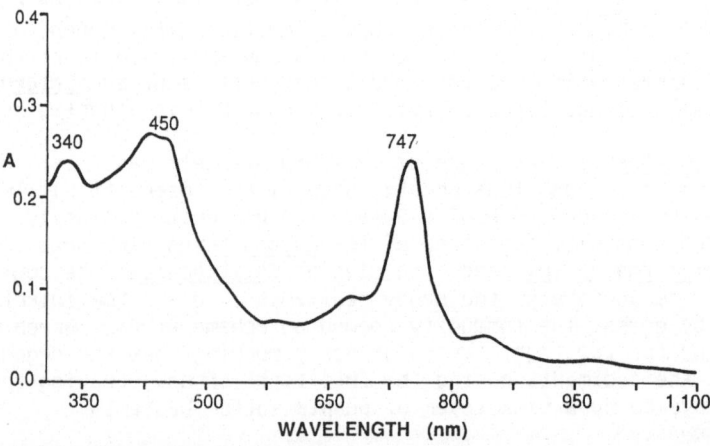

Fig. 1. In vivo absorption spectrum of sonically disrupted Chloroflexus-like filaments in TSM buffer.

Table 1. Photoautotrophic Activity of _Chloroflexus_ with Bicarbonate as Carbon Source

Light Intensity W/m²	Number of Silver Grains/100 μm²	
	Low Sulfide 0.1 mM	High Sulfide 1.0 mM
High: 600-1,000	1.3	0.4
Medium: 100-300	1.1	0.0
Low: 50-100	2.9	0.6
Dark: 0	1.4	0.4
Background	4.2	

Thin sections were cut with a glass knife and post stained in uranyl acetate and lead citrate. The sections were viewed on a Zeiss EM 109 Transmission Electron Microscope. The resulting micrographs showed a filamentous organism with obvious cross-walls and chlorosomes. The multilayered cell wall was about 40 nm thick. The chlorosomes were about 60 nm in diameter and about 120 nm in length and showed no obvious substructure.

An _in vivo_ absorption spectrum of the _Chloroflexus_-like organism from the migrating green layer was prepared as previously described (Pierson et al., 1987). The absorption spectrum (Fig. 1) had maxima at 747 nm, 450 nm, and 340 nm and was similar to that published for _Cf. aurantiacus_ by Pierson and Castenholz (1974).

An initial examination of carbon sources used for phototrophy was done using autoradiographic techniques as described in Pierson et al. (1984). The _Chloroflexus_-like organism was tested for phototrophy, autotrophy and heterotrophy under 4 light conditions and 2 sulfide concentrations (0.1 and 1.0 mM). ^{14}C-labeled sodium bicarbonate (8.4 mCi/mmol, New England Nuclear) was used to test for autotrophy and ^{14}C-labeled sodium acetate (uniformly labeled, 56 mCi/mmol, Amersham) was used to test for heterotrophy. Light conditions were modified with neutral density filters and incident sunlight. Due to variations in light cloud cover, high light intensity ranged from 600 to 1000 W/m², medium from 100 to 300 W/m², and low light intensity from 50 to 100 W/m². Results of the experiments are shown in Tables 1 and 2. These results indicate that the _Chloroflexus_-like organism has a photoheterotrophic physiology with no obvious dependence upon sulfide concentration under the conditions used. Uptake was inhibited at the highest light intensity. No evidence for autotrophy was found.

Table 2. Photoheterotrophic Activity of _Chloroflexus_ with Acetate as Carbon Source

Light Intensity W/m²	Number of Silver Grains/100 μm²	
	Low Sulfide 0.1 mM	High Sulfide 1.0 mM
High: 600-1,000	4.1	20.0
Medium: 100-300	59.9	45.4
Low: 50-100	31.2	42.4
Dark: 0	0.5	1.4
Formalin Control	4.2	
Background	1.0	

The _Chloroflexus_-like organism was maintained in laboratory microcosms with varying degrees of success. The organism did not grow in enrichment or shake cultures in variations of the DG medium used to culture _Cf. aurantiacus_ made up with sea water (Castenholz and Pierson, 1981). Many other species of purple and green phototrophic sulfur bacteria from the same sediments did grow in these media.

Conclusion

The results of this study suggest that the _Chloroflexus_-like organism of the Great Sippewissett Salt Marsh is a member of the family Chloroflexaceae. The organism has a morphology, ultrastructure, pigmentation and carbon physiology similar to _Cf. aurantiacus_.

REFERENCES

Awramik, S. M., 1984, Ancient stromatolites and microbial mats, _in_:
    "Microbial Mats: Stromatolites," Y. Cohen, R. W. Castenholz, and H.
    O. Halvorson, eds., Alan R. Liss, New York.
Castenholz, R. W., and Pierson, B. K., 1981, Isolation of members of the
    family _Chloroflexaceae_, _in_: "The Prokaryotes," M. P. Starr, H. Stolp,
    H. G. Trüper, A. Balows, and H. G. Schlegel, eds., Springer-Verlag,
    Berlin.
Cohen, Y., 1984, The Solar Lake cyanobacterial mats: strategies of
    photosynthetic life under sulfide, _in_: "Microbial Mats:
    Stromatolites," Y. Cohen, R. W. Castenholz, and H. O. Halvorson,
    eds., Alan R. Liss, New York.
D'Amelio, E. D., 1987, Different types of _Chloroflexus_-like microorganisms
    found at the chemocline of hypersaline microbial mats, Poster
    presented at the annual meeting of the Precambrian Paleobiology
    Research Group, Feb. 13-17, University of California, Los Angeles.
D'Amelio, E. D., Cohen, Y., and Des Marais, D. J., 1987, Association of a
    new type of gliding, filamentous, purple phototrophic bacterium
    inside bundles of _Microcoleus chthonoplastes_ in hypersaline
    cyanobacterial mats, _Arch. Microbiol._, 147:213.
Gorlenko, V. M., 1975, Characteristics of filamentous phototrophic bacteria
    from freshwater lakes, _Microbiology_ (English transl. of
    Mikrobiologiya), 44:682.
Olson, J. M., and Pierson, B. K., 1987, Evolution of reaction centers in
    photosynthetic prokaryotes, _Internat. Rev. Cytol._, 108:209.
Pierson, B. K., and Castenholz, R. W., 1974, A phototrophic gliding
    filamentous bacterium of hot springs, _Chloroflexus aurantiacus_, gen.
    and sp. nov., _Arch. Microbiol._, 100:5.
Pierson, B. K., Giovannoni, S. J., and Castenholz, R. W., 1984,
    Physiological ecology of a gliding bacterium containing
    bacteriochlorophyll _a_, _Appl. Environ. Micro._, 47:576.
Pierson, B. K., Oesterle, A., and Murphy, G. L., 1987, Pigments, light
    penetration, and photosynthetic activity in the multi-layered
    microbial mats of Great Sippewissett Salt Marsh, Massachusetts, _FEMS
    Microbiol. Ecol._, submitted (in press)?
Schopf, J. W., and Walter, M. R., 1983, Archean microfossils: New evidence
    of ancient microbes, _in_: "Earth's Earliest Biosphere, its Origin and
    Evolution," J. W. Schopf, ed., Princeton Univ. Press, Princeton, New
    Jersey.
Stackebrandt, E., 1985, Phylogeny and phylogenetic classification of
    prokaryotes, _in_: "Evolution of Prokaryotes," K. H. Schliefer, and E.
    Stackebrandt, eds., Academic Press, New York.
Stolz, J. F., 1984, Fine structure of the stratified microbial community at
    Laguna Figueroa, Baja California, Mexico: II Transmission electron

microscopy as a diagnostic tool in studying microbial communities <u>in</u> <u>situ</u>, <u>in</u>: "Microbial Mats: Stromatolites," Y. Cohen, R. W. Castenholz, and H. O. Halvorson, eds., Alan R. Liss, New York.

THE GREEN SULFUR AND NONSULFUR BACTERIA OF HOT SPRINGS

R.W. Castenholz

Department of Biology
University of Oregon
Eugene, OR 97403, U.S.A.

INTRODUCTION

Hot springs are inhibited primarily by prokaryotes and provide habitats where nearly monospecific populations of phototrophs occur in a recognizable pattern. There is a great diversity of hot springs with regard to chemistry, but within the array, some species of almost all groups of photosynthetic bacteria may be found.

In this article an overall picture of green bacterial distribution will be given in relation to physical and chemical characteristics of hot springs and to the other organisms present. It is of necessity, a summary. The term "green bacteria" here includes members of the Chlorobiaceae and Chloroflexaceae, but also Heliothrix which appears to be related to Chloroflexus although lacking chlorosomes and the bacteriochlorophyll assocaited with these structures.

Hot springs are excellent sources for securing concentrated samples of several species of "green bacteria", which generally occur as discrete layers comprising the major part of many types of compact microbial mats.

Although the green bacteria have not been reviewed solely in context of hot spring habitats before, the habitats of Chloroflexus and related filamentous organisms were described (Castenholz, 1984a) and a recent description of the composition of hot spring microbial mats is available (Castenholz, 1984b).

MATERIALS AND METHODS

Many materials and methods have been used over a number of years to collect the information presented here. Only a few general aspects will be mentioned; many techniques are described in several papers which are referred to.

Field methods in North America, New Zealand, and Iceland where most of these studies were made were faily standard in the use of instruments and of chemical analysis. A Yellow Springs Instrument (Model 42 SC, Yellow Springs, Ohio) thermistor with small tip probe was used in all cases to measure temperatures at the exact boundaries of the organisms in question. This is particularly important, since temperatures can vary dramatically within a few centimeters either vertically or horizontally because of laminar flow patterns or eddies. The measurement of pH was made with a variety of instruments over several years. Free sulfide (H_2S, HS^-, S^{2-}) was measured by either the method of Pachmayr (Castenholz, 1976) or by the similar method of Cline (1969). Both methods were modified at times to accurately quantify sulfide in water samples as small as 0.1 ml.

Diagnostic pigments were used to identify many field populations. Collected material was dispersed in appropriate buffer (TSM: 4.7 g NaCl, 1.2 g $MgSO_4 \cdot 7H_2O$, 2.4 g Tris per liter H_2O, pH adjusted to 7.8) and ultrasonically disrupted. This was followed by centrifugation at ca. 3000 x g before spectroscopic review. Chlorophylls and carotenoids were also extracted in 98% buffered methoanol for quantification.

Microscopy in field locations was with a Cooke-MacArthur field microscope with 40°X and 100°X phase contrast optics. Microelectrode construction and use followed the procedures of Revsbech and Jørgensen (1986). Irradiance was commonly measured with a Lambda LI-185 radiometer.

Fig. 1. The upper temperature-limit of phototrophs in a hot spring of Yellowstone National Park (Spring 74-6, White Creek drainage, Lower Geyser Basin). The 73-74°C upper limit of Synechococcus (S) and the 68 ± 1°C upper limit of the Chloroflexus undermat (C) are indicated by arrows. The flow is toward the foreground. The pH is ca. 8.9. The substrate deposited is siliceous sinter.

RESULTS AND DISCUSSION

Chloroflexus

In hot springs of a pH level higher than 6, the highest temperature-tolerant green bacterium is <u>Chloroflexus aurantiacus</u> (Pierson and Castenholz, 1974a). It is usually supported as an anaerobic photoheterotroph or as an aerobic chemoheterotroph by organic products derived ultimately from one or more species of cyanobacteria which must be in close contact with, or at least, immediately upstream of <u>Chloroflexus</u>. Filaments of <u>Chloroflexus</u> are invariably mixed with the unicells or filaments of cyanobacteria. A strain of <u>Synechococcus</u> cf. <u>lividus</u> grows at temperatures as high as 70-74°C (Castenholz, 1984b; Brock, 1978). Although <u>Chloroflexus</u> will grow in co-culture with this <u>Synechococcus</u> at 70°C and can be isolated from <u>Synechococcus</u> films batched continuously by water in the 74-72°C range, a distinct mat of <u>Chloroflexus</u> underlying the cyanobacterial layer does not develop until waters have cooled in the outlet to about 68-69°C (Fig. 1). Although <u>Chloroflexus</u> is capable of at least minimal growth above 70°C in natural mats, it has not been grown in pure culture above 66°C (Castenholz, unpublished data). In many geographic regions, however, such as Iceland, the highest temperature form of cyanobacterium in

Fig. 2. Co-culture of <u>Synechococcus</u> cf. <u>lividus</u> (OH-53-S, U. of Oregon) and <u>Chloroflexus aurantiacus</u> (J-10-fl, U. of Oregon) on autotrophic DG medium with 15 g agar/liter. The light streaks are <u>Chloroflexus</u> and are yellow to orange in color. Bar indicates 20 µm.

non-sulfide springs is an undescribed species of Chlorogloeopsis, here referred to as "HTF" (high temperature form) and formerly referred to as "HTF Mastigocladus" (Castenholz, 1978). This cyanobacterium grows only up to temperatures of 63-64°C. Consequently, in those areas Chloroflexus is restricted to the temperatures no higher than this, since hot spring flows generally carry very little dissolved organic matter until passing over a mat of primary producers.

The tight coupling between cyanobacteria and Chloroflexus in non-sulfide hot springs, the physiological characteristics of Chloroflexus in culture, and the culturing of Chloroflexus in autotrophic medium in co-culture with a cyanobacterium (Fig. 2) all indicate the dependence of Chloroflexus on an organic substrate derived from a proxiamte organism. In co-culture with Synechococcus (Fig. 2) a direct release product, such as glucose, is involved, but in a hot spring intermediate microorganisms may be involved. Acetate added to hot spring mats is readily assimilated, particularly in the light, and is preferentially taken up by the filaments presumptively identified as Chloroflexus (Ward et al., 1984). Propionate, butyrate, lactate, and ethanol have similar fates. Ward et al. (1984) speculated that fermentative bacteria (including acetogens) may be normal intermediates in the tie between cyanobacteria and Chloroflexus.

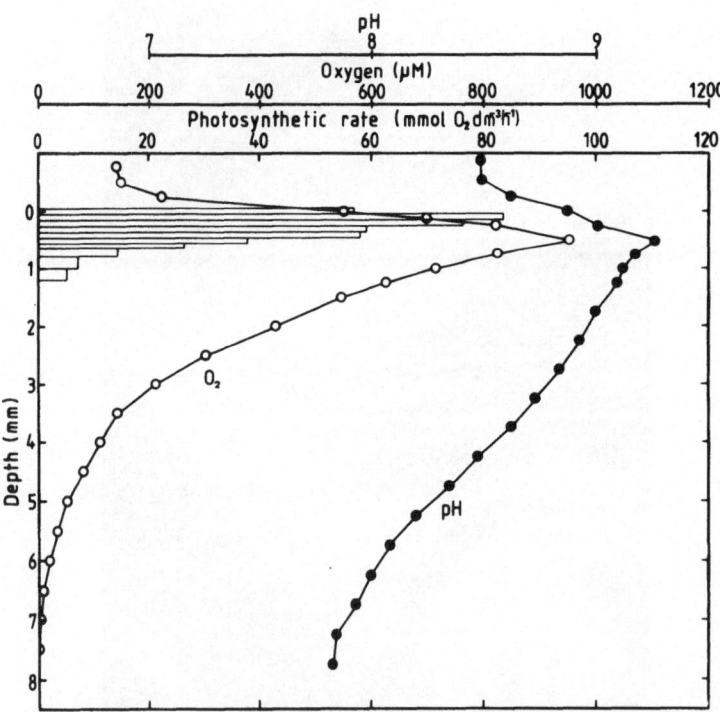

Fig. 3. Profiles of O_2 (open circles), pH (closed circles), and rates of oxygenic photosynthesis in the mat of Octopus Spring, Yellowstone National Park at 55°C. At the surface, the photon flux density was 1,720 μ Einsteins/m² s (11:15 A.M.). The upper 1 mm of mat was predominantly Synechococcus sp.; Chloroflexus was dominant below that. (Revsbech and Ward, 1984; courtesy of Applied and Environmental Microbiology).

The <u>Chloroflexus</u> layers so prominantly displayed and laminated under the top layer of cyanobacteria were once thought to represent an anoxic zone, but are now known to develop O_2 tensions several times the saturation value during daylight hours (Fig. 3). However, during darkness all layers are anoxic. It is assumed that bacteriochlorophyll synthesis occurs during this period, although in culture a low O_2 tension (1%) is required for pigment synthesis in the dark (Redlinger and Fuller, 1985). Undermats of <u>Chloroflexus</u> are a yellow-orange, orange, or reddish color which represents a high content of carotenoid pigments relative to bacteriochlorophylls, particularly to bacteriochlorophyll <u>c</u>. Although a very high light intensity, under anoxic conditions, can produce this state, there is a greater sensitivity of BChl <u>c</u> synthesis to moderate or higher O_2 tensions (Pierson and Castenholz, 1974b; Sprague et al., 1981). O_2 may be the principal agent controlling pigments within microbial mats, since visible wavelengths of light are greatly attenuated by passing through the cyanobacterial topmat.

The descriptions above refer only to hot springs n which free sulfide is very low (< 10 µM) in source water or at least in water in the 60-70°C range. When sulfide is present in higher concentrations two major differences may be seen (Fig. 4). The higher temperature species and strains of cytanobacteria are excluded. Inhibition by sulfide has been demonstrated in at least some of these (Castenholz, 1976). However, sulfide-tolerant cyanobacteria of several genera replace the other types, but not of these is capable of growth above 56°C (Castenholz, 1976; Fig. 2). In some spring systems a cyanobacterium (<u>Spirulina</u>) with an upper limit of 51°C predominates (Castenholz, 1977).Some types of <u>Chloroflexus aurantiacus</u>, however, are quite sulfide-tolerant and even capable of slow photoautotrophic growth with sulfide as reductant (Castenholz, 1973; Madigan and Brock, 1975; Brock, 1978; Giovannoni et al., 1987a). Thus, in sulfide-rich springs with a pH level above 6.0, <u>Chloroflexus</u> mats develop upstream of cyanobacterial mats, and often up to a temperature of about 66°C (Fig. 4). These mats are complately independent of the cyanobacteria. They are known in Iceland, western U.S.A. (mainly Yellowstone National Park) and in New Zealand (Castenholz, 1973). They are not as common as might be expected. High sulfide levels are uncommon in hot springs with a pH above 6.0. In most cases sulfide-rich hot water becomes acid (pH 1-3) as sulfide is oxidized when the water rises to the surface in volcanic areas. In

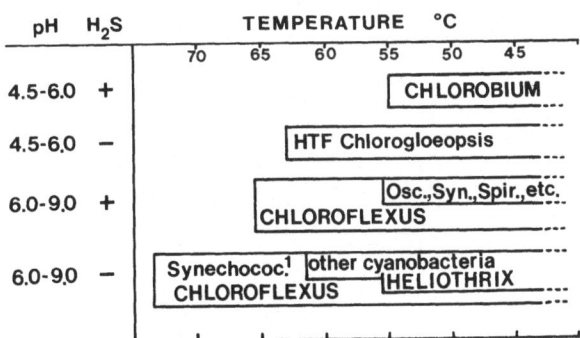

Fig. 4. Summary of the principal distributions of selected "green bacteria" of hot springs in relation to temperature, pH, free sulfide, and cyanobacteria. Geotraphic constraints are not considered. "Green bacteria" are in upper case. [1]A high temperature strain of <u>Synechococcus</u> occurs to temperatures of 73-74°C in western N. America. In non-sulfide springs in Iceland, New Zealand, Europe and many other areas cyanobacteria and <u>Chloroflexus</u> occur only up to temperatures of 63-64°C.

addition, many alkaline springs arising at high source temperatures lose all or most of the sulfide through abiotic or biotic oxidation or through gaseous loss before cooling downstream to temperatures at which <u>Chloroflexus</u> or cyanobacteria can grow.

In some springs the "pure" <u>Chloroflexus</u> mats upstream of cyanobacteria are as orange as they are when in association with cyanobacteria (Castenholz, 1973) (Fig. 5). There is a positive correlation between shallow depth or turbulent flow and orange, independent <u>Chloroflexus</u> mats. The presence of O_2 appear to control the orange coloration of such mats.

In Iceland several sulfide-containing alkaline springs produce orange mats upstream and independent of cyanobacteria (Fig. 5). However, in some of these thermal streams below ca. 63°C, cyanobacterial zones are found a few mm below the top layer of <u>Chloroflexus</u>. Two such streams in southern Iceland were investigated by Jørgensen and Nelson (1987). The cyanobacterial

Fig. 5. Outflow of "Bad.-1" spring, Hveragerdi, Iceland. The direction of the flow is toward the foreground. The area indicated by <u>C</u> is a light orange-red color and occupies the temperature range of 66°C to about 60-62°C. This mat, partly in the form of filamentous streamers, consists of <u>Chloroflexus</u>. No phototrophs occurred above this zone (upper border indicated by arrows). The dark patches on the stream edge are primarily "HTF"-<u>Chlorogloeopsis</u> (below ca. 62°C). The pH was about 8.6; sulfide over the <u>Chloroflexus</u> zone ranged from 60 to 5 µM.

undermat consisted of "HTF" Chlorogloeopsis ("HTF Mastigocladus") which is sensitive to sulfide concentrations over 10 µM. The mat surface of Chloroflexus was bathed by anoxic water with higher sulfide concentrations. Yet, the cyanobacterium at 8-14 mm depth performed oxygenic photosynthesis, apparently protected by the blanket of sulfide-oxidizing Chloroflexus (Jørgensen and Nelson, 1987). In the Mammoth Hot Springs (Yellowstone National Park), several hot springs of the Upper Terraces issue anoxic water at 62-66°C (pH 6.2-6.7) with about 50-120 µM free sulfide. The Chloroflexus mats develop at these sources and are a deep, dull to satiny green on the surface even when exposed to the extremely high light intensities of 1950 m elevation in summer (Fig. 6). Since the sulfide-containing water originates at the appropriate temperature for Chloroflexus, a photoautotrophic Chloroflexus mat develops without any cyanobacteria or other phototrophs (Giovannoni et al., 1987a). The mat is firm and laminated with lower layers of a more yellow to orange color. Mats of this type are possible analogs to Precambrian stromatolite-forming mats that could ahve preceded in time those dominated by oxygenic prokaryotes.

The "green Chloroflexus" (GCF) mats show a significant sulfide-stimulated uptake of HCO_3^- and acetate, and culture isolates behave similarly: sulfide stimulates HCO_3^- incorporation to a much greater degree than in the more studied strains of Cf. aurantiacus (Giovannoni et al., 1987a). Nevertheless, GCF strains also appeared to be photoheterotrophic by "preference". The principal difference between these isolates and those of Cf. aurantiacus was that GCF strains were incapable of aerobic, dark chemoheterotrophic growth, although exposure to O_2 did not irreversibly inhibit growth unless in combination with high light intensity. In this case

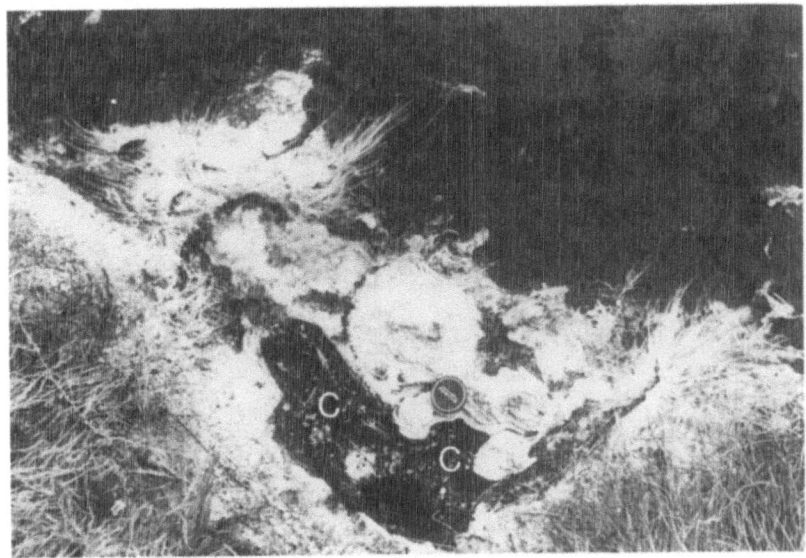

Fig. 6. Hot spring source (64°C) at edge of Painted Pool, Mannoth Springs, Yellowstone National Park. The source opening is the darkest patch below the lens cap. The dark area surrounding this area is the cover of GCF Chloroflexus (C). It is a dull, dark, satiny green. The white area is both a Thermothrix-like bacterium and deposits of travertine. The upper part of the photograph is an orange-red pool mat below about 50°C and consists of a variety of cyanobacteria and other microorganisms. Source pH is 6.4, sulfide about 80 µM. Lens cap is 5.3 cm in diameter.

a rapid degradation of BChl c occurred (Giovannoni et al., 1987a). In field studies using sulfide microelectrodes, a light mediated uptake of sulfide was clearly demonstrated by GCF mats. Also a significant production of sulfide occurred within the mat (Fig. 7). Although it was shown that the uppermost mm of the GCF mat is photoautotrophic, it is possible that the underlying Chloroflexus layers are heterotrophic, relying on organic products from the top layer. It is unlikely that light sufficient for phototrophic metabolism could penetrate much below the green GCF top layer (see below under Chlorobium mats). It is also possible, but only speculative, that at least some of the H_2S produced in the undermat is through the fermentative reduction of elemental sulfur by Chloroflexus itself (Giovannoni et al., 1987a).

Chlorobium

In hot springs where pH levels are generally below 6 and above 4.5, Chlorobium-like green bacteria form essentially monospecific mats when the sulfide concentration is above about 50 µM (Fig. 4). The intolerance of most cyanobacteria to pH levels below about 6 is well known. This factor together with inhibitory sulfide concentrations generally excludes all cyanobacteria from springs in which Chlorobium predominates.

Hot springs with a pH range of 4.5-6.0 and little or no sulfide are generally inhabited by a monospecific population of sulfide-intolerant "HTF" Chlorogloeopsis sp., at least at temperatures in the 63-45°C range (Fig. 4). This cyanobacterium is more tolerant of lower pH values than any other thermophilic cyanobacterium (Castenholz, 1978). Some Chloroflexus may be present.

The pH range 4-6 is relatively uncommon in hot spring waters throughout the world. There is, instead, a bimodal distribution of quite acid waters (pH 1-3) based on sulfuric acid and waters with pH levels above 6 based on the bicarbonate-carbonate buffering system. The common pH 1-3 waters support

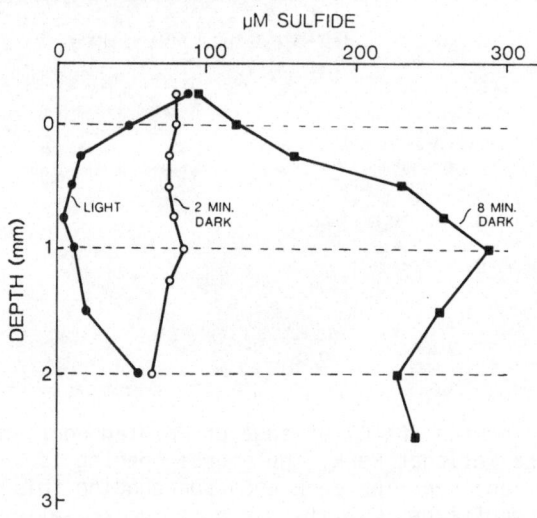

Fig. 7. Profiles of sulfide within a GCF Chloroflexus mat (64 C), Painted Pool, Mannoth Springs, Yellowstone National Park- measured by microelectrode. The photon flux density was 1990 µ Einstein m^{-2} s^{-1}. The pH (6.3) varied little; O_2 was absent throughout. (Giovannoni et al., 1987a; courtesy of Archives of Microbiology).

Syanidium caldarium (a eukaryote) as the only phototroph in the temperature range of 45-56°C. It is especially rare to find pH 4.5-6.0 waters with sufficient sulfide to support Chlorobium and also at an appropriately low temperature.

Such springs were discovered recently in New Zealand in the central volcanic area of North Island (Castenholz, unpublished data) (Fig. 4). In "Travel Lodge Stream" (TLS) (Sulphur Bay, Rotorua) and in other similar springs and thermal streams in the area, Chlorobium-like green bacterial mats cover the substrate from about 55°C to about 45°C or sometimes lower (Fig. 8). The pH varies but is generally in the range 4.5 to 6.0, occasionally higher or lower. In the main stream investigated (TLS), the "Chlorobium" mat was ca. 0.5-3.0 mm thick, unlaminated, compact, but slimy (Fig. 9). The top 0.1-0.2 mm was a yellowish-green color and contained a large amount of elemental sulfur. Below that the color was a dark, dull green (Fig. 9). The sulfide content of water bathing the mat was generally in the 0.2-0.5 mM range.

Like the GCF Chloroflexus mat in sulfide springs of higher pH, microelectrode data showed a rapid light-dependent uptake of sulfide and an equally rapid replenishment in the dark, again partly due to sulfide generated from within (R. W. Castenholz, J. Bauld, and B. B. Jørgensen, unpublished data). Spectroradiometric results show a reduction in wavelengths usable by Chlorobium to less than 0.1% of incident radiation by a vertical depth of 0.7 mm in the mat.

Fig. 8. A small "Chlorobium" spring at Parengarenga on the southwestern shore of Lake Rotoiti, New Zealand. The darker area (arrow) is a covering mat of a Chlorobium-like organism (dull green) from about 50°C (source temperature) to 42°C. The thermometer is near the source. The flow is toward the bottom and right. The white areas are deposits containing much elemental sulfur. The pH was about 5.7 and the sulfide concentration at the source ca. 0.2 mM.

The light-enhanced incorporation of [^{14}C]-HCO$_3^-$ was sulfide-stimulated, and most (but not all) experiments showed stimulation by sulfite (SO$_3^{2-}$) as well. Thiosulfate was ineffective. However, one culture isolate from this stream was able to use thiosulfate (M. Madigan, personal communication). The ability to grow at temperatures above 50°C was previously unknown for any species of <u>Chlorobium</u> or of other genera in the Chlorobiaceae. Other unique characteristics of thermophilic chlorobia may emerge from further work with pure cultures.

Heliothris

<u>Heliothrix oregonensis</u> is a recently described thermophilic, phototrophic bacterium (Pierson et al., 1984, 1985). In appearance it resembles <u>Cf. aurantiacus</u> but is a somewhat wider filament (1.5 µM in diameter), has a bacteriochlorophyll-like pigment with in vivo absorption maxima at 795 and 865 nm, lacks "internal" membranes (except for mesosomes), glides at rates of 0.1-0.4 µm/s, grows in culture (co-culture with <u>Isosphaera pallida</u>, Giovannoni et al, 1987b) under aerobic conditions. In some alkaline pH hot springs it develops bright orange, fluffy surface mats below about 56°C during conditions of high light intensity. The "BChl <u>a</u>" content is ca. 1.0 µg/ml cell dry wt or less, but carotenoid content is very

Fig. 9. A core from "Travel Lodge Stream" taken in a reach of about 45°C, pH 5.9, and a sulfide concentration of 0.25 mM. The diameter of the tube is ca. 3.5 cm. The dark green band on the core top is the <u>Chlorobium</u>-like organism (<u>C</u>) covered by a lighter zone with much elemental sulfur.

high. Although chlorosomes and their pigments are absent, the presence of γ-carotene derivatives as major carotenoids, the filament and cell morphology, and 5S r-RNA nucleotide sequences indicate that this filamentous, gliding organism is related to <u>Chloroflexus</u> more closely than to any other type of photosynthetic prokaryote (Pierson et al., 1985).

It is more difficult to predict its occurrence ore predominance in hot springs than the other "green bacteria". It does not occur in waters of high sulfide (or sulfate); it appears to be restricted to springs of pH above 6.5 or 7.0, but nevertheless there are numerous waters with these conditions and with temperatures between 55 and 40°C that apparently do not harbor populations of <u>Heliothrix</u>.

In Oregon hot springs (e.g., Kahneeta Springs), a bright orange, fluffy mat about 1-5 mm thick develops as a continuous top layer. This is underlain by a deep green zone of cyanobacteria (Fig. 10) which is protected from high intensity short wavelength light by the high carotenoid content of the <u>Heliothrix</u>. The <u>Heliothrix</u> topmat is exposed to saturated levels of O_2 during daytime (Fig. 11). During darkness, it is likely that exposure to some O_2 continues by diffusion through the unconsolidated mat from the water above.

Although <u>Heliothrix</u> appears to be fairly limited in its ecological and geographical distribution, it has been found in some hot springs of Yellowstone National Park and a similar organism occurs as an orange to red layer deep within translucent mats maintained by "Spouting Springs" (Castenholz, 1984b). However, this filamentous organism has a BChl <u>a</u>-like pigment with peaks at 805 and 910 nm instead of at the 795 and 865 nm of the <u>Heliothrix</u> previously described (B. K. Pierson and R. W. Castenholz, unpublished data). This organism is yet to be cultured.

It is apparent from the recentness of many findings that still other members of the green bacterial group are likely to be found in hot springs and that the present summary is merely a progress report.

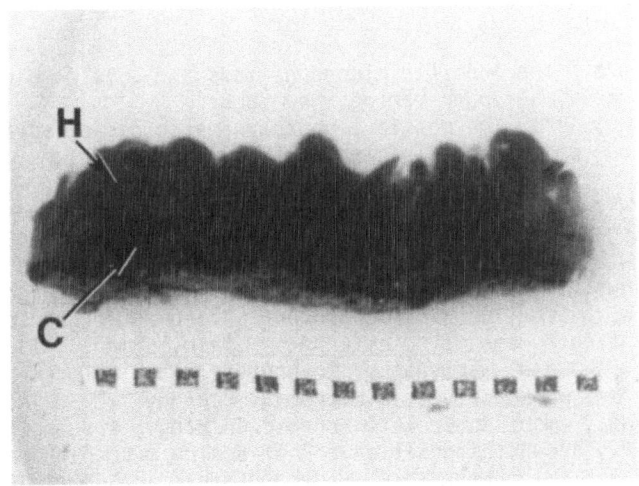

Fig. 10. Section of mat from Kahneeta Hot Springs, Oregon. The lighter colored upper portion (orange) of <u>Heliothrix</u> (<u>H</u>) covers a dense dark (green) layer of cyanobacteria (<u>C</u>). Temperature was ca. 48°C, pH 8.5. The units on the scale are 0.8 mm.

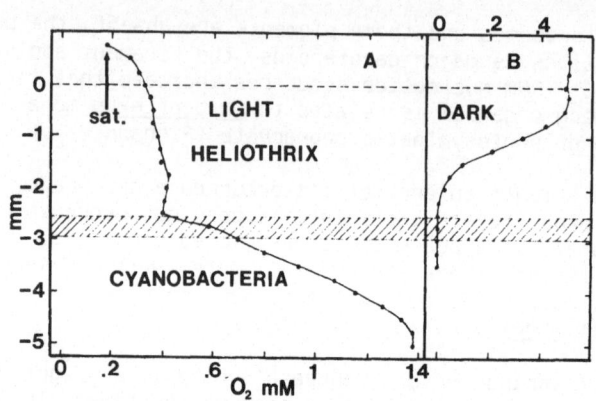

Fig. 11. O₂ profile through a Heliothrix/cyanobacteria mat at Kahneeta Hot Springs, Oregon. The transition between Heliothrix and cyanobacteria in this sample was between 2.5 and 3.0 mm. Temperature 40°C; O₂ saturation value was 0.188 mM. A. The Heliothrix zone was slightly supersaturated with O₂ during daylight hours (irradiance: ca. 790 W m⁻² at mat surface); the cyanobacterial layer was greatly supersaturated. B. In darkness (for 1 h). The O₂ level had fallen to zero in the cyanobacterial layer but some remained in most of the Heliothrix zone. The water moving over the top of the mat from light-exposed areas upstream carried about 2.5 x saturated levels of O₂.

ACKNOWLEDGEMENTS

The author would like to thank the U. S. National Science Foundation for financial support during parts of these studies.

REFERENCES

Brock, T. D., 1978, "Thermophilic Microorganisms and Life at High Temperatures", Springer-Verlag, New York.

Castenholz, R. W., 1973, The possible photosynthetic use of sulfide by the filamentous phototrophic bacteria of hot springs, Limnol. Oceanogr. 18:863.

Castenholz, R. W., 1976, The effect of sulfide on the blue-green algae of hot springs. I. New Zealand and Iceland, J. Phycol., 12:54.

Castenholz, R. W., 1977, The effect of sulfide on the blue-green algae of hot springs. II. Yellowstone National Park, Microbial Ecol., 3:79.

Castenholz, R. W., 1978, The biogeography of hot spring algae through enrichment cultures, Mitt. Internat. Verein. Limnol., 21:296.

Castenholz, R. W., 1984a, Habitats of Chloroflexus and related organisms, in: "Current Perspectives in Microbial Ecology," M. J. Klug and C. A. Reddy, eds., American Society for Microbiology, Washington, D. C.

Castenholz, R. W., 1984b, Composition of hot spring microbial mats: a summary, in: "Microbial Mats: Stromatolites," Y. Cohen, R. W. Castenholz, and H. Halvorson, eds., Alan R. Liss, New York.

Cline, J. D., 1969, Spectrophotometric determination of hydrogensulfide in natural waters, Limnol. Oceanogr., 14:454.

Giovannoni, S. J., Revsbech, N. P., Ward, D. M., and Castenholz, R. W.,

1987a, Obligately phototrophic _Chloroflexus_: primary production in anaerobic hot spring microbial mats, _Arch. Microbiol._, 147:80.

Giovannoni, S. J., Schabtach, E., and Castenholz, R. W., 1987b, _Isosphaera pallida_, gen. and comb. nov., a gliding, budding eubacterium from hot springs, _Arch. Microbiol._, 147:276.

Jørgensen, B. B., and Nelson, D. C., 1987, Bacterial zonation, photosynthesis, and spectral light distribution in hot spring microbial mats of Iceland, _Microbial Ecol._, in press.

Madigan, M. T., and Brock, T. D., 1975, Photosynthetic sulfide oxidation by _Chloroflexus aurantiacus_, a filamentous, photosynthetic, gliding bacterium, _J. Bacteriol._, 122:782.

Pierson, B. K., and Castenholz, R. W., 1974a, A phototrophic gliding filamentous bacterium of hot springs, _Chloroflexus aurantiacus_, gen. and sp. nov., _Arch. Microbiol._, 100:5.

Pierson, B. K., and Castenholz, R. W., 1974b, Studies of pigments and growth in _Chloroflexus aurantiacus_, a phototrophic filamentous bacterium, _Arch. Microbiol._, 100:283.

Pierson, B. K., Giovannoni, S. J., and Castenholz, R. W., 1984, Phynologica ecology of a gliding bacterium containing bacteriochlorophyll _a_, _Appl. Environ. Microbiol._, 47:576.

Pierson, B. K., Giovannoni, S. J., Stahl, D. A., and Castenholz, R. W., 1985, _Heliothrix orgegonensis_, gen. nov., sp. nov., a phototrophic filamentous gliding bacterium containing bacteriochlorophyll _a_, _Arch. Microbiol._, 142:164.

Redlinger, T. E., and Fuller, R. C., 1985, Protein processing as a regulatory mechanism in the synthesis of the photosynthetic antenna in _Chloroflexus_, _Arch. Microbiol._, 141:344.

Revsbech, N. P., and Jørgensen, B. B., 1986, Microelectrodes: their use in microbial ecology, _Adv. Microbial Ecol._, 9:293.

Revsbech, N. P., and Ward, D. M., 1984, Microelectrode studies of interstitial water chemistry and photosynthetic activity in a hot spring microbial mat, _Appl. Environ. Microbiol._, 48:270.

Sprague, S. G., Staehelin, L .A., and Fuller, R. C., 1981, Semiaerobic induction of bacteriochlorophyll synthesis in the green bacterium _Chloroflexus aurantiacus_, _J. Bacteriol._, 147:1032.

Ward, D. M., Beck, E., Revsbech, N. P., Sandbeck, K. A., and Winfrey, M. R., 1984, Decomposition of hot spring microbial mats, _in_ "Microbial Mats: Stromatolites," Y. Cohen, R. W. Castenholz, and H. Halvorson, eds., Alan R. Liss, New York.

ECOLOGICAL NICHES OF GREEN SULFUR AND GLIDING BACTERIA

V.M. Gorlenko

Institute of Microbiology
USSR Academy of Sciences
Moscow 117811, U.S.S.R.

INTRODUCTION

Green bacteria possess unique properties, which make them different from other photosynthetic bacteria (Pfennig, 1967,1977,1978; Ormerod, 1983). Only green bacteria are characterized by the special light-harvesting antenna structures known as chlorosomes and by the specific chlorophylls, BChl c, d, or e, in addition to BChl a.

Modern phylogenetic data show that green bacteria are divided into two branches evolutionarily remote from each other: unicellular green sulfur bacteria and filamentous gliding green bacteria (Fox et al., 1980; Gibson et al., 1985; Woese, 1985). Representatives of these two groups of phototrophic bacteria differ markedly in eco-physiological functions (Pfennig, 1977; Gorlenko et al., 1983).

Evolutionary pathways of unicellular and filamentous green bacteria parted in ancient times, and they are distantly related to other phototrophic and chemotrophic microorganisms. The most deeply isolated is the group of green sulfur bacteria, which includes non-motile representatives of the genera Chlorobium, Pelodictyon, Prosthecochloris and Ancalochloris (Pfennig, 1977; Kondratieva and Gorlenko, 1978) as well as the flexible gliding Chloroherpeton thalassium (Gibson et al., 1984), somewhat similar to filamentous green bacteria. The similarity coefficient S_{AB} of Ch. thalassium with other genera of green sulfur bacteria is 0.4-0.49, whereas non-motile species are less distant; their S_{AB} values are from 0.58 to 0.67. These data indicate, on the one hand, the compactness of the group of the green sulfur bacteria, and on the other hand, its marked divergence, although within certain limits. For comparison it may be noted that among the representatives of the purple sulfur bacteria closely related to each other S_{AB} is not lower than 0.7, and among purple non-sulfur bacteria there are two heterogeneous clusters separated from each other by S_{AB} of 0.3 (Stackebrandt et al., 1984).

Unlike purple bacteria, green sulfur bacteria are not closely related to chemotrophs and they must be a dead-end branch of evolution.

The filamentous gliding green bacterium Chloroflexus aurantiacus is distantly related to morphologically similar heterotrophs of the genus Herpetosyphon (S_{AB} = 0.3-0.4) (Gibson et al., 1985), and evidently they are phylogenetically connected with an unusual filamentous phototrophic bacterium <u>Heliothrix oregonensis</u>, which is devoid of chlorosomes (Pierson et al., 1985).

The evolutionary positions of the unicellular and the filamentous green bacteria correlate closely with their ecological niches. Evolution of microorganisms in the historical past was determined by the evolution of the habitat.

Bacteria must have reacted rapidly to the change of the environment, acquiring new properties and consolidating and improving the old ones, when they remained adapted to the external conditions. Evidently, with the evolution of the biosphere, some ancient forms remained and improved, while other microorganisms with new properties appeared. Qualitative differentiation of the previously monotonous biosphere led to its spatial division into ecological niches which were gradually occupied by various microorganisms. Finally, the conditions of the habitats were preserved more or less, and consequently, microorganisms similar to those which existed at the different stages of biological evolution on the earth, were preserved as well. In other words, prokaryotic microorganisms are living fossils, keeping evolutionary records in their biological code. Therefore, the study of the information macromolecules should result in the reconstruction of the phylogenetic origins of the microbial world.

Analysis of ecological niche distribution shows that evolutionary selection among phototrophic bacteria has resulted in the formation of three groups of microorganisms (Gorlenko, 1981): two extreme groups of specialists - strictly anaerobic, photolithotrophic bacteria and photoorganotrophs; as well as an intermediate group of universalists, performing both anaerobic photo- and aerobic chemotrophic metabolism.

Among nonsulfur purple bacteria there are no obligate photoautotrophs, but their place is occupied by bacteria of the genus <u>Ectothiorhodospira</u>, which are phenotypically similar to Rhodospirillaceae.

There is a vacancy for obligate photoheterotrophs among the purple sulfur bacteria (with the exception of <u>Chromatium purpuratum</u>, which needs organic acids even in the presence of a sulfide). Among filamentous green bacteria, as well as nonsulfur purple bacteria, a dominant branch is that of photo-organoheterotrophs, universalists capable of switching to dark aerobic metabolism.

The development of bacterial universalism is a progressive trend of evolution, leading to the emergence of species via the loss of phototrophic function. Therefore, the greatest number of chemotrophic relatives is observed in the cases of purple bacteria and filamentous green bacteria (Woese, 1985; Gibson et al., 1985). On the other hand, green sulfur bacteria have no close relatives, because they have developed only one evolutionary branch as strict anaerobes, specialists, and autotrophs depending on sulfide.

GREEN SULFUR BACTERIA

The ecological niche of green sulfur bacteria in natural ecosystems is precisely determined. They compete strongly with sulfur purple bacteria, for which photolithautotrophic and photolithoheterotrophic metabolisms are also basic. However, in stratified lakes green sulfur bacteria dominate more often (Gorlenko et al., 1983). We find an explanation for this phenomenon in the structural peculiarities of the photosynthetic apparatus of green bacteria. Purple bacteria have 50-500 light-harvesting molecules of BChl <u>a</u>

or b per reaction centre. These are located in the cytoplasmic membrane or in the chromatophore membrane (Ormerod, 1983). Purple bacteria are able to increase the effectiveness of the photosynthetic apparatus only by increasing the membrane surface of the chromatophores, which are formed by invagination of the cytoplasmic membrane (Fig. 5). Vesicular chromatophores occupy most of the cytoplasm in purple sulfur bacteria.

Unlike purple bacteria, green bacteria possess a highly differentiated photosynthesizing apparatus, including specialized antenna structures, chlorosomes (Staehelin et al., 1978, 1980; Schmidt, 1980; Olson, 1981). Chlorosomes of the green sulfur bacteria look like oval bodies 30-70 x 100-260 nm, enclosed in a thin (2-nm thick) envelope. Each chlorosome consists of 10-30 rod-like elements approximately 10 nm in diameter, which are supposed to contain a BChl-protein complex. Chlorosomes are attached to the internal side of the cytoplasmic membrane, contacting the reaction centres. Chlorosomes contain BChl c, d, or e and most of the carotenoid pigments, whereas BChl a, some carotenoids and a specialized BChl c are localized in the cytoplasmic membrane. There is one reaction centre per 1000-2000 molecules of BChl c, d, or e and 100 molecules of BChl a. Some strains (Chlorobium limicola) contain up to 20% BChl c dry weight.

The photosynthetic apparatus of the green bacteria provides for the high efficiency of solar energy utilization, and they can exist under rather low illumination. The lowest light level for green bacteria is about 20 lux (Biebl and Pfennig, 1978; Parkin and Brock, 1980; Abella et al., 1980). In nature green bacteria monopolize a poorly illuminated anaerobic niche in sulfide-containing lakes (Gorlenko et al., 1983).

Some representatives of the green sulfur bacteria possess adaptive physiological and morphological properties, which allow them to occupy totally the low light ecological niche and also to exceed its limits (Pfennig, 1967; Gorlenko, 1981).

In the course of microevolution there appeared narrowly specialized forms, adapted to the particular conditions of the habitats. These forms are systematized as species. The ability of some species to live under low illumination is increased by the formation of cell appendages, prosthecae, which carry additional chlorosomes (Gorlenko et al., 1983). The existence of prosthecae provides for a higher ratio of the surface area (number of chlorosomes) to cell volume, which significantly improves the energy supply of green bacteria (Table 1). Prosthecae are found in bacteria of the genera Prosthecochloris, Ancalochloris and Pelodictyon (Pfennig, 1977; Gorlenko, 1970; Gorlenko and Lebedeva, 1981; Puchkova and Gorlenko, 1970). These microorganisms occupy the dimmest ecological niches both as benthic and planktonic forms. The prosthecae length of Prosthecochloris varies, depending on the light intensity. Prosthecae are longer in dim light. In this way these green bacteria regulate their energy potential. Prosthecae are no expedient for purple bacteria because they lack chlorosomes.

We suppose that for deep-water species and for some benthic forms of green bacteria gas vesicles, besides flotation, possess an additional function, that of light-trapping. Due to their optical properties central gas vacuoles and gas vacuoles near the cross cell wall reflect the light, redirecting it to the chlorosomes. The role of gas vesicles in the regulation of cell surface to volume ratio is also very important, and consequently, it is important in the regulation of cell energy potential (Table xxx). In such species as Chloronema and Chloroherpeton gas vesicles occupy most of the cytoplasm.

Green bacteria occupy a special light niche, as compared to purple bacteria, not only in respect to the light intensity. Light-harvesting BChls

Table 1. Ecomorphological Characteristics of Green Bacteria.

Species	Cell size μm	V/S	Gas vacuoles	Colour
Chlorobium				
chlorovibrioides	0.3-0.4x	0.06-	absent	green
phaeovibrioides	0.6-0.8	0.08		brown
limicola	0.5-0.7x	0.1-	absent	green
phaeobacteroides	1-2	0.2		brown
vibrioforme	0.5-0.7x	0.160,	absent	green
	1.5	14		
Pelodictyon				
luteolum	0.5-0.7x	0.1-	present	green
phaeum	1.5	0.14		brown
clathratiforme	0.5-1.5x	0.23-	present	green
sp.	1.2-4	0.3		brown
Ancalochloris	0.7-1.0	0.1-	present	green
perfilievii	1.5-2	0.3		
Prosthecochloris	0.5-0.7x	0.04	absent	green
eastuarii	1-1.2			brown
phaeoasteroidea				
Clatrochloris	1x	0.18-	present	green
sulfurica	1.4-1.8	0.2		
Chloroflexus	0.7x1.4	0.18	absent	green
aurantiacus				
Chloronema	2-2.5x	0.37-	present	green
giganteum	2.7-3.6	0.62		
Oscillochloris	4x7	0.78	present	green
chrysea	(0.5x4)*	(0.11)*		
trichoides	1-1.2x	0.25-	present	green
	2-2.4	0.3		

Notes: V = cell volume;   S = surface area; * = incomplete septa.

$\underline{c}$, $\underline{d}$, or $\underline{e}$ have absorption maxima _in vivo_ within the region 760 to 710 nm, which differ from those of Chl $\underline{a}$ (670 nm) and BChls $\underline{a}$ (805 nm) and $\underline{b}$ (1017 nm).

Among the green sulfur bacteria there are two subgroups of morphologically similar bacteria, which differ in pigment compositon (Table 1). The first subgroup includes microorganisms having a green color. They contain BChl $\underline{c}$ or $\underline{d}$ and a small amount of carotenoid (chlorobactene). The second subgroup includes brown-colored bacteria, containing BChl $\underline{e}$ and a great amount of the carotenoid, isorenieratene. At present we know of green and brown pairs for practically all the species of green sulfur bacteria. Symbiotic consortia containing green or brown sulfur bacteria are also known. Enrichment with carotenoids allows "brown" bacteria to occupy a short-wavelength light niche in deep-water stratified lakes independent of green bacteria.

Pfennig (1967) subdivided photosynthetic bacteria into two ecological subgroups depending on their relation to light, sulfide concentration and temperature. The first group, characterized as eurysulfidophylls and

Table 2. Conditions for the Development of Green Sulfur Bacteria in Fresh
Water Dimictic and Meromictic Lakes of the U.S.S.R. (Gorlenko et
al., 1983)

Species[1] or Consortia[2]	t(°C)	H₂S (mg/l)	pH	Eh (mv)	Light* (%)	Depth occur. (m)	Fe²⁺ (mg/l)
[1]Ancalochloris perfilievii	6.3-15	0.75-13 (60)	6.85-7.3 (8.3)	-45-0	0.01-7	0.75-7.5	+ -
[1]Chlorobium phaeobacteroides	5-5.8	0.7-0.5	7.7-7.9	0- -100	0.08-0.13	20-26	2.3
[2]Chlorochromatium sp.	6	traces	-	-	-	3.2	1.8
aggregatum	6-12.5	0.3-9	5.9-7.1	-50-+120	0.3-3.3	1.2-6.5	+
[2]Chloroplana vacuolata	6	traces	-	-	-	3.2	1.8
[2]Pelochromatium roseum	4.6-8.4	0.1-2.7	6.2-7.7	-100-50 (+250)	-	4.3-13.7	+
roseo-viride	4.6-5.2	0.05-1.5	7.2-7.3	+70 (+260)	1.5-2	10.5-10.85	+
[1]Pelodictyon aggregatum	4.5-16	0.15-7	6.7-7.3	-80-+100	1-6	3-11	+
clathratiforme	6.3-9	1.5-12 (60)	6.85-7.3	-150-0	0.01-1	5.5-7.0	+ -

* Light intensity (% of surface intensity).

mesophotophylls, includes purple bacteria of the genera Chromatium,
Thiospirillium, Thiocystis and Thiocapsa as well as green bacteria of the
genus Chlorobium. We assign the green sulfur bacteria Prosthecochloris and
marine Pelodictyon spp. to this group too. These are mainly benthic
microorganisms. Bacteria of the genera Chlorobium and Prosthecochloris
demonstrate the greatest tolerance for sulfide (4-8 mM) in comparison with
purple sulfur bacteria (0.8-4.0 mM). Due to this fact they can develop by
themselves in sulfur springs even under very high insolation conditions. The
second group is oligosulfidophylls and oligophotophylls, which accumulate at
low light intensity, low sulfide content and low temperature. These
conditions are found in fresh meromictic and dimictic lakes (Gorlenko et
al., 1983). It is not accidental that planktonic phototrophic bacteria
containing gas vacuoles including green sulfur bacteria Pelodictyon
clathratiforme and Ancalochloris fall into the second group. Their tolerance
for sulfide is low (0.4-2 mM).

The ecological requirements of some green bacteria simultaneously for
dim light, low sulfide concentration and low temperatures are accounted for
from the energetic point of view. Under low light and low temperature
conditions photooxidation of H_2S occurs slowly but more completely to SO_4^{2-}
without formation of the intermediate $S°$. Four moles of CO_2 are fixed per 1

Table 3. Conditions for the Development of Green Sulfur Bacteria in Saline Lakes (Gorlenko et al., 1983)

Species	Salinity (%)	H$_2$S (mg/l)	t(°C)	pH	E$_n$ (mV)	Depth of occurrence (m)
Chlorobium						
chlorovibrioides	1-2.8	34-320	13	6.4-7.8	-220	0.1-3.25
phaeovibrioides	1-3	30-34	8-13	7.0-7.2	+5- -60	5.7-9.5
vibrioforme	0.7-10.3	119-300	16-28	7.0-7.9	-130- -290	0.15-2.0
Pelodictyon						
luteolum	1-3.5	20-34	12	7.2	-130	2-5.5
phaeum	0.4-3	30-37	3.5-24	7.0-7.7	-40- -170	1.5-10
Prosthecochloris						
aestuarii	0.7-17	42-1600	-4-28	7.2-8.9	-50- -282	
phaeoastheroidea	3-3.5	10-30	13	7.0-7.5	-60- -240	10-13

mole of H$_2$S oxidized. Thus the development of different species of green sulfur bacteria occurs within a wide range of H$_2$S concentration from trace quantities up to 8 mM, the pH being 5.9-7.9 (Tables 2 and 3).

Sulfide is a toxic compound, especially when pH values are low. It is undissociated H$_2$S molecules that are inhibitory. Different sensitivities of phototrophic bacteria to sulfide accounts for the fact that in nature the concentration of hydrogen sulfide becomes the main factor determining the development of this or that species (van Gemerden, 1974; van Gemerden and Beeftin, 1981). When the sulfide content is low, dominance of the species is determined by its saturation constant K(S^{2-}). If the saturation constant (sulfide concentration providing for the highest growth rate of a given microorganism corresponds to or approaches the sulfide content in the medium, this species displaces the competitors with other K(S^{2-}) values (van Gemerden and Jannasch, 1971). If the habitat has a relatively high sulfide concentration, the competition is controlled by the tolerances of the organisms to sulfide. The green sulfur bacteria Cb. limicola at optimal light has a relatively low growth rate on sulfide 0.07 µ max(h^{-1}) and are not able to compete with purple sulfur and even some nonsulfur bacteria 0.12-0.14 µ max(h^{-1}).

Dominance of green bacteria arises form their higher resistance to sulfide and higher affinity for the substrate as well as from its fuller utilization.

Syntrophic associations of green sulfur bacteria with chemo-organotrophic sulfate-reducing and sulfur-reducing bacteria are of great interest. Syntrophic growth of green bacteria takes place when there are trace quantities of sulfide, due to its constant regeneration by the chemo-organotrophic partner. Such associations are well studied in laboratory conditions (Biebl and Pfennig, 1978). No close contact was observed between green and H$_2$S-producing bacteria in mixed cultures. On the other hand in nature there often develop in mass quantities symbiotic consortia containing a central colorless bacterium surrounded with green sulfur bacteria of the genera _Chlorobium_ and _Pelodictyon_ (Gorlenko et al., 1983; Crome and Tyler, 1984). Consortia have a characteristic morphology, and they are traditionally called by binary Latin names: _Chlorochromatium aggregatum_, _Pelochromatium roseum_, _Pelochromatium roseoviride_, _Chlorochromatium glebulum_, _Cylindrogloea bacterifera_, _Chloroplana vacuolata_. The first four types of consortia are formed by a large colorless motile bacterium, to which green bacteria are closely attached, like the grains on an ear of corn. Two layers of phototrophs were revealed in _P. roseoviride_, the internal one being formed from _Chlorobium phaeobacteroides_, and the external one from green _Pelodictyon aggregatum_.

Chloroplana vacuolata consists of a central vacuolated filamentous colorless bacterium and of parallel rows of _P. aggregatum_. The secret of how consortia function has not yet been discovered, because they have not yet been obtained experimentally. However, there is reason to suggest that these are syntrophic associations, similar to the already mentioned ones. The observations show that consortia frequently develop in lakes enriched with ferric oxide and binding H_2S in the form of an unsoluble sulfide. Therefore, the development of green bacteria becomes possible only under conditions of syntrophism and close contact with sulfide-forming bacteria. The whole aggregate is enclosed with slime, which surrounds the consortium with a protective micromedium favourable for ionic exchange. Extremophyllity of green sulfur bacteria is observed only with respect to sulfide and light. There are no extreme thermophiles, halophiles, acidophiles and alkalophiles among them. Therefore the distribution range of green sulfur bacteria in nature is limited to a greater extent than that of purple bacteria.

The highest temperature at which we observed the green bacterium _Prosthecochloris_ was 54°C. It was found in cyanobacterial mats of the hot saline spring Talgi (Dagestan, Caucasus). The isolated culture had a temperature optimum for development of about 30°C and was identified as the mesophyll _Prosthecochloris aestuarii_. The presence of _Prosthecochloris_ was also observed in the heliothermic Solar Lake in the hypolimnion at about 60°C. However, cultures of green bacteria were not isolated from that lake.

Unlike purple bacteria among which there are extreme halophiles, green sulfur bacteria seldom develop in lakes with salinity above 100 o/oo (Table 3) (Gorlenko et al., 1981). As a rule halophility among green bacteria is a special character. A weak halophile is _Chlorobium chlorovibrioides_ (salinity optimum is 0.5-1% NaCl). Moderate halophiles with salinity optimum of 1-4% are _Prosthecochloris aestuarii_, _P. phaeoastheroidea_, _Cb. vibrioforme_, _Pelodictyon luteolum_, _Pd. phaeum_, and _Chloroherpeton thalassium_. The above mentioned species are not observed in fresh ponds. Moderate halophility of green sulfur bacteria correlates with their high tolerance to sulfide.

Green sulfur bacteria are not motile (except for _Ch. thalassium_), and due to their strict anaerobiosis they have to adapt themselves to various conditions of the habitat where they are found.

Planktonic and benthic forms of green sulfur bacteria have appropriate morphological devices. The dominant planktonic species in stratified lakes (_Pelodictyon_ spp., _Ancalochloris_ spp.) have gas vacuoles. Vertical migration is possible by changing the number of gas vacuoles. Gas vacuoles are formed in greater number at low temperatures of about 4°C even in the darkness, and their number decreases at higher temperatures (Pfennig, 1965). Green bacteria therefore float up in the cold water of the hypolimnion and are concentrated in the area of the thermocline at the lower boundary of sunlight penetration. Other planktonic species remain in the zone of water density gradient in the thermocline and halocline due to the small weight and size of individual cells (_Cb. chlorovibrioides_ and _Cb. phaeovibrioides_) (Gorlenko, 1981). Symbiotic consortia of green sulfur bacteria are also mainly planktonic forms. Most of them (Chlorochromatium aggregatum, Chlorochromatium glelum, Pelochromatium roseum, and Pelochromatium roseoviride) are actively motile, due to the central chemotrophic bacterium possessing flagella. The entire aggregate responds to light (positive taxis) and to oxygen (negative taxis), which makes it possible for consortia to migrate into the zone optimal for their existence. Both green bacteria and their chemotrophic partners in Chloroplana vacuolata are supplied with gas vesicles and migrate through the water body like other phototrophic bacteria with gas vacuoles.

Microcolonies of planktonic species have characteristic morphologies ensuring buoyancy and favourable illumination conditions for individual cells. They are often perforated and loose or monolayered and flat.

On the other hand colonies of benthic species of green sulfur bacteria are dense and form much slime by which they are fixed to a substrate. This creates a microenvironment that protects the cells from the changeable conditions of the boundary zone silt-water.

Thus, green sulfur bacteria, a highly specialized physiological group of phototrophic bacteria, have in the course of evolution acquired a number of adaptive morpho-physiological attributes which allow them to occupy completely low light niches of various water ecosystems containing sulfide and possessing certain stability.

FILAMENTOUS GREEN BACTERIA

Multicellular filamentous green bacteria are typically benthic microorganisms which move along the substrate by gliding. Due to photo- and chemotaxis they occupy the most advantageous position among mat-forming microorganisms.

Multicellular prokaryotes have a greater survival potential than one cell of the same species under the same conditions (Wittenbury and Dow, 1981). Interaction between their cells increases, which leads to a complex differentiation and specialized functions. Filamentous green bacteria are multicellular trichome microorganisms, in this respect similar to Oscillatoria-like cyanobacteria. They reproduce by trichome fragmentation into short sections called hormogonia. The functions of hormogonia are those of reproduction and spreading. They are actively motile and often possess gas vacuoles. Cell differentiation of various species of filamentous green bacteria has been insufficiently studied as yet.

Physiology of the thermophilic species, Cf. aurantiacus, has been studied best of all. On the basis of information obtained from a number of strains of this species, it was concluded that the group of filamentous green bacteria consists exclusively of universalists capable of performing different variants of anoxic photosynthesis, as well as of growing in the darkness at the expense of respiration. However, this opinion is beginning to change. Recently ecotypes of the thermophilic Chloroflexus, similar to Cf. aurantiacus but unable to carry out aerobic metabolism, have been described (Giovannoni et al., 1987). The new data show that filamentous green bacteria are rather heterogeneous with respect to both physiology and ecology.

The physiology of mesophilic Chloroflexaceae has been insufficiently studied so far. Only Cf. aurantiacus var. mesophilus, which is similar in physiology and morphology to the thermophilic species (Gorlenko, 1975), has been obtained in pure culture. Cf. aurantiacus var. mesophilus is one of the littoral mat components of the low sulfate fresh water lakes.

Halophilic _Chloroflexus_ sp. was observed in microbial mats of saline shallow ponds having salinity up to 100 o/oo (Gorlenko et al., 1978). It develops in the lower microzones of ponds rich in sulfide and apparently is photoautotrophic and strictly anaerobic. We obtained monocultures of this microorganism from mud samples from shallow lagoons of Lake Sivash and an estuary of the White Sea. They contained BChl _c_ as well as other green bacteria. We failed to purify the isolates and to study their physiology in detail.

Filamentous green bacteria of the genus _Oscillochloris_ contain gas vacuoles. The large species _O. chrysea_ formed rich formations in the bed of a fresh stream where domestic waste waters entered (Gorlenko and Pivovarova, 1978). Terminal cells of these trichome bacteria were differentiated. They have a mucous dome and carry out an attaching function. In _O. chrysea_ an unusual chlorosome arrangement was discovered - on the cytoplasmic membrane of transverse cell septa. Evidently such an arrangement of antenna structures is accounted for by the large size of the trichomes (d = 5 µm), which provides a larger volume of the cytoplasm in comparison with the cell surface. A high energy potential of the cells is maintained by the rational chlorosome arrangement and by the decrease in cytoplasm volume with the help of gas vacuoles.

It is interesting that the trichomes of _O. chrysea_ are stained Gram-positively and the cell wall is similar in structure to that of Gram-positive bacteria. Future studies of pure cultures of this unusual green bacterium will be of great interest for evolutionists and taxonomists.

Another species, _O. trichoides_, has been obtained recently in pure culture. This species prefers to grow anaerobically and photoautotrophically oxidizing H_2S to S^o. The metabolism of _O. trichoides_ is similar to that of _Ch. thalassium_ (Gorlenko and Korotkov, 1977).

The DNA-homology investigation of _O. trichoides_ and _Cf. aurantiacus_, as well as _Cb. limicola_ f. _thiosulfatophilum_ revealed a total lack of homology (Keppen, personal communication).

O. trichoides is a usual component of microbial mats of fresh water ponds with high sulfate content and a sulfide-rich silt.

The area of distribution of filamentous gliding bacteria is not restricted to benthic communities. _Chloromema giganteum_ often develops among the plankton of low-sulfide dimietic mesohumous ponds. The large trichomes of _Cm. giganteum_ (d = 2.5 µm) contain multiple gas vesicles which allow the bacteria to float. The microorganisms have a sheath and actively attach themselves to objects. They glide and show negative aerotaxis and positive phototaxis. It is evident that _Cm. giganteum_ can lead both planktic and benthic ways of life.

According to the above mentioned data filamentous gliding bacteria are a rather divergent group of green bacteria adapted mainly to benthic existence under low light conditions.

Phylogenetically filamentous green bacteria are separated from green sulfur bacteria, but they have in common a highly differentiated photosynthetic apparatus. Specialized antenna structures include chlorosomes and the light-harvesting pigments BChls _c_ or _d_. Filamentous green bacteria containing BChl _e_ have not been discovered yet.

The photosynthetic apparatus of unicellular and filamentous green bacteria is different in some details. Their reaction centres are most different. The reaction centre of green filamentous bacteria is similar to that of purple bacteria. There is a hypothesis that the presence of chlorosomes in phototrophs is determined by plasmids. However, it is evident that the highly differentiated photosynthetic apparatus of green bacteria has emerged in the course of a long evolution and perfection of ecological specialization.

REFERENCES

Abella, C., Montesionos, E., and Guerrero, R., 1980, Field studies on the competition between purple and green sulfur bacteria for available light (Lake Siso, Spain), Dev. Hydrobiol., 3:173.

Biebl, H., and Pfennig, N., 1978, Growth yields of green sulfur bacteria in mixed culture with sulfur and sulfate reducing bacteria, Arch. Microbiol., 117:9.

Croome, R. L., and Tyler, P. A., 1984, The microanatomy and ecology of Chlorochromatium aggregatum in two meromictic lakes in Tasmania, J. Gen. Microbiol., 130:2717.

Dubinina, G. A., and Gorlenko, V. M., 1975, New filamentous photosynthetic green bacteria with gas vacuoles (in Russian), Microbiologiya, 44:511.

Fox, G. E., Stackebrandt, E., Hespell, R. B., Gibson, J., Maniloff, J., Dyer, T. A., Wolfe, R. S., Balch, W. E., Tanner, R. S., Magrum, L. J., Zablen, L. B., Blakemore, R., Gupta, R., Bonen, L., Lewis, B. J., Stahl, D. A., Luchersen, K. R., Chen, K. N., and Woese, C. R., 1980, The phylogeny of the prokaryotes, Science, 209:457.

Gibson, J., Stackebrandt, E., Zablen, L. B., Gupta, R., Woese, C. R., 1979, A phylogenetic analysis of the purple photosynthetic bacteria, Curr. Microbiol., 3:59.

Gibson, J., Ludwig, W., Stackebrandt, E., and Woese, C. R., 1985, The phylogeny of the green photosynthetic bacteria: absence of a close relationship between Chlorobium and Chloroflexus, System. Appl. Microbiol., 6:152.

Giovannoni, S. G., Revsbech, N. P., Ward, D. M., and Castenholz, R. W., 1987, Obligately phototrophic Chloroflexus primary production in anaerobic hot spring microbial mats, Arch. Microbiol., 147:80.

Gorlenko, V. M., 1970, A new phototrophic green sulphur bacterium - Prosthecochloris aestuarii nov. gen., nov. spec. - Z. Allg. Mikrobiol., 10:147.

Gorlenko, V. M., 1975, Characteristics of filamentous phototrophic bacteria from freshwater lakes (in Russian), Microbiologiya 44:756.

Gorlenko, V. M., 1981, Purple and green bacteria and their role in carbon and sulfur cycles, Synopsis of thesis, doctor of biological sciences (in Russian), M.INMI, USSR Academy of Sciences, Moscow.

Gorlenko, V. M., and Lebedeva, E. V., 1971, New green bacteria with autgrowth (in Russian), Microbiologiya, 40:1035.

Gorlenko, V. M., and Pivovarova, T. A., 1977, On the assignment on the blue-green alga Oscillatoria coerulscens Gicklhorn, 1921 to the new genus of chlorobacteria Oscillochloris nov. gen. (in Russian), Izv. Akad. Nauk SSSR, Ser. Biol., 23:396.

Gorlenko, V. M., and Korotkov, S. A., 1979, Morphological and physiological peculiarities of the new filamentous gliding green bacteria Oscillochloris trichoides nov. comb. (in Russian), News of the USSR Academy of Sciences, Ser. Biol., 6:848.

Gorlenko, V. M., Dubinina, G. A., and Kuznetsov, S. I., 1983, The ecology of aquatic microorganisms, in: "Die Binnengewässer," Band 28, W. Ohle, ed., Stuttgart E. Schweizerbartsche Verlagsbuchhandling (Nägele u. Obermiller).

Gorlenko, V. M., Kompantzeva, G. I., Korotkov, S. A., Puchkova, N. N., and Savvichev, A. S., 1984, Development conditions and species composition of phototrophic bacteria in saline shallow ponds of the

Crimea (in Russian), News of the USSR Academy of Sciences, Ser. Biol., 3:362.

Gorlenko, V. M., Kompantzeva, G. I., and Puchkova, N. N., 1985, Temperature influence on the distribution of phototrophic bacteria in thermal springs (in Russian), Microbiologiya, 54:848.

Guerrero, R., Montesionos, E., Estev, I., and Abella, C., 1980, Physiological adaptation and growth of purple and green sulfur bacteria in a meromictic lake (Vila) as compared to a homomictic lake (Siso), Dev. Hydrobiol., 3:161:

Kondrat'eva, E. N., and Gorlenko, V. M., 1978, Purple and green bacteria (in Russian), Achievements of Microbiology, 13:8.

Olson, J. M., 1981, Chlorophyll organisation in green photosynthetic bacteria, Biochim. Biophys. Acta, 594:33.

Parkin, T. B., and Brock, T. D., 1980, Photosynthetic bacterial production in lakes. The effects of light intensity, Limnol. Oceanogr., 25:711.

Pfennig, N., 1967, Photosynthetic bacteria, Ann. Rev. Microbiol., 21:285.

Pfennig, N., 1977, Phototrophic green and purple bacteria: a comparative systematic survey, Ann. Rev. Microbiol., 31:275.

Pfennig, N., 1978, General physiology and ecology of photosynthetic bacteria, in: "The Photosynthetic Bacteria," R. K. Clayton, and W. R. Sistrom, eds., Plenum Press, New York.

Pierson, B. K., and Castenholz, R. W., 1974, A phototrophic gliding filamentous bacterium of hot springs, Chloroflexus aurantiacus, gen. and sp. nov., Arch. Microbiol., 100:5.

Pierson, B. K., and Thornber, J. P., 1983, Isolation and spectral characterisation of photochemical reaction centers from the thermophilic green bacterium Chloroflexus aurantiacus strain J-10 fl., Proc. Natl. Acad. Sci. USA, 80:80.

Puchkova, N. N., and Gorlenko, V. M., 1976, New brown chlorobacteria Prosthecochloris phaeoasteroidea nov. sp. (in Russian) Microbiologiya, 45:655.

Schmidt, K., 1980, A comparative study on composition of chlorosomes (chlorobium vesicles) and cytoplasmic membranes from Chloroflexus aurantiacus strain OK-70-fl and Chlorobium limicola f. thiosulfatophilum strain 6320, Arch. Microbiol., 124:21.

Stackebrandt, E., Folwe, V. J., Schubert, W., and Imhoff, J. F., 1984, Towards a phylogeny of phototrophic purple sulfur bacteria - the genus Ectothiorhodospira, Arch. Microbiol., 137:366.

Staehelin, A., Golecki, J. R., Fuller, R. C., and Drews, G., 1978, Visualization of the supramolecular architecture of chlorosomes in freeze fractured cells of Chloroflexus aurantiacus, Arch. Microbiol., 119:169.

Staehelin, L. A., Golecki, J. R., and Drews, G., 1980, Supramolecular organisation of chlorosomes (Chlorobium vesicles) and their membrane attachment sites in Chlorobium limicola, Biochim. Biophys. Acta, 589:30.

Trüper, H. G., 1976, Higher taxa of the phototrophic bacteria: Chloroflexaceae fam. nov., a family for gliding filamentous, phototrophic "green" bacteria, Intern. J. Syst. Bacteriol., 26:74.

van Gemerden, H., 1974, Coexistence of organisms competing for the same substrate: an example among the purple sulphur bacteria, Microb. Ecol., 1:104.

van Gemerden, H., and Jannasch, H., 1971, Continuous culture of Thiorhodaceae. Sulfide and sulfur limited growth of Chromatium vinosum., Arch. Microbiol., 79:345.

van Gemerden, H., and Beeftink, H., 1981, Coexistence of Chlorobium and Chromatium in a sulfide-limited chemostat, Arch. Microbiol., 129:32.

Woese, C. R., 1984, Why study evolutionary relationships?, in: "Evolution of Prokaryotes," K. M. Schlifer and E. Stackebrandt, eds., Academic Press, London.

DIEL MIGRATION AS A MECHANISM FOR ENRICHMENT OF NATURAL POPULATIONS OF

BRANCHING SPECIES OF PELODICTYON

C. A. Abella and J. Garcia-Gil

Institute of Aquatic Ecology
University College of Girona
(Autonomous University of Barcelona)
Hospital, 6, Girona, E-17071 Spain

INTRODUCTION

Phototrophic bacteria show a wide variety of strategies for survival in planktonic habitats.

Green phototrophic bacteria have no flagella and can move only vertically by means of gas vacuoles. The one known exception to this generalization is Chloronerpeton thalassium, which has gliding-flexing motility (Gibson et al., 1984). Hence Chlorobiaceae must have strategies different from those of Chromatiaceae, which also have many planktonic representatives.

Phototrophic bacteria often form dense populations in the water columns of stratified lakes. Sometimes a single species is dominant, and then a low species diversity results. Many factors are involved in such species distribution: sulfide concentration and affinity, light intensity and quality, assimilation of organic acids, pH, capacity to counteract sedimentation (gas vacuoles, motility, etc.) and formation of different kinds of microcolonies (Parkin and Brock, 1981; van Gemerden, 1983).

Chlorobiaceae are found in the water columns of holomictic and meromictic lakes. Two main pigment/physiological groups become enriched once stable physicochemical conditions allow full competition.

Light quality can explain the selection of brown or green species of Chlorobiaceae in lake water columns (Parkin and Brock, 1980; Montesinos et al., 1983). Almost all species of Chlorobiaceae are known to have green or brown counterparts. This fact shows the importance of light quality in selecting the brown forms in the deepest part of the water column, but this does not explain which species of the group are enriched.

The selective effect of light quality should have a genetic/physiological basis that could be expressed in ecological patterns, but there are no extensive data on this topic. The main pigment differences are isorenieratene and BChl e in the brown forms and chlorobactene and BChl c or d in the green forms. This differentiation should have appeared in the early stages of the group differentiation before any other physiological or morphological adaptation, provided the green or brown counterparts are known to be present beforehand.

The taxonomy of the Chlorobiaceae is mainly based on the following features: pigments, salt dependence, gas vacuoles, appendages, microcolony formation, and formation of consortia.

In the lake water column it is sometimes possible to trace the effect of different physicochemical factors on the specific composition of the population. Frequently a complex pattern is seen based on the influence of a number of different factors which cause the enrichment of a particular community in species of Chlorobiaceae.

In each water column unique ecological conditions are constantly changing in space and time. As a result, development of a distinctive community occurs. Each species occupies its own spatial and temporal ecological niche, which gives it advantages over other forms with analogous physiology. Until now, no model explained accurately how a given species is favoured above others.

The presence of dense populations of brown <u>Pelodictyon "phaeoglomerans"</u> in Buchensee and <u>Pelodictyon clathratiforme</u> in Lake Corominas permits evaluation of the environmental parameters which select the dominant species. Also, the ecological strategies developed by the different species to attain greater fitness can be studied.

In trying to explain the ecology of <u>Pelodictyon</u>, we have compared the physicochemical parameters and the dominant natural populations of the bacteria in the two holomictic lakes.

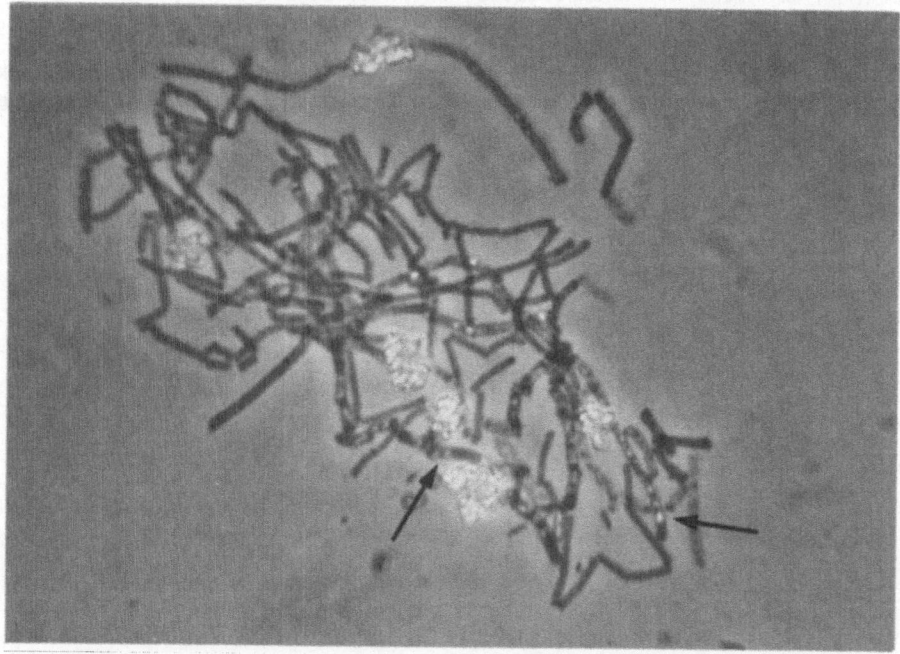

Fig. 1. <u>Pelodictyon clathratiforme</u> strain 7504 in pure culture isolated from Lake Corominas. Note the ternary fission, gas vacuolated cells, net-like microcolony and externally attached sulfur. Phase contrast. See Fig. 2 for approximate size scale.

In this paper we propose a model for the selection of brown or green forms based on the effect of light intensity and quality. The sulfide concentration, and the presence of gas vacuoles are used to explain the selection of species of Pelodictyon over species of Chlorobium. We also introduce a comprehensive model to explain the hierarchy of parameters that select a given species, with special emphasis on Pelodictyon. Further experimental work is needed to quantify the selective importance of each factor.

MATERIALS AND METHODS

Lake Corominas is a small, circular lake located in northeastern Spain in the Karstic area of Banyoles. Buchensee is a small lake near Liggeringen, F.R.G. Samples were taken from the points of maximum depth of the lakes with a Kuenen sampler and were analyzed within 24 hours of sampling.

Pelodictyon clathratiforme 7504 (Fig. 1), and Pelodictyon "phaeoglom-erans" 7502 (Fig. 2) were isolated from Lake Corominas and Buchensee respectively using culture techniques described by Pfennig (1965). Batch culture experiments with these two strains were performed under controlled conditions of light and temperature. Pfennig's medium containing 1 mM H_2S was used for growth. Total carbohydrate content was measured by the anthrone method (Herbert et al., 1975) and proteins by Coomassie Blue (Bradford, 1976). Cell shape and size and numbers of gas vacuoles were monitored continuously. Cell number was determined after first staining the cells with acridine orange, and calculated according to Hobbie et al. (1977).

Pigment concentration was determined by the method of Takahashi and Ichimure (1970). A known volume of lake water was filtered through a $MgCO_3$

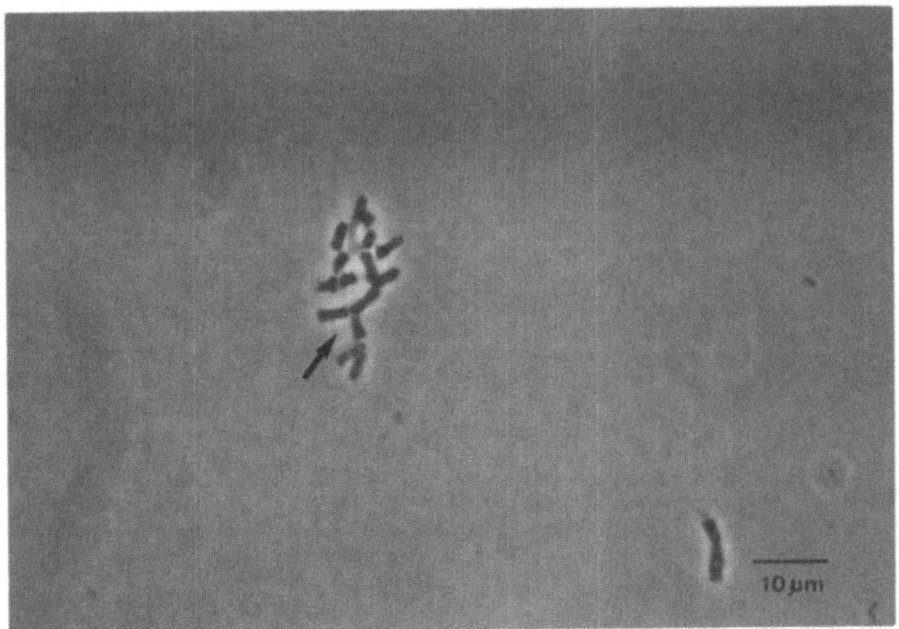

Fig. 2. Pelodictyon "phaeoglomerans" strain 7502 in pure culture isolated from Buchensee. Cells are in exponential growth phase. Note the ternary division at the end of some cells. Epifluorescence micrography with acridine orange.

covered Millipore filter of 0.45-µm pore size (Montesinos et al., 1983). The filtered material was then placed in screw capped tubes containing a known volume of 90% acetone. Calculations of pigment concentration were performed according to Stal et al. (1984). Absorption spectra were determined on a Spectronic 2000 (Bausch and Lomb).

Incident light was measured with a Metrom selenium photoelectric cell Metrom. Water density was calculated from the values obtained for conductivity and temperature (Bührer and Ambühl, 1975).

Conductivity was measured in situ with an YSI conductivimeter. For sulfide concentration an Orion Model 94-16 selective electrode was used. Samples were taken in situ with a sulfide antioxidant buffer (SAOB) composed of 10 M NaOH, ascorbic acid and EDTA.

RESULTS

Populations of branching Pelodictyon (both green and brown species) are present in holomictic lakes. The seasonal circulation of the water mass in such lakes is a common threat in the ecology of these species. No freshwater crenogenic meromictic lake with predominant populations of Pelodictyon has been reported. This fact indicates that physicochemical parameters in the water play a very important role in the establishment of Pelodictyon.

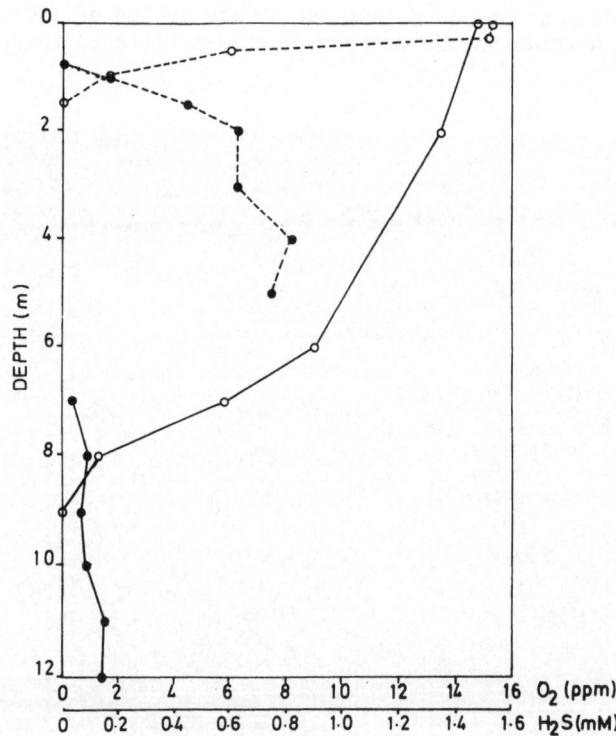

Fig. 3. Vertical distribution of O_2 (o) and H_2S (●) in Lake Corominas (dashed line) and Buchensee (solid line) coinciding with dense populations of Pelodictyon clathratiforme and Pelodictyon phaeoglomerans respectively. Sulfide concentration in Buchensee is plotted ten fold higher in order to obtain comparative profiles.

Fig. 3 shows the vertical distribution of oxygen and sulfide in two holomictic lakes where branching Pelodictyon are found. The sulfide concentration in Buchensee (brown species present) is especially low, reaching as little as 20.5 µM, whereas in Lake Corominas it is much higher (almost 1 mM). In spite of the low sulfide concentration in Buchensee, this concentration can support bacterial growth, at least during the early part of the day.

The low sulfide concentration in Buchensee is also reflected in the BChl e and total cell number profiles (Fig. 4). The population maxima are directly related to the sulfide concentration. Thus, we found that total cell number in Lake Corominas is two orders of magnitude higher than in Buchensee. Nevertheless, the specific content of BChl is higher in the Pelodictyon found in Buchensee, due to the light distribution (Fig. 5). Light penetration is higher in Buchensee but the intensity reaching the bacterial plate is lower (< 0.01%), than in Lake Corominas. This fact, together with changes in light quality reaching bacterial plates, determines the selection of brown species over green.

In Fig. 6 the specific content of total carbohydrate during batch growth of both strains of Pelodictyon is presented. This parameter reaches maximum values at the end of the exponential growth phase. The specific carbohydrate content is 3-4 times higher in P. "phaeoglomerans" 7502 than in the green P. clathratiforme 7504. Specific carbohydrate content drops rapidly (over 2-3 hours) in both cultures to a nearly constant value at the beginning of the stationary phase.

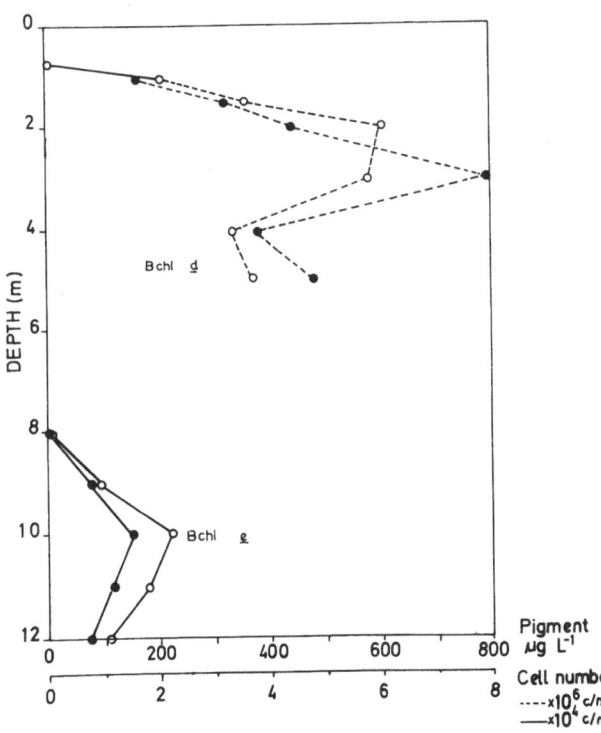

Fig. 4. Vertical distribution of BChl (o) and total cell number (●) in Lake Corominas (dashed line) and Buchensee (solid line). Note that cell number in Lake Corominas is x 10⁶ and in Buchensee is x 10⁴.

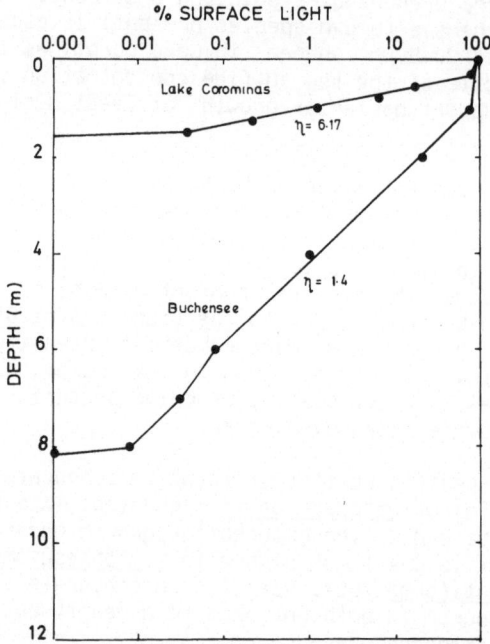

Fig. 5. Light extinction in Lake Corominas and Buchensee water columns
expressed as % of surface light. "η" is the light extinction
coefficient.

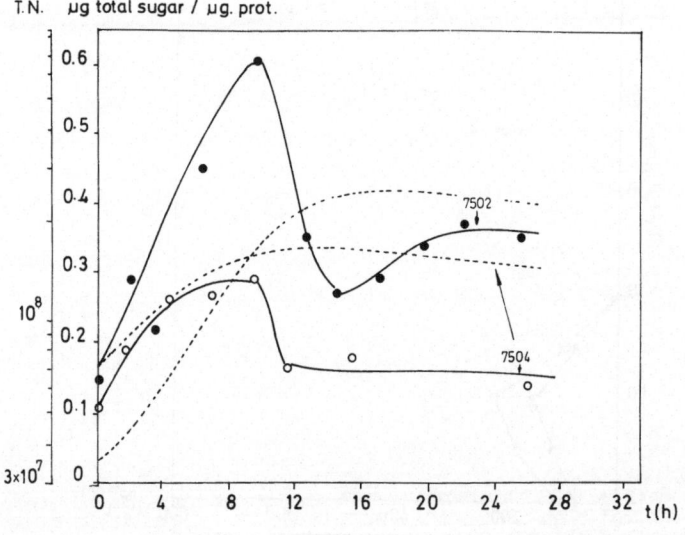

Fig. 6. Batch culture growth on Pfennig's medium of Pelodictyon
"phaeoglomerans" strain 7502 and P. clathratiforme strain 7504.
(---) total number of cells. (●---●; o---o) specific total
carbohydrate ("sugar") content.

Table 1. Comparative Characteristics of Green and Brown Chlorobiaceae. All
green species contain BChl c or d and chlorobactene; all brown
species contain BChl e and isorenieratene.

Species	Shape	G+C(%)	Species	Shape	G+C(%)
Chlorobium limicola	rod	51-58	Chlorobium phaeobacteroides	rod	49-50
Pelodictyon clathratiforme	rod	48.5	Pelodictyon "phaeoglomerans"	rod	49.8
Chlorobium vibrioforme	vibrio	52-57	Chlorobium "phaeovibrioforme"	vibrio	
Chlorobium chlorovibroides	vibrio	-	Chlorobium phaeovibroides	vibrio	52-53
Pelodictyon luteolum	ovoid	53-58	Pelodictyon phaeum	vibrio	-
Prosthecochloris aestuarii	irreg. sphere	50-56	Prosthecochloris phaeoasteroidea	irreg. sphere	52.2
Ancalochloris perfilievii	irreg. sphere	-	not found		
Clathrochloris sulfurica	sphere	-	not found		
"Chlorochromatium aggregatus"	-	-	"Pelochromatium roseum"		
"Chlorochromatium glebulum"	-	-	not found	-	-
"Chloroplana vacuolata"	-	-	not found		

DISCUSSION

Comparative characteristics of brown and green Chlorobiaceae are shown
in Table 1 (Trüper, 1987). The main differences are in pigment composition
and cell structure. G+C (%) ranges between 49 and 58. Some consortia with
names without taxonomic value are also indicated.

In Table 2 a comparison is made between the two main groups of
Pelodictyon. Members of the freshwater group have binary and ternary
division (Pfennig and Cohen-Bazire, 1967), and can develop large three
dimensional microcolonies. Members of the brackish water group have only
binary division and form microcolonies containing only a few cells.

Table 2. Comparison between Fresh and Marine-Brackish Water Species of
Pelodictyon. All species contain gas vacuoles. T.D. = ternary
division.

Species	Shape	T.D.	G+C(%)	Environm.	Clumps (with S°)
P. luteolum	vibrio	-	53-58	brackish w.	-
P. phaeum	vibrio	-	-	brackish w.	-
P. clathratiforme	rod	+	48.5	eutr.fresh w.	+
P. "phaeoglomerans"	rod	+	49.5	eutr.fresh w.	+

Table 3. Groups of Green Phototrophic Bacteria According to
         Indicated Characteristics

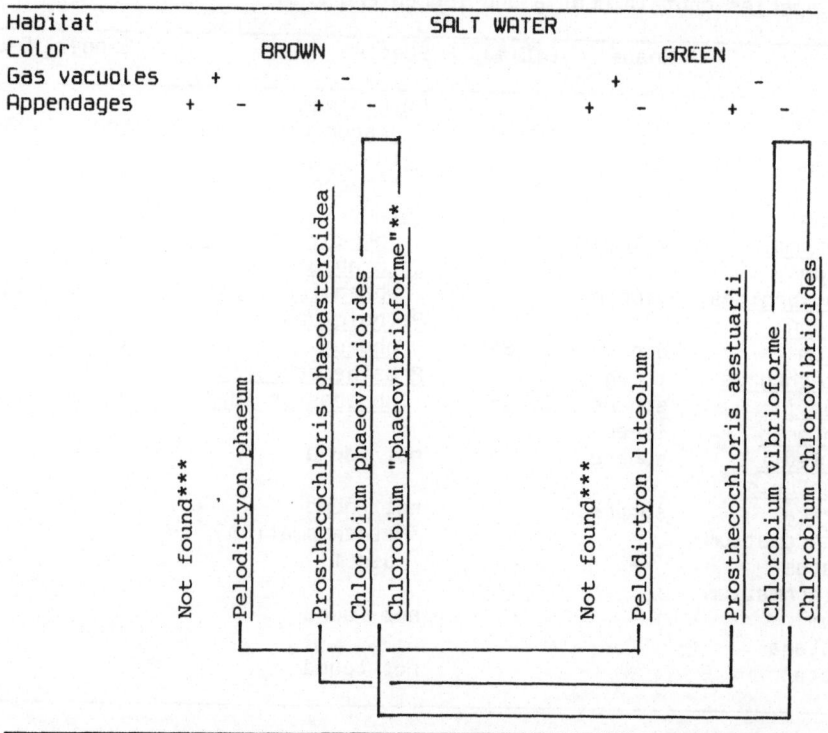

Habitat	SALT WATER								
Color	BROWN					GREEN			
Gas vacuoles	+		−			+		−	
Appendages	+	−	+	−		+	−	+	−

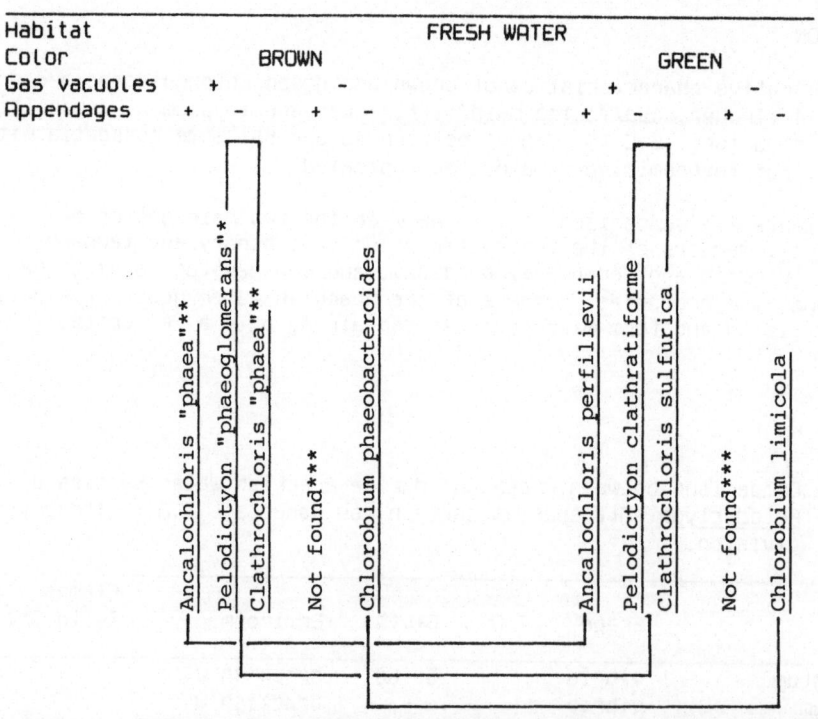

Habitat	FRESH WATER								
Color	BROWN					GREEN			
Gas vacuoles	+		−			+		−	
Appendages	+	−	+	−		+	−	+	−

*In pure culture but not yet described. **Not yet observed, but
 the green form described. ***No brown or green form yet observed.

Table 3 shows an allocation of the different species of Chlorobiaceae (not including the gliding <u>Chloroherpeton thalassium</u>). Two main groups, salt or freshwater, with brown and green genera are present. Further divisions using presence or absence of gas vacuoles and appendages give the final distribution of known species. Some forms are hypothetical but are mentioned because the green form is known. Four of the possible combinations have no representatives. In the salt water group no green or brown species with appendages and gas vacuoles is known. In the fresh water group no green or brown species with appendages is known, that does not have gas vacuoles. At present there is no simple explanation for the missing groups.

A Comprehensive Model for the Enrichment of Different Species of Chlorobiaceae

The flow chart in Fig. 7 (Model A) explains the enrichment of natural populations of photosynthetic bacteria found in holomictic lakes with sulfide in the anoxic hypolimnion, e.g., Corominas and Buchensee. Light intensity and quality and sulfide concentration are used to explain the selection of the phototrophic bacteria.

Light intensity is the first selective factor which can explain the enrichment of Chromatiaceae which are known to require a higher light intensity than Chlorobiaceae. Their plates are frequently located over the latter. The Chlorobiaceae adapt to resulting low light conditions by increased specific BChl content.

The second selective factor is light quality. The light spectrum reaching the bacteria depends on the depth of the water column and on the presence of algae. Light is enriched in the greenish-yellow band (450-650 nm; maximum 540 nm), mainly through differential extinction in the water column, as is the case in deep layers () 8-10 m). Brown Chlorobiaceae will be enriched under these conditions, because they contain isorenieratene as the predominant carotenoid, with maximal absorption at 540 nm.

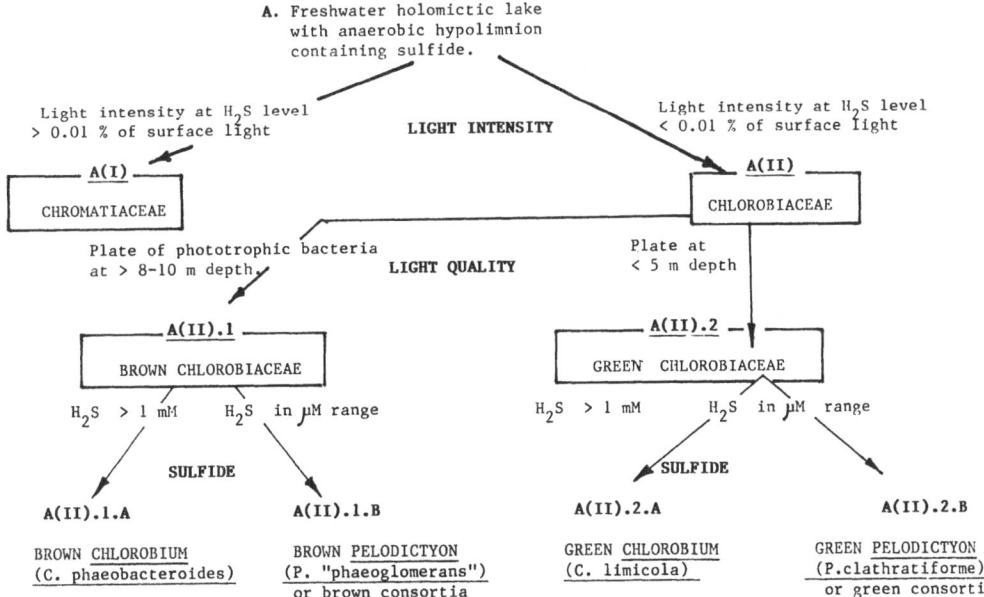

Fig. 7. A model explaining the selective growth of phototrophic bacteria in freshwater holomictic lakes. <u>C.</u> ≡ <u>Cb.</u>

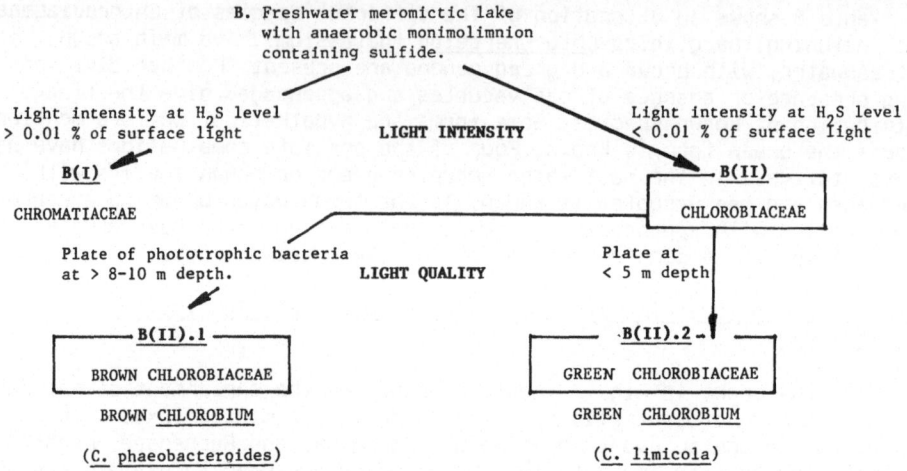

Fig. 8. A model explaining the selective growth of phototrophic bacteria in freshwater meromictic lakes. C. ≡ Cb.

The light spectrum is also enriched in the blue-green area (440-450 nm; maximum 450 nm) through differential extinction by algal populations. This is the case with bacterial plates near the surface. Green Chlorobiaceae develop under these conditions because they have BChl c or d with strong absorption in this spectral band (440-450 nm).

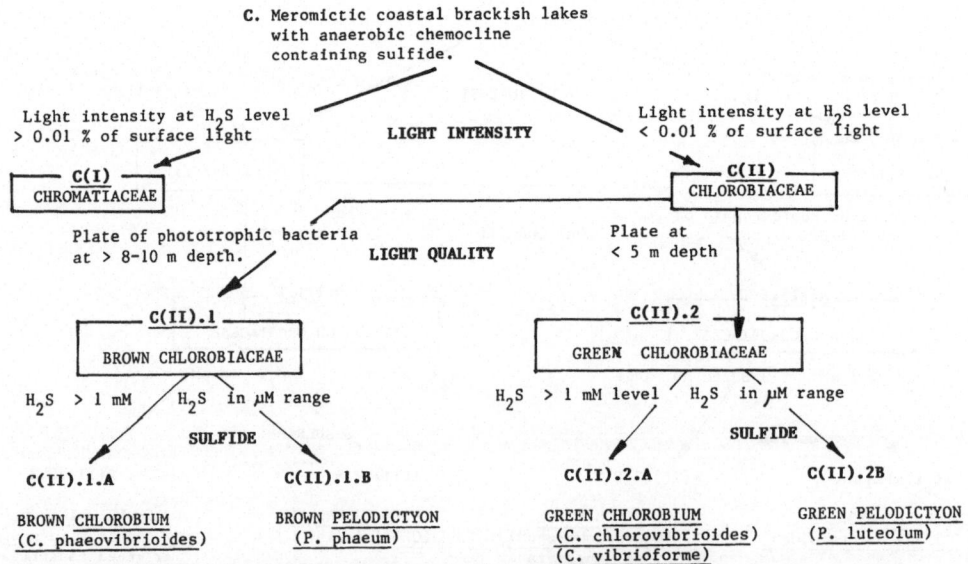

Fig. 9. A model explaining the selective growth of phototrophic bacteria in meromictic coastal brackish lakes. C. ≡ Cb.

Sulfide concentration acts as the third selective factor, enriching either brown Chlorobium or brown Pelodictyon. At concentrations of sulfide exceeding 1 mM, the brown Chlorobium thrives; sulfide depletion by photosynthesis is not important owing to the rapid diffusion of sulfide from water column below. At low sulfide concentrations (around 1 μM), green/brown Pelodictyon or green/brown motile consortia are selected. Sulfide depletion is clearly important and only motile strains with gas vacuoles or flagella have the necessary advantages.

The flow chart in Fig. 8 (Model B) is similar to Model A but applies to freshwater meromictic (crenogenic) lakes with an anaerobic monimolimnion containing sulfide. Light intensity and quality play the same role as in Model A, selecting first between Chromatiaceae and Chlorobiaceae and then between brown and green Chlorobiaceae. Sulfide concentration is not a selective factor for the brown or green genera of Chlorobiaceae at this level, and usually only the brown Chlorobium phaeobacteroides or green Chlorobium limicola are enriched depending on the light quality.

The flow chart in Fig. 9 (Model C) is also similar to Model A but applies to meromictic coastal brackish waters with anaerobic monimolimnion containing sulfide. Light intensity and quality again play the same roles as in Model A. Sulfide concentration is the selective factor enriching for brown Chlorobium or brown Pelodictyon. Above 1 mM sulfide the brown Chlorobium thrives; sulfide depeltion by photosynthesis is unimportant owing to the rapid diffusion of sulfide from the water column (Jørgensen et al., 1979).

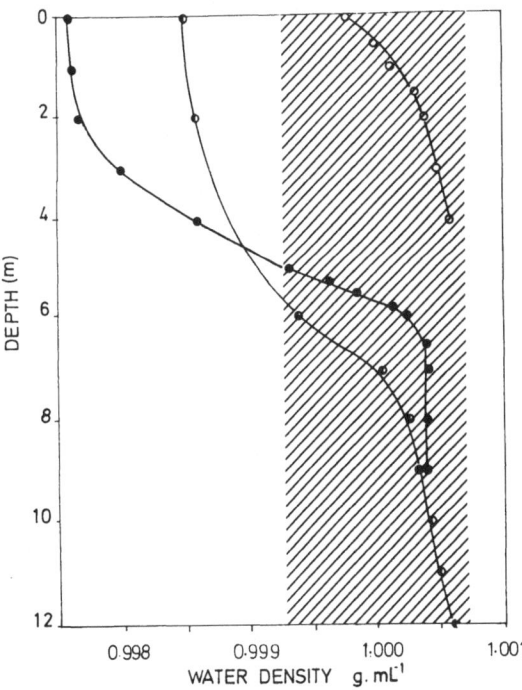

Fig. 10. Density profiles of two holomictic lakes (Corominas, o---o and Buchensee, ●---●) and one meromictic freshwater lake (Vilar, Spain, o---o=. Shaded area represents the density range in which bacterial populations (Chlorobiaceae) are placed.

At a low sulfide concentration (around 1 µM), brown <u>Pelodictyon</u> is selected. Sulfide depletion is clearly an important factor, and only strains which can keep their position in the water column by means of gas vacuoles (neutral buoyance), or flagella have the necessary advantage. <u>Pelodictyon</u> can reach the top of the sulfide layer and thrive at a low sulfide concentration. Positive buoyancy is no longer useful because the water density gradient is very steep at the chemocline and salt-adapted cells find lower density waters.

Gradients of Water Density at the Thermocline or Chemocline of Lakes with Chlorobiaceae

Fig. 10 shows density profiles calculated from conductivity and temperature values for three different freshwater lakes. A smooth gradient of 0.5×10^{-3} density units per meter of depth is present in Lake Corominas and Buchensee and 2×10^{-3} units per meter in the meromictic Lake Vilar. This gradient is very steep in the case of meromictic brackish coastal lakes, and we found a chemocline gradient of 25×10^{-3} units per meter in Massona coastal lagoon in northwest Spain.

Although all freshwater Chlorobiaceae grow in a relatively narrow range of densities (Fig. 6, shaded area), the presence of an intense gradient, as in meromictic fresh and brackish waters, limits the upward diffusion of sulfide, since chemical and biological oxidation can be very intense at the chemocline level. However, if the gradient is sufficiently smooth (as in holomictic lakes), the sulfide level moves nearer to the lake surface (Fig. 11).

Microcolony Formation

Microcolony formation is crucial to explain vertical movement in the lake water column. Buoyant or sinking particles follow Stokes' law: $v = 2 \cdot g \cdot r^2 \cdot (p-p')/9\eta$, where g is the gravitational acceleration, r is the

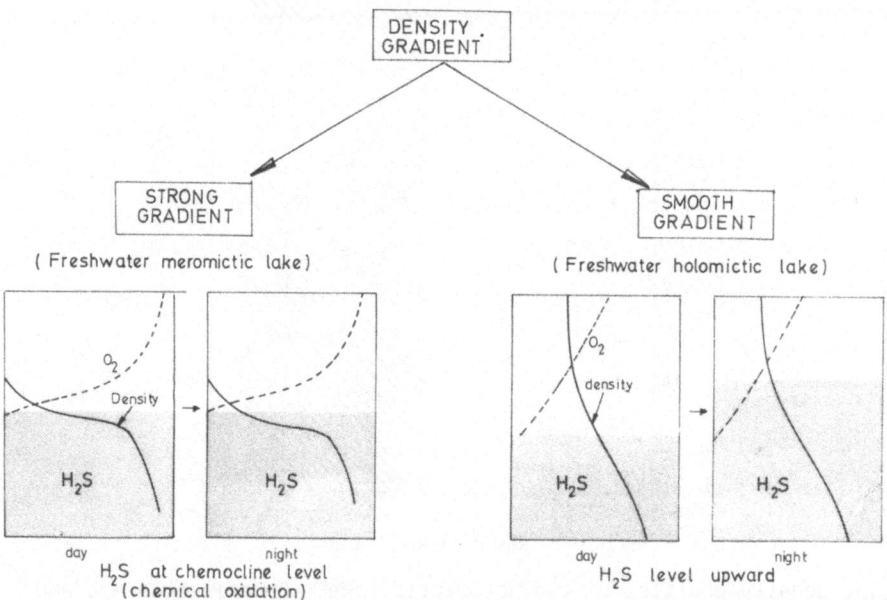

Fig. 11. Schematic diel behaviour of H₂S level in freshwater lakes related to density gradients.

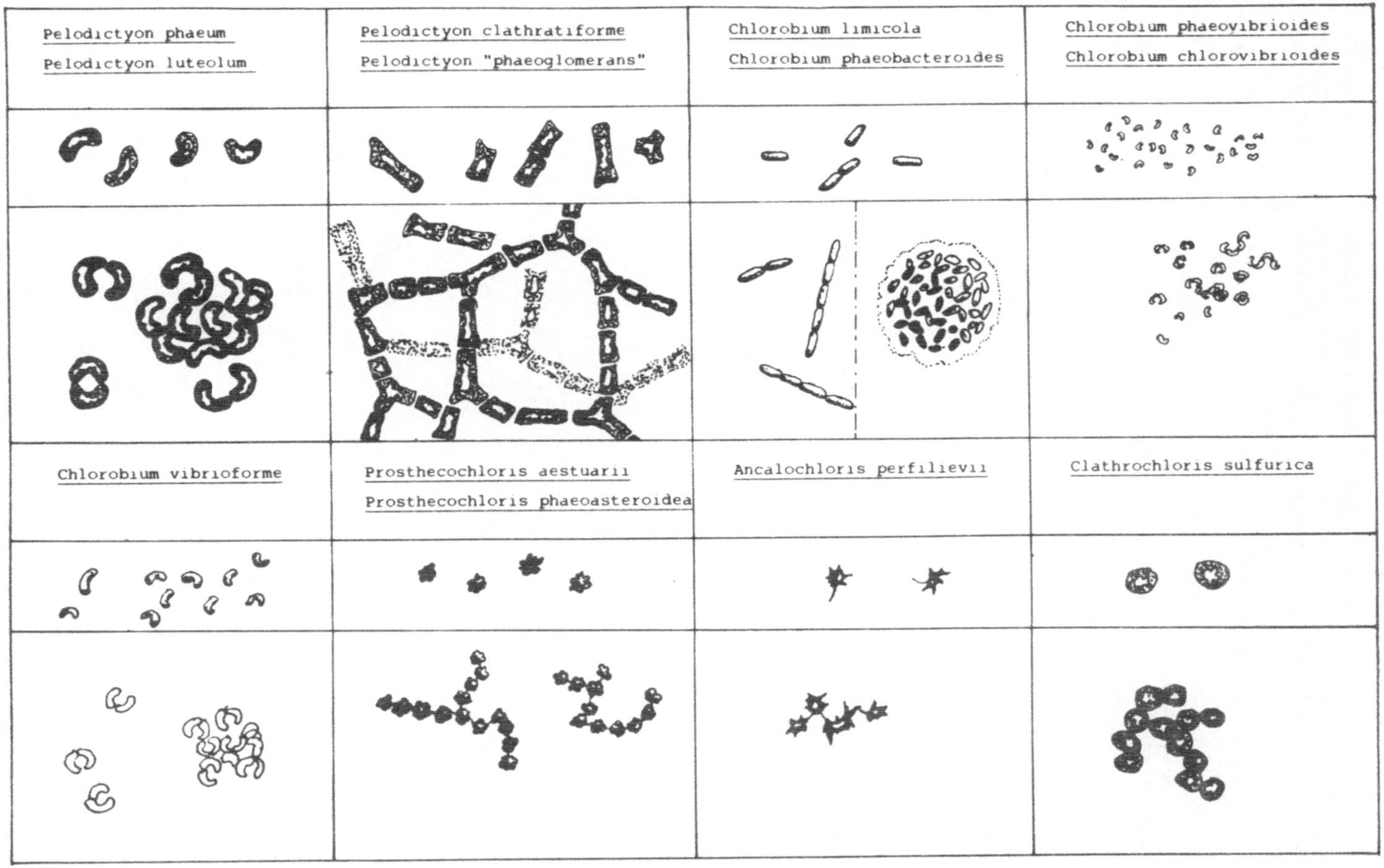

Fig. 12. Different kinds of microcolony formation in Chlorobiaceae. Individual cells and microcolonies are indicated.

radius, p is the particle density, p' the medium density and the viscosity of the medium. In a uniform medium sufficiently rapid movement by buoyancy can be achieved only by increasing $\underline{r}$, the microcolony size, and lowering the density of the cells by gas vacuoles.

Microcolony formation is an almost constant feature in natural populations of planktonic Chlorobiaceae. Different kinds of microcolonies are known (Fig. 12). The simplest kind is due to filamentation caused by formation of chains of cells, as is frequent in <u>Chlorobium</u> species, and in marine <u>Prosthecochloris</u>. Another way to form microcolonies is by surrounding the cells with slime. This kind of microcolony is known in some strains of <u>Chlorobium limicola</u> (Brugada, 1986) and marine <u>Pelodictyon</u> (Gorlenko, 1972). The most complex way to construct microcolonies is by the formation of three-dimensional nets as in branching <u>Pelodictyon</u> species. A complex net is formed by a combination of binary and tertiary cell division (see Fig. 1).

Neutral Buoyancy Versus Positive Buoyancy

Cell movement by positive buoyancy has been described in cyanobacteria (Konopka and Schnur, 1980; Okada and Aiba, 1983; van Rijn and Shilo, 1985) and in heterotrophic bacteria (Konopka, 1977). There are occasional references concerning phototrophic bacteria (Clark and Walsby, 1978a,b). Kholer et al. (1984) first described a diel migration in a natural population of <u>Thiopedia rosea</u> in the water column of Rotsee.

Clark and Walsby (1978b) found that only <u>Thiopedia rosea</u> and <u>Pelodictyon clathratiforme</u> showed vertical migration. The apparent microcolony radius was high enough to have significant buoyancy velocity to migrate. In fact it could be assumed that the distance of migration in the water column was directly proportional to the radius of the microcolony, provided the other parameters (gas vacuoles and ballast) are constant.

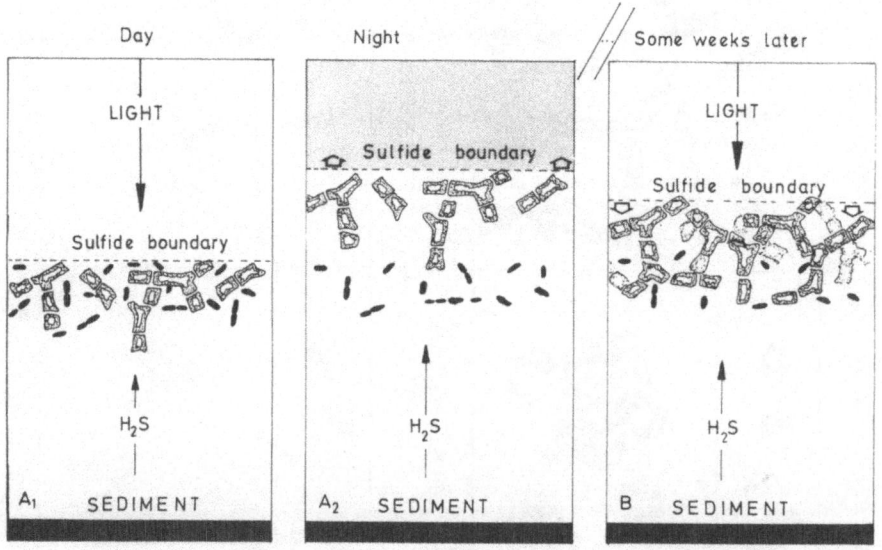

Fig. 13. Schematic model for the evolution of <u>Pelodictyon</u> and <u>Chlorobium</u> populations in holomictic lakes. A_1 and A_2: diel cycle; <u>Pelodictyon</u> migrates upward (development of vacuolation in darkness). B: Some weeks later the <u>Pelodictyon</u> population is enriched over the <u>Chlorobium</u> population.

Fig. 13 presents a schematic representation in which the relative size of microcolony is compared with the buoyancy and sinking of two populations of Pelodictyon and Chlorobium. In individual cells or small microcolonies gas vacuoles could be important to avoid sinking, by providing neutral buoyancy. This would allow the development of a population maximum even if growth is relatively slow (Clark and Walsby 1978a). In this way the organism could coexist with species without gas vacuoles but with higher growth rates, such as Chlorobium or Prosthecochloris.

Diel Migration as a Selective System: Physiocochemical Parameters Required

The branching species of Pelodictyon, P. clathratiforme and P. "phaeoglomerans", have a selective advantage over non-vacuolated forms since they can migrate in the water column following sulfide depletion and diffusion. Sulfide limitation plays an important role in the ecological behaviour of the branching Pelodictyon species in both lakes studied. The

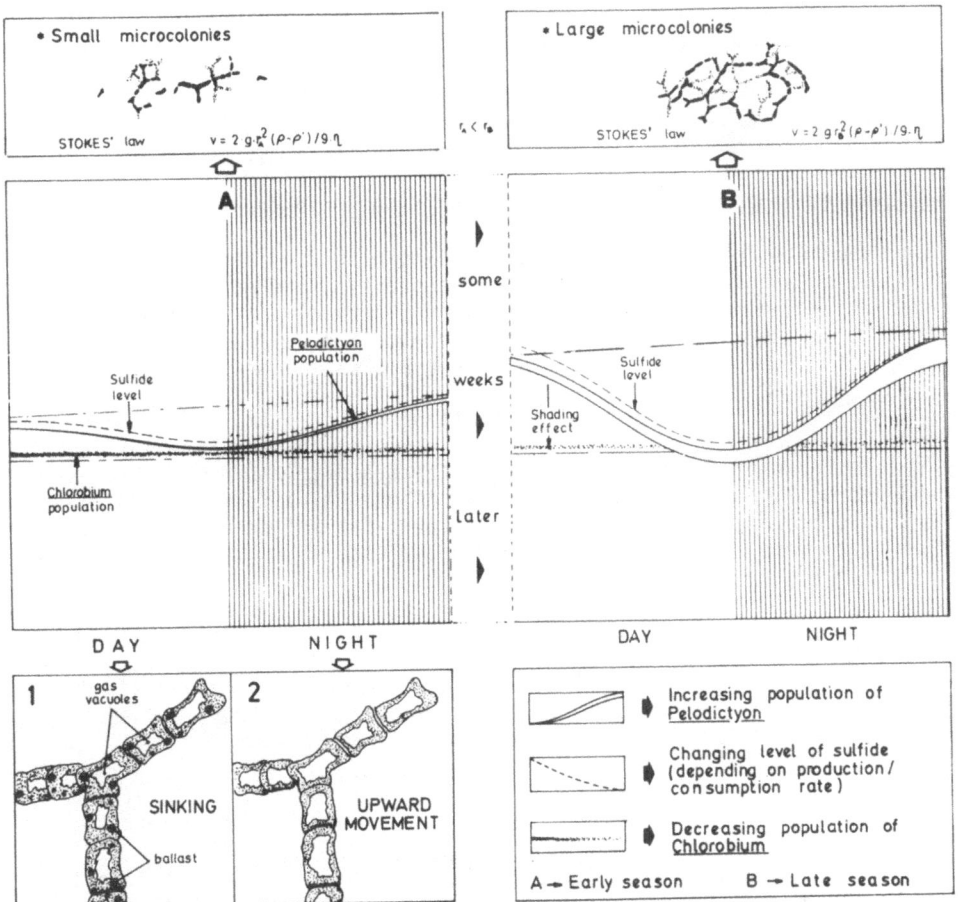

Fig. 14. General scheme for explanation of Pelodictyon diel migration. The Chlorobium population present together with Pelodictyon in the water column gradually decreases owing to the imposed biological filter of the Pelodictyon population. The size of the Pelodictyon microcolony is proportional to the range movement according to Stoke's law. The "undulating" pattern expresses the complex behaviour of Pelodictyon natural populations.

continuous oxidation of sulfide at the H_2S/O_2 interface during the day by phototrophic bacteria causes branching <u>Pelodictyon</u> species to migrate to lower regions with higher sulfide concentration.

In Fig. 14 is given a graphic explanation of the diel movements of branching <u>Pelodictyon</u> species. <u>Pelodictyon</u> can move up during the night following sulfide diffusion, by controlling the ballast and turgor pressure inside the cells. In the early morning they reach the top, just at the border of the illuminated sulfide layer. At dawn <u>Pelodictyon</u> start sinking following sulfide depletion, and the migration starts again. This could be explained by a higher ballast and turgor pressure inside the cell, which causes the weaker gas vacuoles to collapse. The changes in specific content of total carbohydrates in both strains of <u>Pelodictyon</u> studied (see Fig. 6), corroborates the importance of cell content as a result of the physiological state of cells. When cells are growing the carbohydrate content increases but drops when the stationary phase is entered as a result of sulfide limitation. Carbohydrate probably has an important role as cell ballast (van Rijn and Shilo, 1985).

The diel migration starts in the spring months with a narrow range proportional to microcolony size. During summer and until the fall mixing, the migration distance grows, as the sulfide level approaches the lake sur-face as shown schematically in Fig. 14. These "undulating" movements always give a selective advantage over non-motile or non-vacuolated species, which are shaded by the biological filter due to the <u>Pelodictyon</u> population.

Other Strategies such as Syntrophic Consortia or Appendages

Only motile consortia of <u>Chlorobium</u> with heterotrophic cells can coexist with <u>Pelodictyon</u>. In synthrophic consortia, the non-motile cells of Chlorobiaceae can move with the microcolony (Pfennig, 1980) and hence reach the best conditions of light and sulfide at the top of gradients in the water column. No marine or brackish water examples of consortia are known.

Another kind of strategy for maintaining the position in the water column is the presence of appendages. These structures are frequent both in marine and freshwater Chlorobiaceae. <u>Prosthecochloris</u> and <u>Ancalochloris</u> are appendaged non-vacuolated species. The appendages give to the individual cell or microcolony a high ratio of surface to volume, preventing rapid sedimentation and allowing population development.

ACKNOWLEDGEMENTS

We wish to thank Dr. E. Montesinos, R. C. Brunet, D. Brugada, and the people of the Laboratory of Microbiology, Faculty of Biology, University of Constance, F. R. G. for help in different parts of this work.

REFERENCES

Bradford, M. M., 1976, A rapid and sensitive method for the quantification of microgram quantities of protein utilizing the principle of protein-dye binding, <u>Anal. Biochem.</u>, 27:248.
Brugada, D., 1986, Sulfur metabolism and detoxifying role of phototrophic bacteria of Vilar Lake (Banyoles), Master Science Thesis, <u>Autonomous University of Barcelona</u>, Girona, Spain.
Bührer, H., and Ambühl, H., 1975, Einleitung von Abwasser in Seen. <u>Schweiz. Z. Hydrol.</u>, 37:347.
Clark, A. E., and Walsby, A. E., 1978a, The occurrence of gas-vacuolate bacteria in lakes, <u>Arch. Microbiol.</u>, 118:223.

Clark, A. E., and Walsby, A. E., 1978b, The development and vertical distribution of populations of gas-vacuolate bacteria in a eutrophic, monomictic lake, Arch. Microbiol., 118:229.

Gibson, J., Pfennig, N., and Waterbury, J. B., 1984, Chloroherpeton thalassium gen. nov. et spec. nov., a non-filamentous, flexing and gliding green sulfur bacterium, Arch. Microbiol., 138:96.

Gorlenko, V. M., 1972, A new species of phototrophic brown sulfur bacteria, (in Russian), Mikrobiologiya, 41:370.

Herbert, D., Phipps, P. J., and Strange, R. E., 1975, Chemical analysis of microbial cells, Methods in Microbiology, Vol. 5, Academic Press, New York.

Hobbie, J. E., Daley, J. R., and Jasper, S., 1977, Use of Nuclepore filters for counting bacteria by flourescence microscopy, Appl. Environ. Microbiol., 33:1225.

Jørgensen, B. B., Kuenen, J. G., and Cohen, Y., 1979, Microbial transformations of sulfur compounds in a stratified lake (Solar Lake, Sinai), Limnol. Oceanogr., 24:799.

Kholer, H. P., Ahring, B., Abella, C. A., Ingvorsen, K., Keweloh, H., Laczko, E., Stupperich, E., and Tomei, F., 1984, Bacteriological studies on the sulfur cycle in the anaerobic part of the hypolimnion and in the surface sediments of Rotsee in Switzerland, FEMS Microbiol. Lett., 21:279.

Konopka, A. E., 1977, Inhibition of gas vesicle production in Microcyclus aquaticus by 1-lysine, Can. J. Microbiol., 23:363.

Konopka, A., and Schnur, M., 1980, Effect of light intensity on macromolecular synthesis in cyanobacteria, Microbial Ecol., 6:291.

Montesinos, E., Guerrero, R., Abella, C. A., and Esteve, I., 1983, Ecology and physiology of competition for light between Chlorobium limicola and Chlorobium phaeobacteroides in natural habitats, Appl. Environ. Microbiol., 46:1007.

Okada, M., and Aiba, S., 1983, Simulation of water-bloom in a eutrophic lake. II. Reassessment of buoyancy, gas vacuole and turgor pressure of Microcystis aeruginosa, Water Res., 17:877.

Parkin, T. B., and Brock, T. D., 1980, The effects of light quality on the growth of photosynthetic bacteria in lakes, Arch. Microbiol., 125:19.

Parkin, T. B., and Brock, T. D., 1981, The role of phototrophic bacteria in the sulfur cycle of a meromictic lake, Limnol. Oceanogr., 26:880.

Pfennig, N., 1965, Anreicherungskulturen für Röte und Grüne Schwefelbakteria. Zentralbl. Bacteriol. Parasiten. Infektionskr. Hyg., Abt. 1, Suppl. 1:179.

Pfennig, N., 1980, Syntrophic mixed cultures and symbiotic consortia with phototrophic bacteria, in: "Anaerobes and Anaerobic Infections," G. Gottschalk, N. Pfennig and H. Werner, eds., Gustav Fischer Verlag, Stuttgart.

Pfennig, N., and Cohen-Bazire, G., 1967, Some properties of the green bacterium Pelodictyon clathratiforme, Arch. Microbiol., 59:226.

Stal, L. J., van Gemerden, H., and Krumbein, W. E., 1984, The simultaneous assay of chlorophyll and bacteriochlorophyll in natural microbial communities, J. Microbiol. Meth., 2:295.

Takahashi, M., and Ichimura, S., 1970, Photosynthetic properties and growth of photosynthetic sulfur bacteria in lakes, Limnol. Oceanogr., 15:929.

Trüper, H. G., 1987, Phototrophic bacteria (an incoherent group of prokaryotes). A taxonomy versus phylogenetic survey, Microbiologia SEM, 3:71.

van Gemerden, H., 1983, Physiological ecology of purple and green bacteria, Annal. Microbiol. (Inst. Pasteur), 134:73.

van Rijn, J., and Shilo, M., 1985, Carbohydrate fluctuations, gas vacuolation, and vertical migration of scum-forming cyanobacteria in fishponds, Limnol. Oceanogr., 30:1219.

GROWTH OF <u>CHLOROBIUM LIMICOLA</u> F. <u>THIOSULFATOPHILUM</u> ON POLYSULFIDES

P.T. Visscher[a,b] and H. van Gemerden[a]

[a]Department of Microbiology, University of Groningen
 Kerklaan 30, NL-9751 NN Haren, The Netherlands
[b]Netherlands Institute for Sea Research
 P.O.Box 59, NL-1790 AB Den Burg, Texel, The Netherlands

INTRODUCTION

Elemental sulfur is commonly found as an intermediate product in the oxidation of sulfide by green and purple bacteria. In the early log phase, when growing on sulfide, batch cultures of these anoxygenic phototrophic bacteria have a milky appearance due to the often massive accumulation of elemental sulfur, either stored intracellularly (Chromatiaceae) or extracellularly (Chlorobiaceae and Ectothiorhodospiraceae). The amount of sulfur decreases after sulfide depletion, since its oxidation to sulfate continues, whereas its formation from sulfide has stopped. However, in some species sulfur oxidation does not start before the sulfide supply is exhausted (Trüper, 1978). In batch cultures of <u>Chromatium vinosum</u> it has been observed that, prior to sulfide depletion, the specific content of sulfur (mmol mg protein^{-1}) shows little variation (van Gemerden, 1968, 1984). The maximum specific content of sulfur in different species of purple sulfur bacteria is very similar, i.e., 30-35% of the ash-free dry weight (van Niel, 1931; Trüper and Schlegel, 1964; van Gemerden, 1968; Mas and van Gemerden, 1987). <u>Chlorobium</u> species have been much less studied in this respect, but generally marine species deposit less sulfur than fresh water strains do. The accumulation of elemental sulfur is the result of the nature of the initial substrate, but the specific content of sulfur is very much influenced by the growth rate. The constant content of sulfur in batch cultures, described above, can be explained by the fact that during the exponential growth phase the organisms are growing at μ_{max}. Exponential growth at lower specific growth rates can be obtained in continuous cultures.

In the simultaneous presence of extracellular elemental sulfur (S^o) and hydrogen sulfide (H_2S), polysulfides are formed abiotically. Experiments have been performed on the utilization of such polysulfides by <u>Cb. limicola</u>.

MATERIALS AND METHODS

Organism and Cultivation

<u>Cb. limicola</u> f. <u>thiosulfatophilum</u> DSM 249 was originally kindly provided by Prof. Norbert Pfennig. Composition of culture media, growth

287

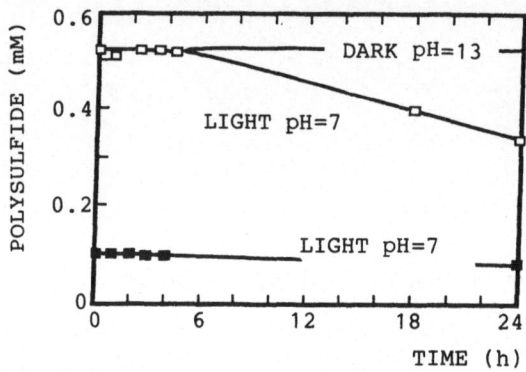

Fig. 1. Polysulfide stability at pH = 7 and pH = 13 during a 24 h period.
Two concentrations (0.1 and 0.5 mM, respectively) were tested under
dark or light conditions.

conditions and equipment for continuous culture have been described
previously (van Gemerden and Beeftink, 1978; Beeftink and van Gemerden,
1979).

Analysis

Sulfide was determined either by the colorimetric methylene blue method
after Pachmayr (1960; Trüper and Schlegel, 1964), or when sulfide
concentrations were below 0.2 mM, with an ion-specific electrode (Berner,
1963; van Gemerden, 1987). Elemental sulfur and BChl $\underline{c}$ were measured
consecutively in methanol extracts (Stal et al., 1984). Polysulfides and
thiosulfate were assayed with the cyanolytic method described by Schedel
(1978) and Then (1984). Polysulfide synthesis was carried out according to
Feher and Laue (1956). The stability of these polysulfides was tested at
several pH values, both under light and dark conditions (Fig. 1). Protein
was determined with the Folin-phenol reagent (Lowry et al., 1951) after
extraction of elemental sulfur. Glycogen was assayed with the anthrone

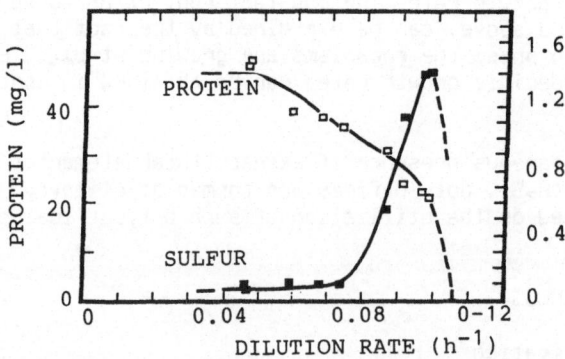

Fig. 2. Steady-state data on protein and elemental sulfur for Chlorobium
limicola f. thiosulfatophilum at various dilution rates, growing on
sulfide as the limiting nutrient (S_R = 1.97 mM).

reagent and calculated using a conversion factor of 7.4 mg mmol^{-1} RP (reducing power) according to van Gemerden and Beeftink (1978).

RESULTS AND DISCUSSION

Steady States

In continuous cultures of Cb. limicola f. thiosulfatophilum, growing with sulfide as the limiting substrate, the maximum concentration of elemental sulfur was found at high dilution rates (Fig. 2). At a dilution rate close to the maximum specific growth rate, the concentration of elemental sulfur was about 1.5 mM, i.e., 70-80% of the sulfide supplied from the reservoir solution S_R = 2 mM) was not oxidized beyond the state of elemental sulfur. With increasing dilution rates decreasing steady-state protein concentration were observed and thus the specific content of sulfur at high dilution rates can reach very high levels (Fig. 2). The deposition of elemental sulfur decreased with a decreasing dilution rate but appeared to level off. At dilution rates below approximately 0.06 h^{-1} very little sulfur was found (Fig. 2). During sulfide limitation not all of the sulfide is oxidized. However, the affinity of Cb. limicola f. thiosulfatophilum for this substrate is very high. Consequently, the concentration of sulfide in the culture was very low even at high dilution rates. The maximum concentration observed was 20 µM (D = 0.099 h^{-1}), while the concentration of sulfide at a dilution rate of 0.046 h^{-1} was about 1 µM. These concentrations are far below the inhibitory level generally observed for anoxygenic phototrophic bacteria. For this Chlorobium strain the inhibition constant K_i, at which μ = 0.5 μ_{max}, was reported to be 3 mM (van Gemerden, 1984). Using the direct linear plot (Eisenthal and Cornish-Bowden, 1974) the maximum specific growth rate on sulfide was determined to be 0.105 h^{-1} and the affinity constant K_s 1.5 µM. These data agree with earlier observations and show that green sulfur bacteria have very high affinities for sulfide. This could imply that in the simultaneous presence of hydrogen sulfide and elemental sulfur the sulfide would be used by Chlorobium for CO_2 fixation rather than react abiotically with elemental sulfur to form polysulfides.

However, in a situation of either continuous supply or excess of hydrogen sulfide a certain fraction of this might react with elemental sulfur to form polysulfides. The concentration of these polysulfides is a function of the concentration of the reacting compounds. Polysulfides (estimated as S_3^{2-}) were indeed detected at all dilution rates tested, the maximum concentration observed being 70 µM (D = 0.99 h^{-1}). When comparing the molar concentrations of sulfide and polysulfide it should be noted that in the oxidation of 1 mol of trisulfide (S_3^{2-}) to sulfate, 3.5 times more electrons are released than in the oxidation of 1 mol of hydrogen sulfide (20 and 8, respectively). As will be discussed below, polysulfide utilization by Chlorobium is blocked in the presence of hydrogen sulfide. Also, de novo protein synthesis is required before polysulfides can be oxidized. Once Cb. limicola f. thiosulfatophilum is induced to grow on polysulfide, this sulfur compound is an excellent substrate, provided sulfide is absent. Continuous cultures were grown with polysulfides (S_R = 0.499 mM S_3^{2-}) as the only limiting nutrient. Data were collected during steady conditions at dilution rates of 0.049 h^{-1}, 0.057 h^{-1}, and 0.059 h^{-1}. The values of the kinetics parameters K_s and μ_{max} were graphically derived with the direct linear plot (Eisenthal and Cornish-Bowden, 1974). The three lines intersected at one point; apparently S_3^{2-} is a non-inhibitory substrate. Therefore, the kinetics could be described according to Monod. The saturation constant was found to be 5.86 µM S_3^{2-} and the maximum specific growth rate 0.071 h^{-1}. The most striking difference, compared to cultures limited by sulfide, was the low concentration of elemental sulfur. At a dilution rate of 0.059 h^{-1} (82% of μ_{max}) only 4% of the reducing

equivalents supplied with the metering pump were recovered as elemental
sulfur. During growth on sulfide the concentration of elemental sulfur at
the corresponding dilution rate (i.e., 0.09 h⁻¹, which is 82% of the maximum
attainable specific growth rate on sulfide) was 1.2 mM, or 43% of the
reducing equivalents present in the reservoir bottles.

As could be expected, the yield per mmol reducing equivalent from
polysulfides was about the same as found for sulfide. At a dilution rate of
0.049 h⁻¹ the values for protein and BChl $\underline{c}$ were 2.95 mg mmol⁻¹ and 0.24 mg
mmol⁻¹ reducing power, respectively.

It has been reported that thiosulfate is formed in batch cultures of
green sulfur bacteria (Schedel, 1978). The production of thiosulfate during
sulfide limitation appeared to be marginal, however. The maximum
concentration of thiosulfate (1.1 µM) was found at the highest dilution rate
tested (D = 0.99 h⁻¹). With respect to the potential molar yield of
thiosulfate as electron donor, thiosulfate and sulfide are directly
comparable. In view of the increasing concentration of reduced forms of
sulfur with increasing dilution rate, and the fact that electron donors are
used virtually exclusively for growth, the concentration of structural cell
material can be expected to decrease with increasing dilution rate, as was
observed. Both the concentration of protein and of the photopigment BChl $\underline{c}$
decreased with increasing dilution rate. The yield of structural cell
material in terms of protein appeared to be 3.28 mg mmol⁻¹ reducing power,
irrespective of the dilution rate (standard deviation = 0.22, n = 9). With
BChl $\underline{c}$ as the parameter for structural cell material, similar data were
observed, i.e., 0.25 mg mmol⁻¹ reducing power (s.d. = 0.03, n = 9).

Transient States

The presence of sulfide and elemental sulfur in steady-state cultures
of $\underline{Chlorobium}$ (growing on sulfide) is the consequence of the growth-rate
limiting nature of these electron donors. Upon deprivation of sulfide from
the reservoir bottle (metering pump stopped) a rapid depletion of sulfide
would be expected. If polysulfides were actively oxidized during the
preceding steady state, the same would be expected for this reduced form of
sulfur. The time course of sulfide and polysulfides during a shift-down from
D = 0.091 h⁻¹ to D = 0 is shown in Fig. 3. The sulfide present during the
steady-state condition (21 µM) was depleted in less than 10 min, at a rate
much higher than abiotic formation of polysulfides. However, it took about
40 min before the concentration of polysulfides started to decrease, and

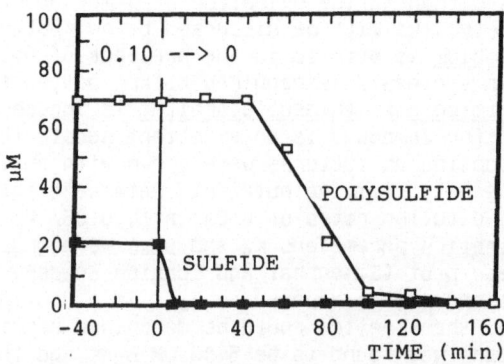

Fig. 3. Sulfide and polysulfide oxidation by $\underline{Cb.\ limicola\ f.}$
$\underline{thiosulfatophilum}$ during a shift-down experiment from D = 0.10 h⁻¹
to D = 0 (metering pump stopped at t = 0).

complete depletion did not occur within 90 min. Apparently at time zero the cells were not equipped to oxidize polysulfides, despite their continuous confrontation with this form of reduced sulfur during the preceding steady state.

It appeared that _de novo_ protein synthesis is required in sulfide-grown cells before polysulfides can be oxidized. In a similar experiment to that described above, (dilution rate = 0.068 h^{-1}, S^{2-} = 3.1 µM, S$_3^{2-}$ = 44.4 µM), the addition of chloramphenicol (CAP, final concentration 114 mg l^{-1}) immediately after the metering pump was stopped resulted in complete incapacity to oxidize polysulfides. Identical results were obtained with puromycin (final concentration 54 mg l^{-1}). In addition to having a requirement for protein synthesis, the oxidation of polysulfides is competitively inhibited by hydrogen sulfide. This was demonstrated by the addition of sulfide (final concentration 14.7 µM) to a culture actively oxidizing polysulfides (Fig. 4). In the preceding steady state (D = 0.091 h^{-1}) the concentrations of sulfide and polysulfides were 9.8 µM and 67.4 µM, respectively. After sulfide had been depleted as the result of stopping the pump and the concentration of polysulfides had been decreased to approximately 35 µM, sulfide was added to a final concentration of 14.7 µM. The concentration of polysulfides increased within 10 min from 35 µM to about 70 µM (Fig. 4). After the complete oxidation of the added sulfide, the cells resumed the oxidation of polysulfides at approximately the same rate as before addition of sulfide. The final result was the complete oxidation of both sulfide and polysulfide prior to the depletion of elemental sulfur (data not shown). However, when the production of cell material is compared with substrate utilization, the actual reactions are by no means clear.

A complicating factor is the possible participation of polysulfides with different chain lengths. These molecules cannot be distinguished, since all polysulfides are measured as thiocyanate. However, the average chain length can be calculated from the sulfide produced during the assay. Reaction equilibria demonstrate that trisulfide (S$_3^{2-}$) is more likely to be formed than disulfide (S$_2^{2-}$; Feher and Berthold, 1953; Feher and Laue, 1956). Disulfide either dissipates to form hydrogen sulfide and sulfur or reacts with sulfur to form trisulfide. Therefore, the concentration of disulfide will be very low. During the steady state of the experiment shown in Fig. 4, the average chain length was determined as 2.85 and the CNS$^-$ concentration as 134.8 µM. From these data it was calculated that the dominant polysulfide present was trisulfide.

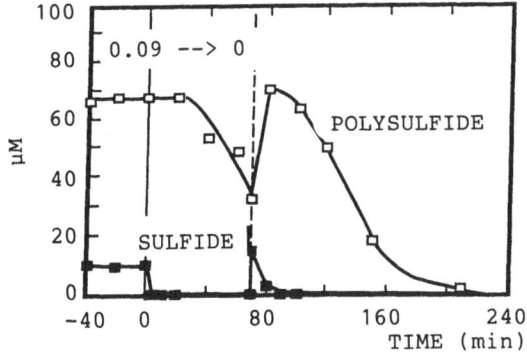

Fig. 4. Sulfide and polysulfide oxidation by _Cb. limicola_ f. _thiosulfatophilum_ during a shift-down from D = 0.09 h^{-1} to D = 0 (metering pump stopped at t = 0). Sulfide was replenished at t = 70 min to a final concentration of 14 µM.

At time zero the organisms were deprived of substrate addition from the reservoir bottle (pump stopped). Within minutes the specific growth rate must have declined from 0.099 h^{-1} to virtually zero because of lack of substrate and inability to use polysulfides. After 40 min polysulfide utilization started. If it is tentatively assumed that S_3^{2-} becomes fully oxidized to sulfate, the production of structural cell material would have been equivalent to $S_3^{2-} + 4\ OH^- + 8\ H_2O \longrightarrow 3\ SO_4^{2-} + 20\ H$ or, (67-35) x 20/1000 x 3.28 = 2.1 mg protein 1^{-1}, or, more likely, (67-35) x 20/1000 x 7.4 = 4.7 mg glycogen l^{-1}. Neither of these increases was found. Also, if trisulfide were formed according to $S^{2-} + 2\ S^\circ \longrightarrow S_3^{2-}$, the increase in trisulfide after the addition of sulfide at t = 40 min could have been 16 µM at the most, leading in total to 51 µM; in fact, 70 µM was found 10 min after the addition.

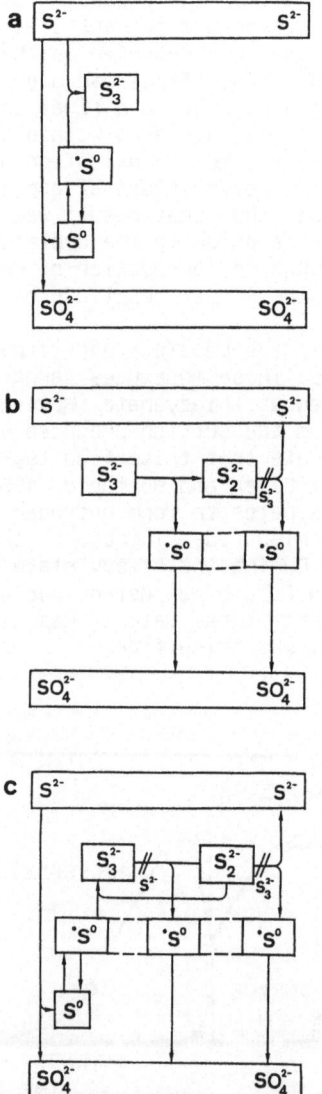

Fig. 5. Hypothetical representation of electron donor utilization by <u>Cb. limicola</u> f. <u>thilsulfatophilum</u>: (a) growth on sulfide alone, (b) growth on polysulfide alone, and (c) sudden appearance of sulfide during growth on polysulfide.

Therefore, it appears likely that trisulfide is oxidized to another form of sulfur. The hypothesis is put forward that this reduced form is disulfide. In this case the reaction would be $S_3^{2-} \longrightarrow S_2^{2-} + {}^*S^o$ in which ${}^*S^o$ designates an activated configuration of elemental sulfur, which is rapidly oxidized by Chlorobium to form sulfate. In this way the oxidation of 35 μM S_3^{2-} would result in the production of only 0.9 mg protein l^{-1} or 1.9 mg glycogen l^{-1}.

Subsequently, when hydrogen sulfide is present the disulfide produced reacts abiotically with activated elemental sulfur, which originates from the excess of elemental sulfur (1000 μM), to form trisulfide (Fig. 5). It is anticipated that biotic oxidation of disulfide is inhibited by trisulfide. Consequently, the concentration of trisulfide would increase by 35 to a final 70 μM, which was the actual concentration observed.

Chlorobium has an extremely high affinity for sulfide (K_s approximately 1 μM). Therefore, virtually all of the sulfide added will be oxidized by the bacteria instantaneously, and none will result in the formation of trisulfide according to the equation $S^{2-} + 2\ S^o \longrightarrow S_3^{2-}$.

To summarize, two different situations can be distinguished. During the first, Chlorobium is growing with hydrogen sulfide as the limiting substrate (Fig. 5a). Sulfide is being oxidized either to elemental sulfur which is deposited outside the cell, or to sulfate. Under steady-state conditions of continuous culture of Chlorobium on hydrogen sulfide large amounts of elemental sulfur appear, a part of which might be present in some kind of activated configuration generated as a result of, for example, crystallization enthalpy of the major fraction of this sulfur. The activated form is available for abiotic reaction with the residual hydrogen sulfide to form polysulfides. Though being a potential electron donor for CO_2 fixation, this sulfur compound would not be oxidized as long as hydrogen sulfide is present.

In the other situation Chlorobium is induced to grow on polysulfides (Fig. 5b), and no hydrogen sulfide is present. Though the mechanism of polysulfide utilization is not yet elucidated, we hypothesize that trisulfide is split into disulfide and an active form of elemental sulfur (${}^*S^o$). The latter is immediately oxidized to sulfate, whereas the disulfide remains unused as long as trisulfide is present. When growing at lower growth rates in continuous culture on trisulfide, a part of the disulfide is further dissipated and thus trisulfide and disulfide are in equilibrium.

In the case of a sudden appearance of hydrogen sulfide, as mimicked during the shift-down experiment, the organisms immediately stop using polysulfide and begin to utilize hydrogen sulfide (Fig. 5c). Obviously, hydrogen sulfide is preferred by Cb. limicola at all times, even if it is present at lower concentrations than other potential electron donors.

As long as hydrogen sulfide is present, the disulfide pool can react with the activated form of elemental sulfur. In between sulfide depletion and reappearance (Fig. 4) the activated sulfur is used by the organisms, together with the activated sulfur obtained from trisulfide. It is not possible for disulfide to react with activated elemental sulfur to produce trisulfide again, during growth on the latter. Only when hydrogen sulfide is available does trisulfide dissipation stop, and also activated sulfur oxidation. Hence abiotic formation of trisulfide is possible, as was observed (Fig. 4).

In conclusion, we hypothesize that hydrogen sulfide acts as a "super" substrate, i.e., hydrogen sulfide is preferred at all times, and at all concentrations above all other reduced sulfur compounds. When hydrogen sulfide is not present, an activated configuration of elemental sulfur appears to be the preferred substitute.

REFERENCES

Beefting, H. H., and van Gemerden, H., 1979, Actual and potential rates of substrate oxidation and product formation in continuous cultures of Chromatium vinosum, Arch. Microbiol., 121:161.

Berner, R. A., 1963, Electrode studies of hydrogen sulfide in marine sediments, Geochim. Cosmochem. Acta, 27:563.

Eisenthal, R., and Cornish-Bowden, A., 1974, The direct linear plot. A new graphic procedure for estimating enzyme kinetic parameters, Biochem. J., 139:715.

Feher, F., and Berthold, D., 1953, Über das System des Natrium-Schwefels, Z. Anorg. Allg. Chem., 273:144.

Feher, F., and Laue, W., 1956, Beitrage zur Chemie des Schwefels. XXIX. Über die Darstellung von Rohsulfanen, Z. Anorg. Allg. Chem., 288:103.

Lowry, O. H., Rosenbrough, N. J., Farr, A. L., and Randall, R. J., 1951, Protein measurements with the Folin-phenol reagent, J. Biol. Chem., 193:265.

Mas, J., and van Gemerden, H., 1987, Influence of sulfur accumulation and composition of sulfur globule on cell volume and buoyant density of Chromatium vinosum, Arch. Microbiol., 146:362.

Pachmayr, F., 1960, Vorkommen und Bestimmung von Schwefelverbindungen in Mineralwasser, Ph.D. Thesis, Univ. München, F.R.G.

Schedel, M., 1978, Untersuchungen zur anaeroben Oxidation reduzierter Schwefelverbindungen durch Thiobacillus denitrificans, Chromatium vinosum und Chlorobium limicola, Ph.D. Thesis, Univ. Bonn, F.R.G.

Stal, L. J., van Gemerden, H., and Krumbein, W., 1984, The simultaneous assay of chlorophyll and bacteriochlorophyll in microbial communities, J. Microbial Meth., 2:295.

Then, J., 1984, Beitrage zur Sulfidoxidation durch Ectothiorhodospira abdelmalekii und Ectothiorhodospira halochloris, Ph.D. Thesis, Univ. Bonn, F.R.G.

Trüper, H. G., 1978, Sulfur metabolism, in: "The Photosynthetic Bacteria," R. K. Clayton and W. R. Sistrom, eds., Plenum, New York.

Trüper, H. G., and Schlegel, H. G., 1964, Sulphur metabolism in Thiorhodaceae. I. Quantitative measurements on growing cells of Chromatium okenii, Anthonie van Leeuwenhoek J. Microbiol. Serol., 30:225.

van Gemerden, H., 1968, Growth measurements of Chromatium cultures, Arch. Microbiol., 64:103.

van Gemerden, H., 1984, The sulfide affinity of phototrophic bacteria in relation to the location of elemental sulfur, Arch. Microbiol., 139:289.

van Gemerden, H., 1987, Competition between purple sulfur bacteria and green sulfur bacteria: role of sulfide, sulfur and polysulfides, Acta Acad. Abo., 47:13.

van Gemerden, H., and Beeftink, H. H., 1978, Specific rates of substrate oxidation and product formation in autotrophically growing Chromatium vinosum cultures, Arch. Microbiol., 119:135.

van Niel, C. B., 1931, On the morphology and physiology of purple and green sulphur bacteria, Arch. Mikrobiol., 3:1.

PHYSIOLOGICAL ASPECTS OF HIGH SULFIDE TOLERANCE IN A PHOTOSYNTHETIC

BACTERIUM

T. Bergstein-Ben Dan

The Yigal Alon Kinneret Limnological Laboratory
Israel Oceanographic & Limnological Research
P.O. Box 345, Tiberias, Israel

INTRODUCTION

Photosynthetic bacteria are found in the hypolimnion of stratified lakes, where sulfide is present. H_2S concentrations in the thermocline range from 3-12 µM in Lake Vechten (Steenbergen et al., 1987) to 0.19 mM in Lake Belovod (Sorokin, 1970), 0.062-0.125 mM in Lake Kolksee (Olah et al., 1973) and 0.09-0.13 mM in Lake Kinneret (Serruya et al., 1974).

Growth of photosynthetic sulfur bacteria is affected mainly by light intensity and by the presence of reduced sulfur compounds (Bergstein et al., 1979; Bergstein and Cavari, 1983; van Gemerden, 1967). Cell yield as well as photosynthetic activity is influenced by light intensity. Maximal cell yield of Chlorobium phaeoacteroides is achieved at all light intensities used (5-50 µE m^{-2} s^{-1}). Photosynthetic activity is maximal at 10 µE m^{-2} s^{-1} (Bergstein et al., 1979). Above and below this intensity photosynthetic activity is relatively low. Sulfide consumption is influenced by light intensity as well. At higher light intensity (in the range of 0.5-25 µE m^{-2} s^{-1}) sulfide consumption is increased (Bergstein and Cavari, 1983).

Cell growth as well as photosynthetic activity is also influenced by sulfide concentration. Increasing the sulfide concentration from 0.53 mM to 1.99 mM causes an increase of cell yield of Chlorobium limicola by about 3 fold (Bieble and Pfennig, 1978). Van Gemerden (1967) reported that an increase of sulfide concentration from 0.35 to 3.15 mM increased the cell yield of Chromatium by about 7 fold. For Cb. phaeobacteroides both sulfide and sulfur consumption were found to be affected by sulfide concentration. Raising the sulfide concentration from 0.28 mM to 5.05 mM caused an increase in the amount of $S^=$ utilized per growth unit (1 OD at 715 nm) from 0.58 mM to 2.32 mM. Growth and photosynthetic activity were maximal at 2 mM H_2S and sharply decreased at concentrations above 5 mM (Bergstein and Cavari, 1983).

MATERIALS AND METHODS

Bacterial Isolation

Bacteria were isolated from samples taken from the organic detritus layer (~0.2-0.3 M H_2S) sampled from a water pond on Aldabra atoll in the

Table 1. ATP Content of Cells after
Incubation for Five Days
under Different Light
Intensities at S_t = 3 mM

Light intensity ($\mu E\ m^{-2}\ s^{-1}$)	ATP/10^6 cells (pg)
75	0.99
40	1.58
15	1.91
7	3.92

Seychelles Islands (Indian Ocean). About 1 ml of the organic detritus layer was inoculated into 10-ml Pfennig medium-agar shakes (Pfennig, 1965) with modifications described previously (Bergstein et al., 1979) and incubated at 30-50 $\mu E\ m^{-2}\ s^{-1}$ light intensity. After about 10 days green colonies developed, and they were separately transferred into liquid Pfennig medium (pH 6.5). Each green colony grew into a green culture. Subsamples (0.5 ml) from these cultures were filtered on 25 mm GF/C filters (Whatman) and the filters extracted with 90% acetone. Bacterial pigments in the extracts, as well as living cells, were determined spectrophotometrically. According to these results and by microscopic inspection we identified the bacterium as Clathrochloris sulfuricon (Pfennig and Trüper, 1969, 1974).

Clathrochloris cultures were grown in Pfennig medium at 26°C with slight modifications, as described previously (Bergstein et al., 1979). The absorption at 755 nm (in vivo maximum of the bacterial chlorophyll) was taken as a measure of growth of the culture, and of cell yields.

Total Reduced Sulfur (S_t) Concentrations in Laboratory Experiments

S_t utilization by the culture was measured with a pH$_2$S electrode in the subsamples taken from the growth vessels. S_t concentrations were calculated from the measured H_2S activities by the protolysis constant α_0 (Stumm and Morgan, 1981). For pH < 9 S_t is equal to H_2S + HS^-, therefore [Sred] = $[H_2S]/\alpha_0$. Previous investigators have measured sulfide concentrations by the methylene blue method (Trüper and Schlegel, 1964), which detects all the reduced forms of S (H_2S + HS^-) defined in this paper as S_t.

Photosynthetic Activity

Carbon dioxide uptake was measured by a variation of the basic

Table 2. Photosynthetic Activity per 10^6 Cells at Different Light
Intensities and S_t Concentrations

Light intensity ($\mu E\ m^{-2}\ s^{-1}$)	[S_t] (mM)	Incubation time (days)	Maximal photosynthetic activity after incubation (μg C-CO_2)
22.0	6.0	7	1.20
22.0	8.7	7	1.14
5.6	6.0	9	4.52
5.6	8.7	9	4.60

Steeman-Nielsen (1952) technique. After addition of 5 μC_i NaH^{14}CO$_3$ (60 mC$_i$/mmol; The Radiochemical Center, Amersham) to 100 ml <u>Clathrochloris</u> culture, 1 ml culture was filtered under mild vacuum (100 mm Hg) on 25 mm membrane filters (pore size, 0.45 μm, Millipore Corporation, Bedford MA), rinsed with 10 ml distilled water, and fumed in HCl vapour to eliminate

remaining traces of inorganic ^{14}C. The filters were placed in scintillation liquid containing 70% toluene and 30% Lumax, and the radioactivity counted in a Kontron MR 300 scintillation counter.

Light Intensity and ATP

Light intensity was measured with a Lamda L1-185 quantum meter, and ATP was measured by the luciferin-luciferase assay (Cavari, 1976).

Direct Counts

Direct counts were performed by acridine orange staining, and the bacteria were counted using epifluorescent microscopy (Hobbie et al., 1977).

RESULTS AND DISCUSSION

Bacteria Description

A typical absorption spectrum (Fig. 1) of a suspension of living cells shows maxima at 460 and 755 nm. An acetone extract shows peaks at 437 and 668 nm (Fig. 2). The cells are spherical (1-1.5 μm in diameter) and arranged in chains which are held together in loose, trellis-shaped aggregates.

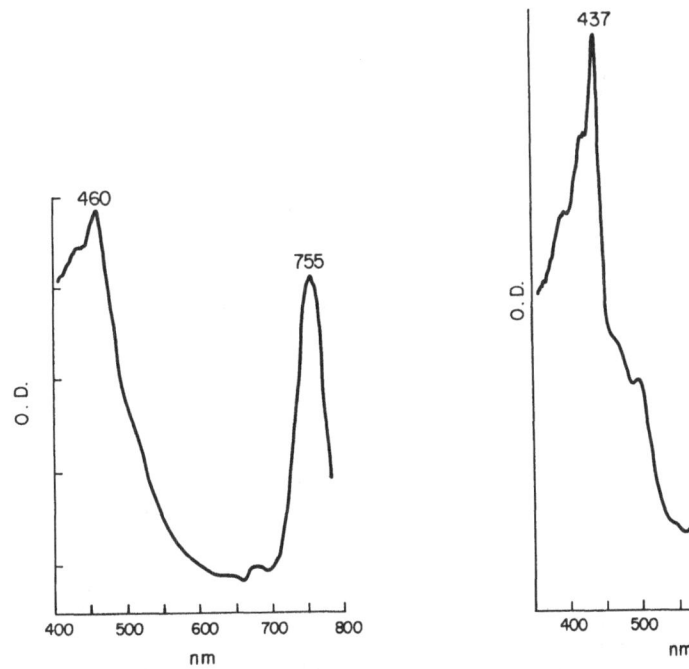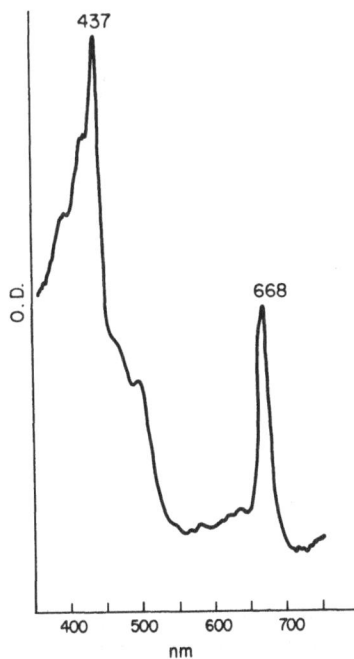

Fig. 1. Absorption spectrum of living cells of <u>Clathrochloris sulfurica</u>.
Fig. 2. Absorption spectrum of acetone extract of <u>C. sulfurica</u> culture.

Factors Affecting Growth and Photosynthetic Activity

Light intensity. Growth of Clathrochloris as a function of light intensity was studied at several S_t concentrations. At all S_t concentrations used (3, 6, 8.7 mM) cell numbers were similar in all cultures incubated under the light intensities used (5.6-75 µE m^{-2} s^{-1}). ATP production was maximal at 7 µE m^{-2} s^{-1} (Fig. 3 and 4). Table 1 shows that as light intensity decreases there is a corresponding increase in ATP concentration within the cells. When initial S_t in the medium was 6 mM, bacterial chlorophyll formation was similar at light intensities of 5.6, and 22 µE m^{-2} s-1. At an initial S_t of 8.7 mM bacterial chlorophyll formation was lowest at a light intensity of 22 µE m^{-2} s^{-1} and reached higher values when light intensity decreased. Cell numbers were similar at all light intensities (Fig. 5). Photosynthetic activity per cell was similar at the two S_t concentrations and at the two light intensities (Table 2). The above results suggest that cell number and photosynthetic activity are similar in cultures incubated at different light intensities and are independent of intracellular ATP.

Reduced sulfur compounds (S_t). Utilization of different S_t concentrations as well as pH values were measured during cell growth at light intensities of 10.3 and 5.6 µE m^{-2} s^{-1} (Fig. 6 and 7). Fig. 6 shows that at the highest S_t concentration the greatest growth was achieved (as

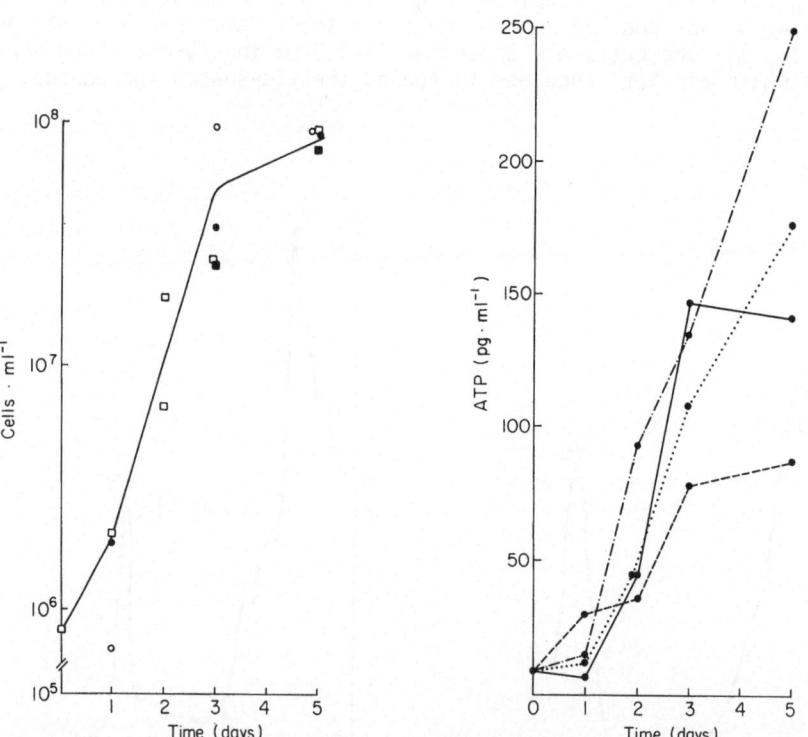

Fig. 3. Effect of light intensity on cell number direct count of Clathrochloris incubated with initial $[S_t]$ = 3 mM: 7 (o), 15 (□), 40 (●), and 75 (■) µE m^{-2} s^{-1}.

Fig. 4. Effect of light intensity on ATP concentration of Clathrochloris cultures incubated with initial $[S_t]$ = 3 mM: 7 (-·-), 15 (···), 40 (---), and 75 (——) µE m^{-2} s^{-1}.

expressed by OD at 755 nm). With an S_t of 12.7 mM maximal OD was 1.76. During growth the pH remained constant at about 6.6. With 19 mM S_t maximal growth was 2.00 OD, pH increased from 6.5 to 6.7 after 3 days, and then dropped to 5.7. With 34 mM S_t, the OD reached 2.24, pH increased to 6.8 and after three days decreased to 5.4.

A similar pattern could be observed when bacteria were incubated under a light intensity of 5.6 µE m⁻² s⁻¹ (Fig. 7a, b, c), but in this case there is a smaller reduction in the pH as can be seen from Fig. 7b, c. The pH dropped from 6.56 to 6.2 at an initial S_t concentration of 34 mM, and from 6.5 to 6.2 when the initial S_t was 19 mM. After three days (in all the treatments) most of the S_t was taken up, even though bacterial growth was not completed. (At the lowest light intensity the log phase had only started.) The continuation of growth was probably due to globular sulfur consumption. As a result of $S°$ oxidation, SO_4^{2-} was produced, which caused a pH drop at the high S_t concentrations.

The rate of S_t consumption in the first three days of incubation increased as the S_t concentration was raised (Table 3). Furthermore with increasing S_t concentrations in the medium (from 12.7 to 34 mM) the amount of S_t consumed per growth unit at 755 nm rose from 16.8 to 75.6 mM at a light intensity of 10.3 µE m⁻² s⁻¹ and from about 80 to 154.6 mM at 5.6 µE m⁻² s⁻¹. These values are very high compared to a maximal value of 2.32 mM obtained by Cb. phaeobacteroides (Bergstein and Cavari, 1983).

The results of this work suggest that even though light intensity has an affect on the ATP content of Clathrochloris cells, there is no influence

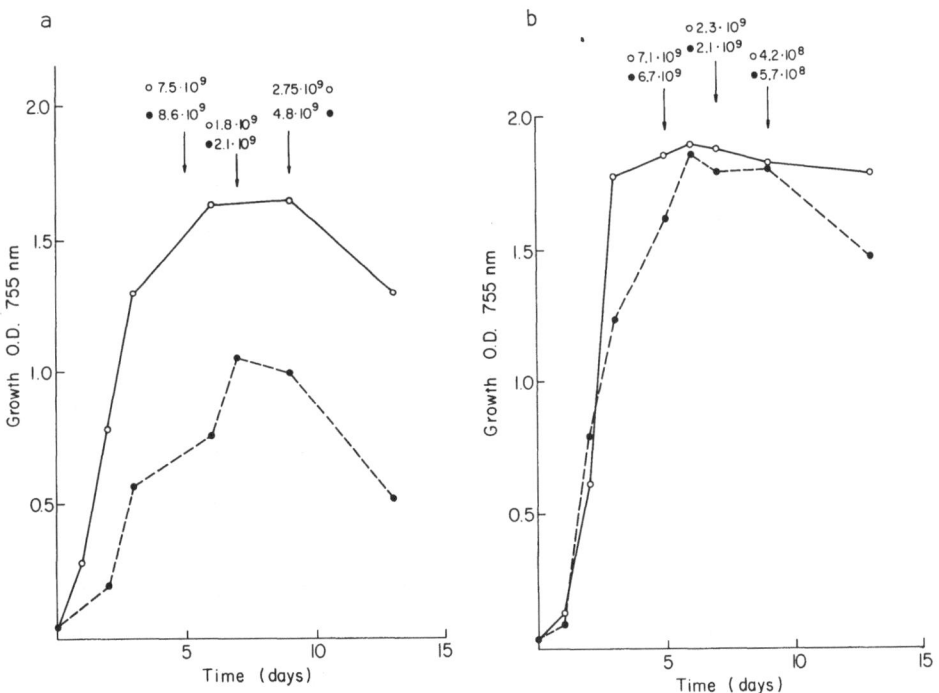

Fig. 5. Effect of light intensity and initial S_t concentration on Clathrochloris growth measured as bacterial chlorophyll OD at 755 nm and as cell numbers per 1 ml direct count. Cultures were incubated under 22 µE m⁻² s⁻¹ (a) and 5.6 µE m⁻² s⁻¹ (b) with [S_t] = 0.6 mM (o) and 8.7 mM (○).

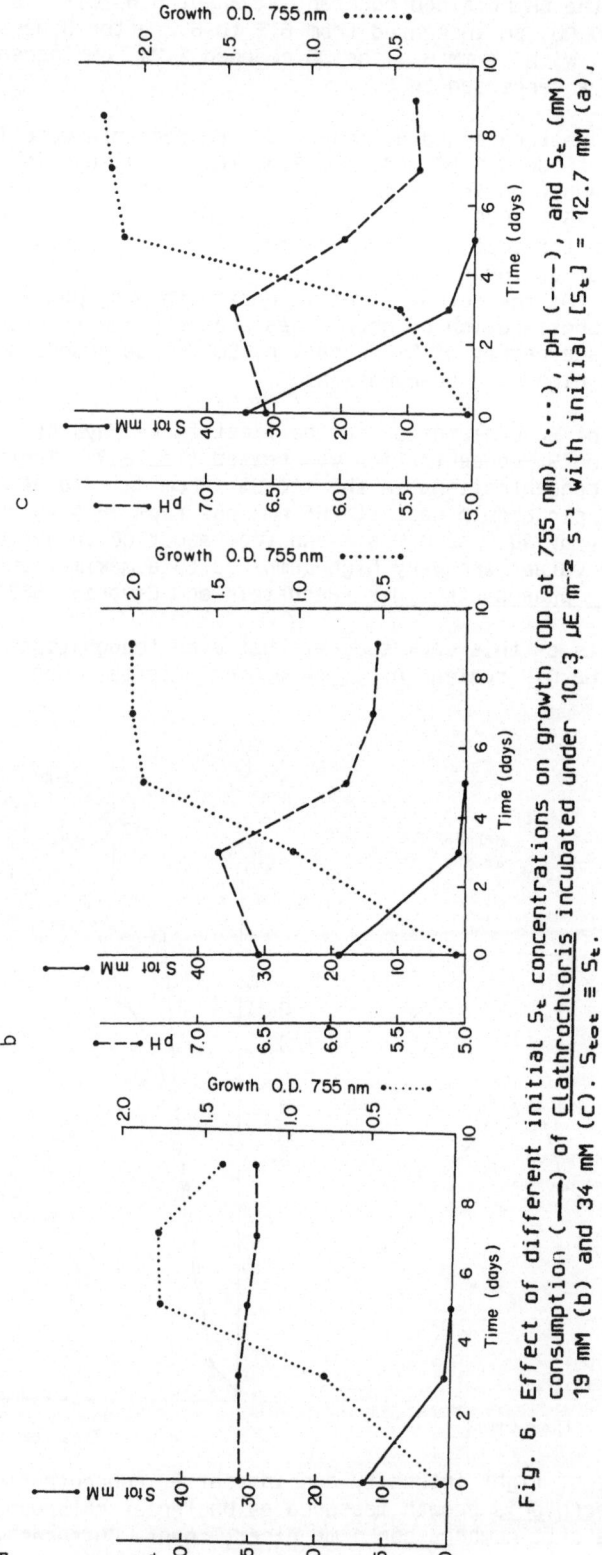

Fig. 6. Effect of different initial S_t concentrations on growth (OD at 755 nm) (····), pH (---), and S_t (mM) consumption (——) of <u>Clathrochloris</u> incubated under 10.3 µE m^{-2} s^{-1} with initial $[S_t]$ = 12.7 mM (a), 19 mM (b) and 34 mM (c). $S_{tot} \equiv S_t$.

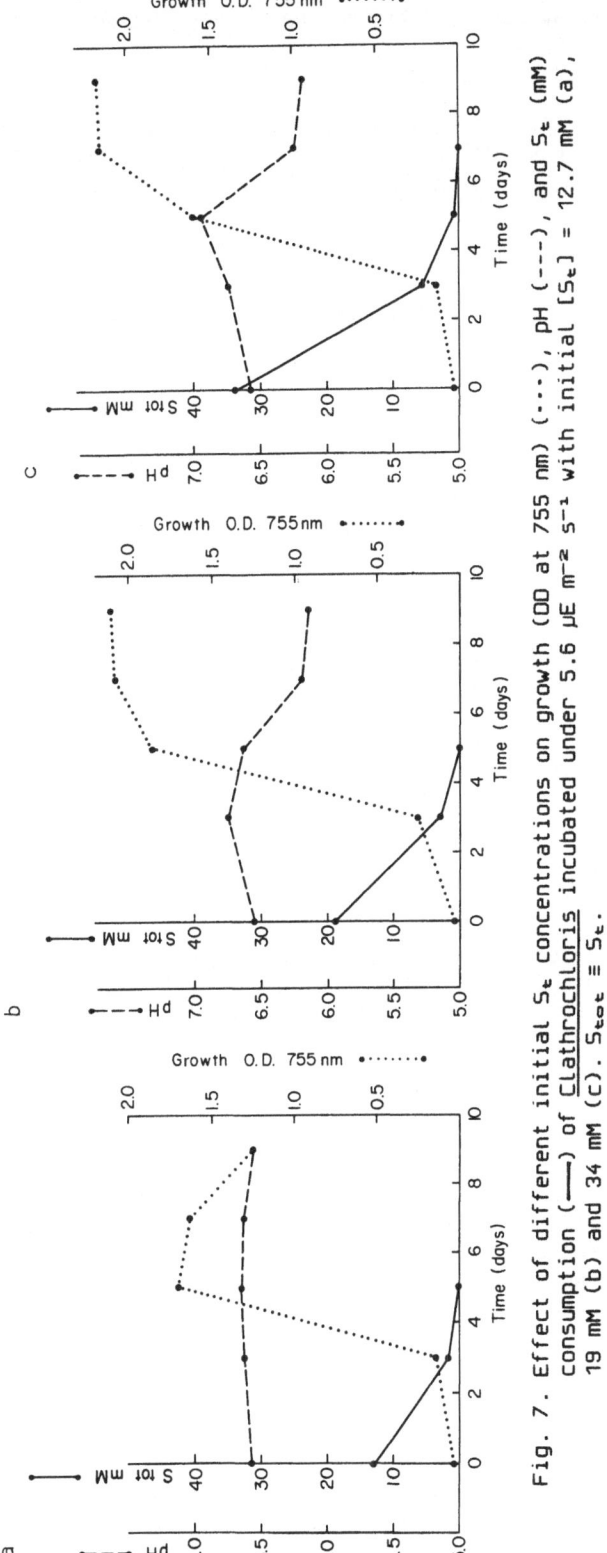

Fig. 7. Effect of different initial S_t concentrations on growth (OD at 755 nm) ($\cdots$), pH (---), and S_t (mM) consumption (——) of <u>Clathrochloris</u> incubated under 5.6 μE m^{-2} s^{-1} with initial [S_t] = 12.7 mM (a), 19 mM (b) and 34 mM (c). $S_{tot} = S_t$.

Table 3. S_t Utilization Rate and S_t Required per Unit of Growth at Different Initial S_t Concentrations. The S_t required was calculated for the first 3 days.

Light intensity ($\mu E \ m^{-2} \ s^{-1}$)	Initial [S_t] (mM)	S_t utilization rate (mM day^{-1})	OD at 755 nm after 3 days	S_t required per OD unit at 755 nm (mM)
10.3	12.7	4.04	0.76	16.8
10.3	19.0	5.77	1.03	18.4
10.3	34.0	10.0	0.45	75.6
5.6	12.7	3.66	0.14	91.1
5.6	19.0	5.39	0.25	75.8
5.6	34.0	9.53	0.22	154.6

on cell number or photosynthetic activity, in contrast to results with other photosynthetic bacteria (Bergstein et al., 1979; Takahaski and Ichimura, 1970). The natural environment of Clathrochloris contains sulfide concentrations several orders of magnitude higher (~0.2-0.3 M H_2S) than the natural habitats of most other photosynthetic sulfur bacteria, which might explain its extremely high use of reduced sulfur ($H_2S + HS^-$).

Sulfide concentrations in the range of 2-3 mM were found to be optimal for the growth and photosynthetic activity of photosynthetic sulfur bacteria (Biebl and Pfennig, 1978; van Gemerden, 1967; Bergstein and Cavari, 1983). No effect of sulfide concentration on photosynthetic activity per cell was found when S_t in the medium increased from 6 to 8.7 mM. Since C. sulfurica has a very high S_t utilization our results suggest that the bacterium might be useful in removing sulfide from aquatic ecosystem.

ACKNOWLEDGEMENTS

The author wishes to thank Dr. Tonnies Frevert for supplying the organic detritus sample, Dr. D. Wynne for helpful comments and reviewing the manuscript, and Dr. P. Walline for taking the microscopic pictures.

SUMMARY

A photosynthetic sulfur bacterium isolated from a high sulfide-containing organic detritus layer was identified as Clathrochloris sulfurica. Cell numbers and photosynthetic activities of these bacteria were similar at light intensities of 6.5-75 $\mu E \ m^{-2} \ s^{-1}$ and S_t concentrations of 6-8.7 mM, independent of intracellular ATP. C. sulfurica utilized S_t up to 34 mM and its consumption rate and the amount consumed per growth unit increased as the initial concentration of S_t was raised.

REFERENCES

Bergstein, T., Henis, Y., and Cavari, B. Z., 1979, Investigations on the photosynthetic sulfur bacterium Chlorobium phaeobacteroides causing seasonal blooms in Lake Kinneret, Can. J. Microbiol., 25:999.
Bergstein, T., and Cavari, B. Z., 1983, Sulfide utilization by the photosynthetic bacterium Chlorobium phaeobacteroides, Hydrobiol., 106:241.
Biebl, H., and Pfennig, N., 1978, Growth yield of green sulfur bacteria in

mixed cultures with sulfur and sulfate reducing bacteria, Arch. Microbiol., 117:9.

Cavari, B. Z., 1976, ATP in Lake Kinneret: indicator of microbial biomass or of phosphorus deficiency? Limnol. Oceanogr., 21:231.

Hobbie, J. E., Daley, R. J., and Jasper, S., 1977, Use of nuclepore filters for counting bacteria by fluorescence microscopy, Appl. Environ. Microbiol., 33:1225.

Olah, J., Biebl, H., and Overbeck, J., 1973, Photoorganotrophic utilization of acetate in a stratified eutrophic lake, Hydrologiai Kozlony, 1:21.

Pfennig, N., 1965, Anveicherungskulturen für Rote and Grüne Schwefelbakterium, Zentralbl. Bakteriol. Parasitenkd. Infektionskr. Hyg., Abt. 1., Suppl. 1:179.

Pfennig, N., and Trüper, H. G., 1969, Phototrophic bacteria (in German) GSF Report M 32, pp. 117-118.

Pfennig, N., and Trüper, H. G., 1974, The phototrophic bacteria, in: "Bergey's Manual of Determinative Bacteriology", 8th ed., R. E. Buchanan and N. E. Gibbons, eds., Williams & Wilkins, Baltimore.

Serruya, C., Edelstein, M., Pollingher, U., and Serruya, S., 1974, Lake Kinneret sediments: nutrient composition of the pore water exchanges, Limnol. Oceanogr., 19:489.

Sorokin, Yu. I., 1970, Interrelations between sulfur and carbon turnover in meromictic lakes, Arch. Hydrobiol., 66:391.

Steeman-Nielsen, E., 1952, The use of radioactive carbon (^{14}C) for measuring organic production in the sea, J. Cons. Int. Explor. Mer., 18:117.

Steenbergen, C. L. M., Korthols, H. J., and Van Nes, M., 1987, Ecological observations on phototrophic sulfur bacteria and the role of these bacteria in the sulfur cycle of monomictic Lake Vechten, Acta Academiae Aboensis, 47:97.

Stumm, W., and Morgan, J. J., 1981, "Aquatic Chemistry", 2nd ed., Wiley, New York.

Takahashi, M., and Ichimura, S., 1970, Photosynthetic properties and growth of photosynthetic sulfur bacteria in lakes, Limnol. Oceanogr., 15:929.

Trüper, H. G., and Schlegel, N. G., 1964, Sulfur metabolism in Thiorhodaceae. I. Quantitative measurements on growing cells of Chromatium okenii, Antonie van Leeuwenhoek J. Microbiol. Serol., 30:225.

van Gemerden, H., 1967, On the bacterial sulfur cycle of inland waters, Ph.D. thesis, Rijksuniversiteit te Leiden, The Netherlands.

DIFFERENCES BETWEEN *IN SITU* AND *IN VITRO* REDOX CONDITIONS DUE TO THE

ACTIVITY OF THE GREEN SULFUR BACTERIUM *CHLOROBIUM PHAEOBACTEROIDES*

W. Eckert

Department of Microbiology, University of Bonn
Meckenheimer Allee 168, D-5300 Bonn 1, F.R.G.
Present address: Kinneret Limnological Laboratory
P.O.Box 345, 14-102 Tiberias, Israel

SUMMARY

In warm monomictic Lake Kinneret a bloom of the green sulfur bacterial
species *Chlorobium phaeobacteroides* can be observed from July until
September when the thermocline is stabilized at a depth of 15-20 m (Cavari
et al., 1973; Bergstein et al., 1979; Eckert et al., 1986). In mesomictic
lakes the habitat of phototrophic sulfur bacteria is established when during
the process of stratification the so-called sulfuretum reaches water layers
within the zone of light penetration (Pfennig, 1967). This situation
represents the final state in the succession of different hypolimnetic
redox-systems. Various descriptions of the changes of microbial communities
with time and the corresponding hydrochemical conditions are given in the
literature (Hutchinson, 1957; Wetzel, 1975). But only in a few works such as
those of Baas-Becking et al. (1960) or Sorokin (1970) have the prevailing
hydrochemical conditions in sulfureta been investigated in detail.

According to Stumm and Morgan (1981) and Frevert (1979) aquatic redox
systems can be characterized by means of the master variables, pH and pe (pe
= -log a_e-). There is an ongoing discussion concerning the information value
of measured redox levels (Harrison, 1972; Jacobs, 1974; Wimpenny, 1976;
Whitfield, 1974; Kjaergaard, 1977; Morris and Stumm, 1966; Eckert, 1987;
Boulegue and Michard, 1979). Proceeding from the assumption that the
intracellular microbial metabolism reflects the extracellular conditions,
Frevert (1984) introduced a sensor-specific redox concept which is
independent of the internal redox equilibrium in the test solution but
depends on the equilibrium between the test solution and an appropriate
sensor. This consideration enables a systematic identification of certain
predominant redox systems as exemplified by the $S(+VI)/S(-II)-H_2O$ system of
the sulfuretum. A problem in the past has been to obtain *in situ* pH and pe
data of sufficient reproducibility from the H_2S-bearing hypolimnion, a
requirement which is not fulfilled by conventional AgCl//glass and
AgCl//Pt-metal electrode combinations. In this work *in situ* profiles of
pH_2S, pH and pe values were measured with a recently developed probe which
by-passes former technical problems. The presence of phototrophic bacteria
was examined by measuring the BChl absorption peaks in acetone extracts of
water samples.

Fig. 1 shows a profile as obtained during the _Chlorobium_ bloom. The metalimnetic pe, values ranged from 0.16 to -0.70 at a pH range from 7.58 to 7.53. The corresponding pS(-II) values were 4.62 and 3.69.

In order to investigate the redox range covered by sulfur bacteria interaction, a sediment core experiment was carried out using the arrangement shown by Fig. 2. Anaerobic sterifiltered lake water from the thermocline layer served as medium. The pH value was maintained constant at pH 7.0 ± 0.1. By alternate incubation in light and dark the activity of phototrophic bacteria was stimulated and inhibited, respectively. The results are shown in Fig. 3. Note the quick response of the system when switched from light to dark and vice versa. The measured pe values covered a range from pe -2.58 to 4.04 at pS(-II) values ranging from 4.47 to 16.3. A comparison with the findings of Baas-Becking et al. (1960) who defined a range of pe from -5.05 to 0.12 for the green phototrophic sulfur bacteria shows that these bacteria, due to the oxidation of hydrogen sulfide, are establishing more positive pe values than previously reported.

In Fig. 4 and 5 the pe and pS(-II) values from the experiment and from the first 5 m of the metalimnion, respectively, are plotted against each other and compared to the thermodynamic equilibrium curve as calculated from the measured pH_2S values:

$$pe = 2.4 - pH + 0.5\ pH_2S \quad \text{(Stumm and Morgan, 1981)}.$$

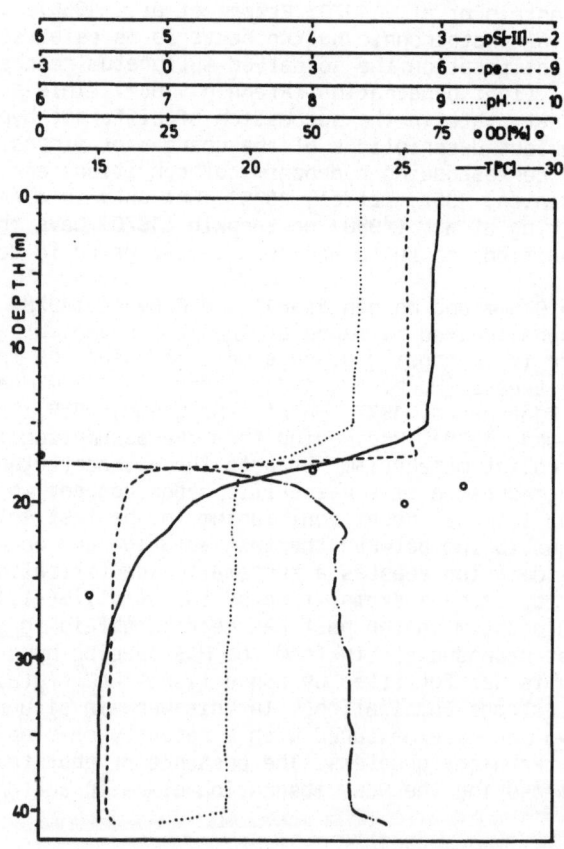

Fig. 1. _In situ_ profile from 6.7.86.

Apparently the experimental pe data show the same tendency as the theoretical curve, whereas the lake data are more positive despite lower pH_2S values - a phenomenon which is in need of further research. One possible reason could be an indirect pH effect, due to a different response of the Pt sensor to H_2S and HS-fractions. This would make uncertain the use of pe_7 values, as used currently for comparison of redox data from different pH regimes.

ACKNOWLEDGEMENTS

This work was supported by the MINERVA foundation (Heidelberg/Rehovot), by the DAAD (Deutscher Akademischer Austauschdienst), and by the DFG (Deutsche Forschungsgemeinschaft).

REFERENCES

Baas-Becking, L. G., Kaplan, I. R., and Moore, D., 1960, Limits of the natural environment in terms of pH and oxidation-reduction potential, J. Geol., 68:243.

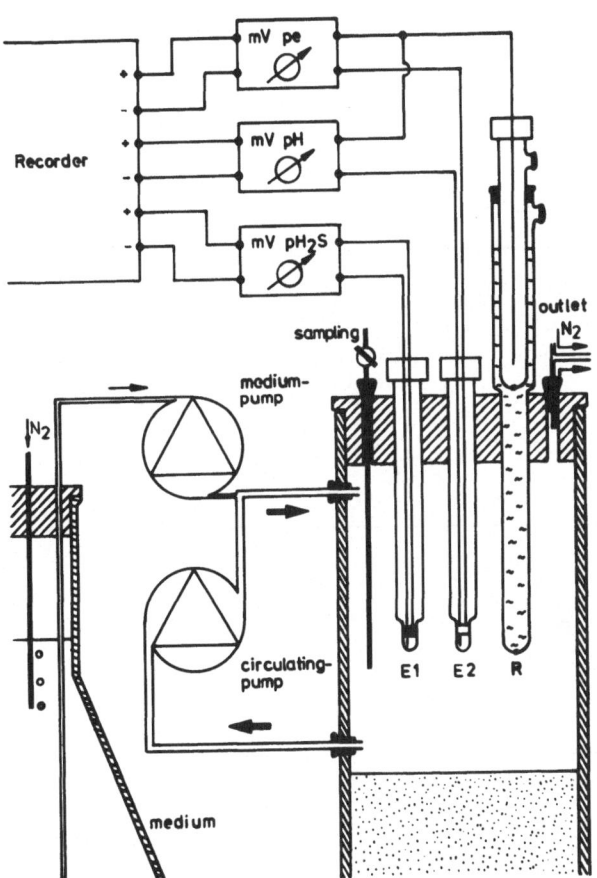

Fig. 2. Experimental arrangement: E1 = pH_2S electrode, E2 = Pt-glass electrode, R = double junction reference electrode.

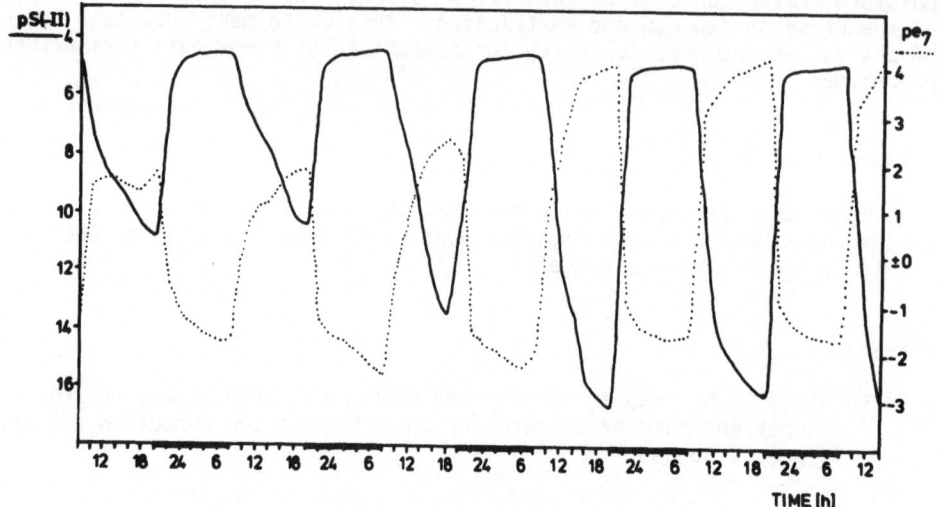

Fig. 3. pS(-II)-pe₇ data versus time.

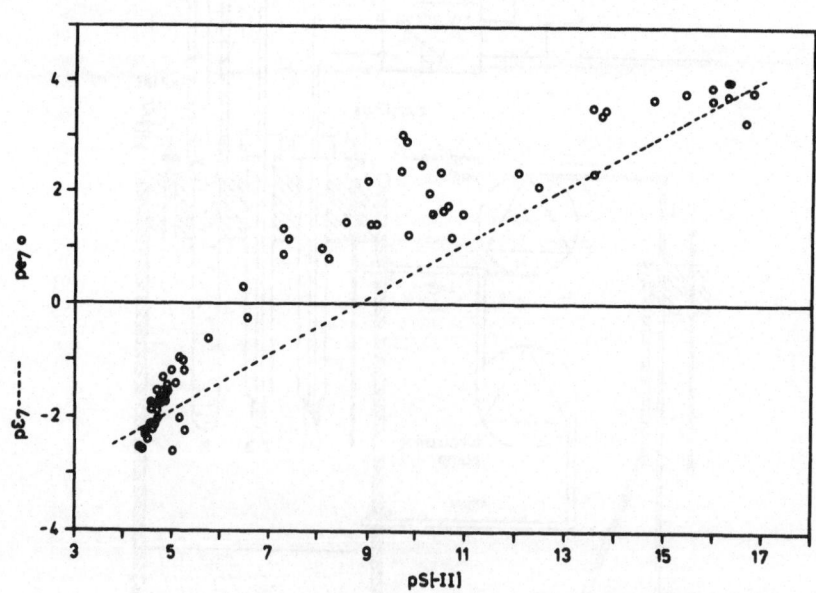

Fig. 4. pe₇-pS(-II) Diagram (experimental data).

Bergstein, T., Henis, Y., and Cavari, B. Z., 1979, Investigations on the photosynthetic sulfur bacterium Chlorobium phaeobacteroides causing seasonal blooms in Lake Kinneret, Can. J. Microbiol., 25:999.

Boulegue, J., and Michard, G., 1979, Sulfur speciations and redox processes in reducing environments, in: "Chemical Modeling in Aqueous Systems," E. A. Jenne, ed., ACI Symposium Series, No. 93:25.

Cavari, B. Z., Uriel, O., Gophen, M., and Berman, T., 1973, Physiological characteristics of Chlorobium and Chromatium strains found in Lake Kinneret, 1st Int. Congr. Bacteriol. Jerusalem (Israel), 11:147.

Eckert, W., Frevert, T., Bergstein-BenDan, T., and Cavari, B.Z., 1986, Competitive development of Thiocapsa roeseopersicina and Chlorobium phaeobacteroides in Lake Kinneret, Can. J. Microbiol., 32:917.

Eckert, W., 1987, Electrochemical identification of microbially mediated hydrogen sulfide oxidation, Biogeochemistry, in press.

Frevert, T., 1979, The pe redox concept in natural sediment-water systems: its role in controlling phosphorous release from lake sediments, Arch. Hydrobiol./Suppl., 55:278.

Frevert, T., 1984, Can the redox conditions in natural waters be predicted by a single parameter?, Schweiz. Z. Hydrol., 46:269.

Harrison, D. E. F., 1972, Physiological effects of dissolved oxygen tension and redox potential on growing populations of microorganisms, J. Appl. Chem. Biotechnol., 22:417.

Hutchinson, G. E., 1957, A Treatise on Limnology Vol. 1, John Whiley, London.

Jacob, H. E., 1974, Reasons for the redox potential in microbial cultures, Biotechnol. Bioeng. Symp., 4:781.

Kjaergaard, L., 1977, The redox potential: its use and control in biotechnology, Adv. Biochem. Eng., 7:131.

Morris, J. C. and Stumm, W., 1967, Redox equilibria and measurements of potentials in the aquatic environment, Adv. Chem. Ser., 67:270.

Pfennig, N., 1967, Photosynthetic bacteria, Ann. Rev. Microbiol., 21:464.

Sorokin, Y. I., 1970, Interrelations between sulfur and carbon turnover in meromictic lakes, Arch. Hydrobiol., 66:391.

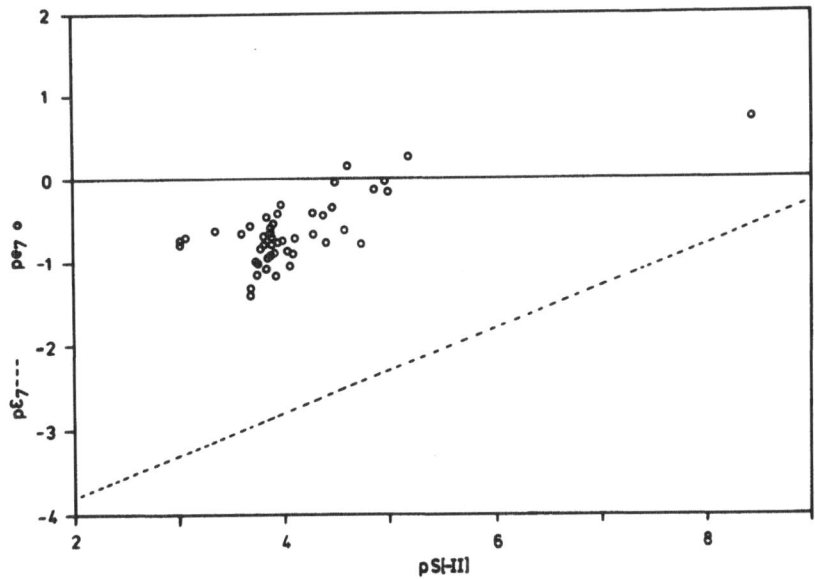

Fig. 5. pe$_7$-pS(-II) Diagram (metalimnion data '86).

Stumm, W., and Morgen, J. J., 1981, Aquatic Chemistry (2nd ed.), Wiley
    Interscience, New York.
Wetzel, R. G., 1975, Limnology, Saunders, Philadelphia.
Whitfield, M., 1974, Thermodynamic limitations on the use of the platinum
    electrode in Eh measurements, Limnol. Oceanogr., 19:857.
Wimpenny, J. W.T., 1976, Can culture redox potential be a useful indicator
    of oxygen metabolism by microorganisms?, J. Appl. Cem. Biotechnol.,
    26:48.

A STUDY OF GREEN PHOTOSYNTHETIC BACTERIA FROM A THERMAL SULFUR SPRING

N.J. Khinvasara

Department of Microbiology
MACS Research Institute
Law College Road, Pune, India

INTRODUCTION

The green photosynthetic bacteria are a physiological-ecological group of anaerobic phototrophic bacteria with anoxygenic photosynthesis (Pfennig and Trüper, 1981). The most important environmental factors affecting the growth of these organisms are anaerobic conditions, the presence of hydrogen sulfide, and illumination. The green photosynthetic bacteria have a wide ecological distribution ranging from freshwater habitats such as ponds, ditches, stagnant bodies of water, lakes, etc.; marine and saline habitats such as seawater pools, salt marshes, closed bays and estuaries and thermal habitats such as sulfur springs. Extensive reviews have appeared in the literature on the distribution of green sulfur bacteria in nature (Kondratieva, 1965; Gorlenko et al., 1977).

Sulfur springs create aquatic habitats with relatively constant sulfide concentrations. Szafer (1910) and Strzeszewski (1913) in Poland first reported visible mass accumulations of green and purple sulfur bacteria in the effluent of such habitats exposed to light (40-100 mg H_2S/litre). Green sulfur bacteria occurred as the predominant forms at highest sulfide concentrations, followed downstream by different species of purple sulfur bacteria. Miyoshi (1897) reported the presence of purple sulfur bacteria in warm sulfur springs (35°-44°C) in Japan. Another green bacterium, Chloroflexus, has often been isolated from alkaline hot springs in most regions of the globe (Pierson and Castenholz, 1974).

In our microbiological studies of different ecosystems rich in sulfur or its compounds such as stagnant ponds, estuarine areas, acid mine drainage, thermal springs, etc., we decided to take a look at photosynthetic bacteria since they are also known to interact with sulfur and bring about its transformation in sulfur cycle. In order to study the occurrence of green photosynthetic bacteria in thermal sulfur springs, a spring in Western Maharashtra (India) was selected for microbiological studies.

MATERIALS AND METHODS

The thermal sulfur spring selected for our study consisted of a single point source of eruption (temperature at source was 70°C), where a small pool was formed. The hot water gushed out of this pool and flowed out in the form of a stream, which meandered through the vegetation, finally emptying into a stagnant pond, at a distance of about 1 km from the source of eruption. A strong smell of sulfur pervaded the atmosphere around the spring, indicating a high sulfur content of the water.

Sample Collection

For microbiological studies, water, sediment and mat samples were collected at various points selected randomly along the course of the stream. The collected samples were placed in sterile glass bottles which were stored in heatproof containers. The temperature, pH, and sulfide content of the water at different points of sampling were noted. The samples were transported to the laboratory within 2 h and immediately processed for the enrichment of different types of microorganisms.

Enrichment

Microscopic examination of the different samples revealed the presence of a vast array of microbial shapes and sizes. Since it was not possible nor desirable to enrich for the different types present, it was decided to set up enrichments to more or less select the following types:
- Chloroflexus spp. in samples collected at a point where temperature was between 50-60°C (50 m downstream);
- thermophilic Thiobacillus spp. in samples collected at a point where temperature was between 40-50°C (100 m downstream);
- green sulfur bacteria in samples collected at a point where the temperature was between 30-40°C (200 m downstream).
The above bias was to simplify this preliminary study by looking for selected groups of microorganisms.

The enrichments for green sulfur bacteria were set up in Medium 1 used for cultivation of green and purple sulfur bacteria according to the method described by Pfennig and Trüper (1981). The enrichments for Chloroflexus spp. were set up in Medium D as described by Castenholz and Pierson (1981) and that for sulfur-utilizing Thiobacillus spp. was set up using different media described by Kuenen and Tuovinen (1981).

Isolation and Identification

The above enrichment procedures were used to select out the predominant species in the samples and the representative predominant cultures were purified and identified up to genus level, using the methods described for each of them in the reference quoted above.

RESULTS AND DISCUSSION

As mentioned earlier, it was decided to look for only three groups of microorganisms, viz. Chloroflexus, Thiobacillus and green sulfur bacteria. Consequently, the different enrichments were scanned for the presence of these microorganisms.

It was observed that the enrichments from samples collected at points of high temperature (50-60°C) yielded predominantly cultures of Chloroflexus spp. This observation further confirmed the fact that Chloroflexus was most dominant in that region of the spring as evidenced by the formation of thick orange-coloured gel-like mats. This particular region of the stream was found to have a high sulfide concentration (8-10 mM sulfide) and a pH of 6.0.

From midstream points, a few isolates of _Chloroflexus_ were obtained, probably due to the lower temperature of this region (40-50°C). However, several isolates of _Thiobacillus_ spp. capable of utilizing sulfur as energy source were obtained. These cultures were moderately thermophilic, having temperature optima around 45°C and being capable of slow oxidation of sulfur to sulfuric acid. The reason for the occurrence of these organisms was probably due to the fact that the spring water was rich in elemental sulfur, the energy substrate for these microorganisms.

At downstream points, green sulfur bacteria were observed because of lower temperatures (30-40°C) and slightly acid pH (6.6-6.8). Most of the isolates from this region were _Prosthecochloris_ spp. and a few _Chlorobium_ spp. The _Prosthecochloris_ cultures were mostly found to be contaminated with a Gram-negative rod-shaped organism capable of reducing elemental sulfur and later identified as _Desulfuromonas_ spp. Since acetate was added in Medium 1 used for enrichment of _Prosthecochloris_, _Desulfuromonas_ formed a dense culture in syntrophy with the former. The syntrophy of these two cultures was further proved by adding small quantities of sulfide to the syntrophic mixture growing on ethanol or acetate. This gave very dense cultures.

The above study does not claim to reveal the entire microbiological picture of the stream. An ecological niche such as a sulfur spring is known to harbour a complex group of microorganisms interacting with each other and varying in composition, depending upon both internal and external factors. A clear picture can only emerge after extensive studies. The intent of the present study was to highlight the presence of photosynthetic bacteria in the stream, with some idea about their distribution.

REFERENCES

Castenholz, R. W., and Pierson, B. K., 1981, Isolation of the members of the family Chloroflexaceae, in: "The Prokaryotes," M. P. Starr, H. Stolp, H. G. Trüper, A. Balows and H. G. Schlegel, eds., Springer-Verlag, Berlin.

Gorlenko, V. M., Dubinina, G. A., and Kusnetsov, S. I., 1977, Ecology of aquatic microorganisms (Russian), Izdatel'stvo Nauka, Moscow.

Kondratieva, E. N., 1965, Photosynthetic Bacteria, Israel Program for Scientific Translations, Jerusalem.

Kuenen, J. C., and Tuovinen, O. H., 1981, The genera _Thiobacillus_ and _Thiomicrospira_, in: "The Prokaryotes," M. P. Starr, H. Stolp, H. G. Trüper, A. Balows and H. G. Schlegel, eds., Springer-Verlag, Berlin.

Miyoshi, M., 1897, Studien über die Schwefelrasenbildung und die Schwefelbakterien der Thermen von Yumoto bei Nikko, _Centralblatt für Bakteriologie, Parasitenkunde u. Infektionskrankheiten Abt. 2, 3:526._

Pierson, B. K., and Castenholz, R. W., 1974, A phototrophic gliding filamentous bacterium of hot springs, _Chloroflexus aurantiacus_, gen. and sp. nov., _Arch. Mikrobiol._, 100:5.

Pfennig, N., and Trüper, H. G., 1981, Isolation of members of the families Chromatiaceae and Chlorobiaceae, in: "The Prokaryotes," M. P. Starr, H. Stolp, H. G. Trüper, A. Balows, and H. G. Schlegel, eds., Springer-Verlag, Berlin.

Strzeszewski, B., 1913, Beitrag zur Kenntnis der Schwefelflora in der Umgebung von Krakau, _Bulletin de l'Academie des Sciences de Cracovie_, Serie B, 209.

Szafer, W., 1910, Zur Kenntnis der Schwefelflora in der Umgebung von Lemberg, _Bulletin de l'Academie des Sciences de Cracovie_, Serie B, 161.

NATURAL GENETIC TRANSFORMATION IN <u>CHLOROBIUM</u>

J.G Ormerod

Department of Biology
University of Oslo
N-0316 Oslo 3, Norway

INTRODUCTION

The availability of a system for the transfer of genes is a prerequisite for genetic mapping and manipulation of an organism. Among phototrophic microorganisms, only the cyanobacteria have been shown to be transformable without special treatment (Shestakov and Khuyen, 1970). Purple bacteria can be transformed with plasmid DNA after treatment with $CaCl_2$ (Fornari and Kaplan, 1982).

It has now been established that <u>Chlorobium</u> cultures exhibit a natural competence for genetic transformation and the purpose of this communication is to describe some of the properties of the system and to indicate its potential in the study of green sulfur bacteria.

MATERIALS AND METHODS

Bacterial Strains and Growth Media

<u>Chlorobium limicola</u> strains 8327 and Tassajara were grown on the medium of Pfennig (1961) as modified by Sirevåg and Ormerod (1970). This medium, designated TA, contains thiosulfate and acetate, and was solidified as required by mixing 1 vol. with 0.66 vol. of molten 3.5% agar solution containing 0.05% Na thioglycollate and resazurin (1 mg/l) as redox indicator. Where required, a filter-sterilised solution of streptomycin was added to give 20 μg streptomycin(ml (TASm agar). Pouring of plates and, as far as possible all other operations, were performed in an anaerobic chamber containing N_2, CO_2, H_2 (85:10:5, vol/vol). Liquid cultures were grown in the light in 8 ml screw cap tubes and growth was followed by measuring A_{650} in a Spectronic colorimeter. All incubations were made at 30°C.

A spontaneous, streptomycin resistant mutant of strain 8327, isolated by Mrs. Gerd Asklund, which grew well on agar containing 100 μg streptomycin/ml, was designated 8327 Sm^r. Growth of the wild type is inhibited by 2 μg/streptomycin/ml.

Transformation Procedure

A modification of the convenient and simple agar plate technique of

315

Juni and Heym (1980) was used when only qualitative or semiquantitative data were required. A colony of the donor strain suspended in dilute saline-sodium citrate-SDS solution was heated for 1 h at 60°C and a loopful of the resulting DNA extract mixed with a loopful of the donor cells on a TA agar plate. After incubation for 10-12 h in the light, cells were streaked on TASm agar and incubation was continued for 2-6 days. The concentration of SDS in the extracting solution in the published method is 0.05%. Experiment showed that transformation of <u>Chlorobium</u> was just as efficient when 0.01% SDS was used. In fact transformation occurred when the cells were heated in saline citrate without SDS.

For quantitative measurements of transformation, 100-200 µl of a culture of strain 8327 in TA medium was mixed in a glass tube with 10 µl of a solution of DNA purified from 8327SM^r by the method of Marmur (1961). After incubation for 2 h in the light, 10 µl DNase (1 mg/ml) was added and incubation continued for 10-12 h, after which samples (100 µl) were plated out on TASm agar and incubated for 2-6 days.

RESULTS

Conditions Required for Uptake of DNA and Transformation

Since streptomycin kills bacterial cells, it was necessary to ensure that the gene for resistance had been expressed in the transformed cells before they were exposed to streptomycin. Using the agar plate transformation technique, it was found that when cells were transferred to TASm agar immediately after addition of Sm^r DNA, no transformants appeared. If cells were incubated for 4 h after DNA addition, a few colonies appeared after plating out on TASm agar but when the transfer was made after 10 h or more, hundreds of transformants appeared.

The plate technique was also used to determine the time required for uptake of DNA. A loopful of DNase solution (1 mg/ml) was mixed with the cells on the agar surface at various intervals after the addition of the 8327SM^r DNA. When DNase was added at the same time as the DNA, no transformants appeared. Transformation could be demonstrated when DNase was added 30 min after the DNA but the frequency was much greater after 1 h and particularly after 2 h, after which it was difficult to demonstrate further increase.

The requirement for light during DNA uptake and transformation was investigated in liquid medium. Cells mixed with DNA were incubated for 4 h in the light or dark, DNase was added, and the tubes were further incubated in the light or dark for 12 h, at which time samples were plated out on TASm agar. The results are shown in Table 1.

Table 1. Effect of Light on Transformation of
         <u>Cb. Limicola</u> 8327. DNA was added at
         0 h, DNase at 4 h and the cells were
         plated out at 20 h.

Incubation conditions		No. of transformants
0-4 h	4-20 h	
light	light	920
light	dark	0
dark	light	1408
dark	dark	0

316

Table 2. Dependence of Transformation on DNA
Concentration

DNA concentration (ng/ml)	Transformants/ml
0.006	3.5×10^2
0.064	1.6×10^3
0.64	1.7×10^4
64	1.8×10^4

Time and Duration of Competence

Preliminary results with the agar plate technique had indicated that
cells were competent at all stages of the growth and stationary phases, and
this was confirmed in the following experiment: A screw cap tube containing
8 ml TA medium was inoculated with Cb. limicola 8327 and incubated at 30°C
in the light. Growth was measured colorimetrically and samples (150 µl) were
removed at intervals and treated with DNA and DNase as described under
MATERIALS AND METHODS. The results are shown in Fig. 1.

DNA Concentration Dependence

Cells from a mid-log phase culture were transformed in liquid medium
with various amounts of DNA. The results are shown in Table 2.

Interstrain Transformation

The agar plate method was used to determine whether a crude DNA extract
from the streptomycin resistant mutant of strain 8327 could transform other
Chlorobium strains. The bacteriochlorophyll c-containing mutant of strain
8327 (Broch-Due and Ormerod, 1978) was transformed with high efficiency, but
the Tassajara strain gave no transformants. Attempts to transform strains
8327 and Tassajara to kanamycin resistance with crude DNA preparations from
cells of Rhodobacter capsulatus containing plasmids R6845 and R1822

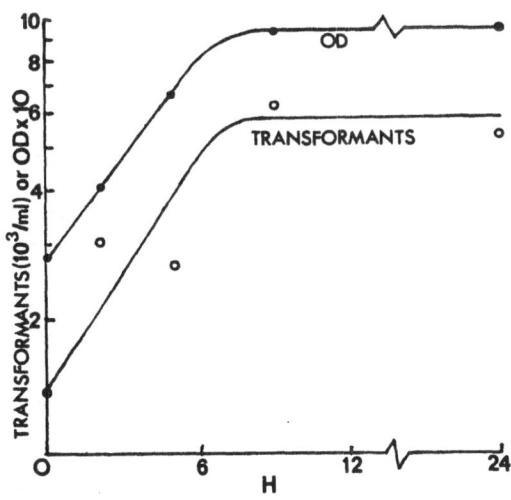

Fig. 1. Competence during growth cycle.

respectively were consistently negative. Attempts to transfer these plasmids by conjugation with Rb. capsulatus or plasmid RP 4 with Escherichia coli J53 were also negative (L. Sveen and J. G. Ormerod, unpublished work).

DISCUSSION

Ideally, an auxotrophic mutant should have been used in this study in view of the toxicity of streptomycin to cells in which the resistance gene has not been expressed. However, attempts to isolate auxotrophic mutants, including several cycles of penicillin enrichment after mutagenisation, gave negative results.

The requirement for incubation in the light before plating out on streptomycin agar results in some inaccuracy because growth of the transformed cells occurs. Nevertheless, the results permit a definition of the properties of this genetic system.

The absolute sensitivity of the system to DNase shows that this is a true transformation. Contact of Chlorobium cells with DNA for only 30 min was sufficient to allow some transformation, showing that uptake must have taken place in this period, although much more transformation occurred after 2 or more hours of contact. The uptake process was not light dependent (Table 1); in fact darkness seemed to enhance the process. On the other hand, light was essential in the subsequent incubation period before exposure to streptomycin. This is not surprising, since Chlorobium is an obligate phototroph and protein synthesis would be required.

The fact that Chlorobium cultures were competent at all stages in the growth cycle (Fig. 1) makes this a very convenient experimental system. The number of transformants in relation to DNA concentration was relatively high and the system became saturated at less than 1 ng DNA/ml. Culture cell counts were not made in these experiments, but a rough estimate of transformation frequency would indicate that one cell in about 10^5 was transformed.

Transformation is a useful tool for investigating the taxonomic relationships of bacteria. The BChl c-containing mutant of strain 8327 was efficiently transformed by DNA from 8327. This mutant was originally isolated from cultures of strain 8327 that had been grown for prolonged periods at low light intensity, and the ease of transformation observed confirms that it really is a mutant of 8327. On the other hand, the Tassajara strain was not transformable by 8327 DNA. Obviously more work of this kind is required to establish the relationships of the various Chlorobium strains.

Transformation of Chlorobium with DNA from a plasmid-containing purple bacterium did not occur, possibly because a crude DNA preparation was used. However, other bacteria which exhibit natural transformation have given similar results (Carlson et al., 1983) and it was found necessary to incorporate chromosomal DNA from the recipient into the plasmid to make it acceptable (Carlson et al., 1984). This has not yet been tested with Chlorobium but if the technique does prove to be successful, this would provide a useful connection with the purple bacteria, where so much elegant work on the genetics of photosynthesis has been carried out (Marrs, 1983). This is rather important, since it was not possible to transfer plasmid-borne kanamycin resistance to Chlorobium by conjugation.

The possible significance of transformation in nature is an interesting aspect of the ecology of bacteria. The fact that detergent was not necessary in the DNA extraction procedure and the very small amounts of DNA required

318

make it seem likely that transformation of <u>Chlorobium</u> could occur in nature.

ACKNOWLEDGEMENTS

Thanks are due to Dr. Judy Wall for the cultures of <u>Rhodobacter capsulatus</u> and to Dr. Bärbel Friedrich for the culture of <u>E. coli</u> J53.

REFERENCES

Broch-Due, M., and Ormerod, J. G., 1978, Isolation of a BChl <u>c</u> mutant from <u>Chlorobium</u> with BChl <u>d</u> by cultivation at low light intensity, <u>FEMS Microbiol. Lett</u>., 3:305.

Carlson, C. A., Pierson, L. S., Rosen, J. J., and Ingraham, J. L., 1983, <u>Pseudomonas stutzeri</u> and related species undergo natural transformation, <u>J. Bacteriol</u>., 153:93.

Carlson, C. A., Steenbergen, S. M., and Ingraham, J. L., 1984, Natural transformation of <u>Pseudomonas stutzeri</u> by plasmids that contain cloned fragments of chromosomal deoxyribonucleic acid, <u>Arch. Microbiol</u>., 140:134.

Fornari, C. S., and Kaplan, S., 1982, Genetic transformation of <u>Rhodopseudomonas sphaeroides</u> by plasmid DNA, <u>J. Bacteriol</u>., 152:89.

Juni, E., and Heym, G. A., 1980, Transformation assay for identification of psychotrophic Achromobacters, <u>Appl. Env. Microbiol</u>., 40:1106.

Marmur, J., 1961, A procedure for the isolation of deoxyribonucleic acid from microorganisms, <u>J. Mol. Biol</u>., 3:208.

Marrs, B., 1983, Genetics and molecular biology, <u>in</u>: "The Phototrophic Bacteria," J. G. Ormerod, ed., Blackwell Scientific, Oxford.

Pfennig, N., 1961, Eine vollsynthetische Nährlösung zur selektiven Anreicherung einiger Schwefelpurpurbakterien, <u>Naturwiss</u>., 48:136.

Shestakov, S. V., and Khuyen, N. T., 1970, Evidence for genetic transformation in blue-green alga <u>Anacystis nidulans</u>, <u>Mol. Gen. Genet</u>., 107:372.

Sirevåg, R., and Ormerod, J. G., 1970, Carbon dioxide fixation in green sulphur bacteria, <u>Biochem. J</u>., 120:339.

CONTRIBUTORS INDEX